# METHODS IN MOLECULAR BIOLOGY



For further volumes:
http://www.springer.com/series/7651

For over 35 years, biological scientists have come to rely on the research protocols and methodologies in the critically acclaimed *Methods in Molecular Biology* series. The series was the first to introduce the step-by-step protocols approach that has become the standard in all biomedical protocol publishing. Each protocol is provided in readily-reproducible step-by-step fashion, opening with an introductory overview, a list of the materials and reagents needed to complete the experiment, and followed by a detailed procedure that is supported with a helpful notes section offering tips and tricks of the trade as well as troubleshooting advice. These hallmark features were introduced by series editor Dr. John Walker and constitute the key ingredient in each and every volume of the *Methods in Molecular Biology* series. Tested and trusted, comprehensive and reliable, all protocols from the series are indexed in PubMed.

# Large Language Models (LLMs) in Protein Bioinformatics

Edited by

**Dukka B. KC**

*Department of Computer Science, Golisano College of Computing and Information Sciences, Rochester Institute of Technology, Rochester, NY, USA*

Humana Press

*Editor*
Dukka B. KC
Department of Computer Science,
Golisano College of Computing and
Information Sciences
Rochester Institute of Technology
Rochester, NY, USA

ISSN 1064-3745 ISSN 1940-6029 (electronic)
Methods in Molecular Biology
ISBN 978-1-0716-4622-9 ISBN 978-1-0716-4623-6 (eBook)
https://doi.org/10.1007/978-1-0716-4623-6



This Humana imprint is published by the registered company Springer Science+Business Media, LLC, part of Springer Nature.
The registered company address is: 1 New York Plaza, New York, NY 10004, U.S.A.

# Dedication

To my late parents, Ram Prasad KC and Tika Devi KC; wife, Bidya KC; daughters, Diva and Dika KC; and my dog, Kancha.

# Preface

The recent wave in AI is to replace the task-specific models with *foundation models* that are trained on a broad set of unlabeled data that can be used for different tasks with minimal fine-tuning. These models are called *foundation models* as they serve as the foundation for many applications of the AI model. Large Language Models (LLMs) are a class of foundation models that are (pre)trained on enormous amounts of data to provide the foundational capabilities needed to drive multiple use cases and applications. LLMs are typically based on a transformer architecture and involve training on a massive corpus of data (e.g., text). The transformer architecture of LLMs allows LLMs to effectively handle long context and sequential information. LLMs represent a significant breakthrough in natural language processing (NLP) and are designed to *understand and generate* texts/contents. LLMs have found application in text/content generation, content summarization, AI assistants, code generation, and language translation, among others.

LLMs have shown significant promise in various research fields including *protein bioinformatics.* Thanks to advances in LLMs, the field of protein bioinformatics has also witnessed a lot of advances in various areas including but not limited to protein structure prediction, protein function prediction, and others. Starting with training of Protein Language Models (PLMs, LLMs that are trained on protein sequence/structure) and the subsequent application of these PLMs, the field has seen a plethora of approaches for various protein bioinformatics tasks.

The application of LLMs in protein bioinformatics can be broadly classified into two categories: representation (understanding) and protein design/engineering (generation). In the representation category, the protein language model is used to extract embeddings typically from the last layer and then these embeddings are used for downstream prediction/classification tasks. Most people refer to this as static, pretrained embeddings, and it has been one the most common approaches in protein bioinformatics. Recently, in NLP, we have seen a few works that use task-specific supervised fine-tuning and training for both the PLM encoder and the prediction head. The field of LLMs is developing at a much faster pace, and new topics like AI agents, etc. are already becoming more popular. We hope to see the application of AI agents and other new emerging themes in the area of protein bioinformatics as well. Additionally, there are newer concepts like large context models (LCMs), and we also expect to see their application in the field.

The following chapters are included in this volume of Methods in Molecular Biology. The book begins (Chap. 1) by setting up the premise for the development of LLM-based approaches in protein bioinformatics, specifically, surveying recent pretrained protein language models. This chapter gives an excellent overview of various types of protein language models based on architecture, aka encoder-only, encoder-decoder, and decoder-only. The chapter also briefly summarizes the most recent trends in the field: fine-tuning protein language models and multimodal protein language models, among other things.

In Chap. 2, Dong Xu's group from the University of Missouri-Columbia describes S-PLM, a 3D structure-aware protein language model. Motivated by the recent advances in protein structure prediction approaches, S-PLM allows one to obtain structural embeddings by leveraging a Geometric Vector Perceptron (GVP) model to process the 3D coordinates of protein.

Protein language models have high demand for computational resources. In Chap. 3, Yan Wang, Zhidong Xue, and colleagues describe their *lightweight* protein language model called ProtFlash. ProtFlash uses several key technological breakthroughs including mixed-chunk attention, among other things. The authors also describe step-by-step instructions for utilizing the ProtFlash library.

Bonvin's group from Utrecht University describes an approach called DeepRank-GNN-esm based on protein language models for protein-protein interaction prediction to rank protein-protein models (scoring problem) in Chap. 4. As the ranking of good models from the large pool of generated models in protein-protein interaction prediction is an important step, this chapter describes the use of protein language model (ESM-2) features to improve protein-protein interaction prediction.

In Chap. 5, Daisuke Kihara from Purdue University and colleagues describe GO2Sum, gene ontology (GO) terms summarizer that uses a protein-language model. Essentially, GO2Sum takes a list of GO terms as input and converts them into a summary that describes various GO aspects of a protein. The chapter also describes the web server of GO2Sum.

Jinalin Cheng's group from University of Missouri-Columbia describes their recently developed protein function annotation tool called TransFun in Chap. 6. Recognizing the lack of functional annotations for many proteins, TransFun leverages embeddings from ESM-1b and predicted structures from AlphaFold to predict function for a given protein. The authors also describe in detail how to get started with TransFun.

In Chap. 7, Lydia Fredollino's group at University of Michigan describes InterLabelGO +, a top-performing model to predict GO term in CAFA5. InterLabelGO+ is an approach for prediction of protein functions in the form of gene ontology that uses the ESM2 protein language model to extract sequence features. Additionally, the group also describes the procedure to perform protein GO term prediction with InterLabelGO+ webserver and the standalone package in details.

In Chap. 8, Ana Rojas and collaborators from Centro Andaluz De Biologia Del Desarrollo discuss the application of a protein language model (ProtTrans) for protein function annotation. Additionally, they also describe the FANTASIA tool for large-scale annotation of uncharacterized proteomes.

Debswapna Bhattacharya from Virginia Tech and collaborators summarize the recent advances in protein-nucleic acid binding site prediction approaches that harness protein language models in Chap. 9. Additionally, the chapter also presents their own approach called EquiPNAS that integrates pLM with equivariant deep graph neural networks for protein-DNA and protein-RNA binding site prediction.

In Chap. 10, Henrik Nielsen from the Technical University of Denmark describes three important tools that his group recently developed related to what proteins belong to which compartments, making use of protein language models. Specifically, the chapter describes SignalP6.0 for prediction of signal peptides, DeepLoc2.1 for prediction of subcellular location and membrane association in eukaryotes, and DeepLocPro1.0 for prediction of subcellular location in prokaryotes.

Iman Dehzangi and colleagues from Rutgers University-Camden discuss their tool CNN-Meth for predicting lysine methylation sites that uses evolutionary information and structural features in Chap. 11. Although their method does not directly use a protein language model, the Position-Specific Scoring Matrix (PSSM) features in their approach can readily be replaced by protein language model-based embeddings.

Pier Luigi Martelli and colleagues from the University of Bologna describe their approach for characterizing proteins and for predicting the pathogenicity of human protein variants. Their approach uses embeddings from protein language models. Additionally, they describe their Bioinformatics Sweeties, a web portal, that has a list of bioinformatics tools for characterizing proteins and different aspects of pathogenic variants with examples in Chap. 12.

In Chap. 13, Shandar Ahmad's group from Jawaharlal Nehru University discusses various existing approaches for prediction of biological function that leverages protein language models and NLP-based techniques. Additionally, the survey highlights the major advances in the field and possible future directions for the research in the field.

Shanfeng Zhu and Jianyi Yang's group discusses their recent approach, in Chap. 14, for homologous protein search and sequence alignment that uses protein language models. PLMSearch is their protein language model-based tools for searching homologous sequences, and PLMAlign is their tool for aligning remote homologous sequences. The chapter also describes in detail how to use these tools.

In Chap. 15, Siwei Chen's group at the Broad Institute of MIT and Harvard summarizes the recent advances in protein-protein interaction analysis that leverages protein language models. Essentially, the computational tools for predicting protein-protein interactions and protein-protein interaction site prediction are discussed in detail. The chapter also highlights some of the other promising areas of PPI prediction, including PPI hotspots among others.

Identifying protein-peptide binding residue is important for understanding the mechanisms of protein functions and drug discovery. In Chap. 16, Leyi Wei and colleagues describe their PepBCL tool to predict protein-peptide binding site. PepBCL uses a pre-trained BERT model called ProtBert-BFD to generate the encoding vector.

Bioactive peptide discovery is another important field across food, nutraceuticals, cosmetics, and pharmaceuticals. In Chap. 17, Yonghui Li and colleagues from Kansas State University describe their tool that leverages a protein language model for predicting peptide bioactivity. Their approach, called UniDL4BioPep, uses the ESM protein language model.

In Chap. 18, Boxue Tian's group at Tsinghua University discusses a new tool called CLAPE for the prediction of protein-ligand binding site. CLAPE uses contrastive learning and the pretrained protein language model ProtBERT. The authors describe in detail the architecture, model performance, and datasets utilized in the training of CLAPE as well as how to use CLAPE.

Finally, in Chap. 19, My group from the Rochester Institute of Technology focus our chapter on a survey of recent advances in the prediction of post-translational modification sites in proteins that leverage large language models. We also identify emerging trends in the field and outline some of the challenges and future research directions in the field.

I hope readers receive this book as a comprehensive collection of methods, resources, and studies that use LLMs in protein bioinformatics. In addition to the description of these approaches, I believe the book will also serve as a practical guide for using these LLM-based tools in relation to various protein bioinformatics tasks. I am hopeful that this book exhibits a state of the art of the current research field in the arena and provides future trends in the field regarding the use of LLMs for protein bioinformatics.

This book would not have been possible without the generous contributions from all the contributing authors in this book. I sincerely thank them for their contributions. I am also thankful to the series editor, Dr. John M. Walker, for patiently working with me throughout the process. In addition, I would like to thank all the lab members of my lab, my collaborators and my colleagues in the Department of Computer Science at Rochester Institute of Technology (RIT). Most importantly, I would like to sincerely acknowledge the inspiration, encouragement, and guidance I received from my mentors Dr. Tatsuya Akutsu, Dr. Dennis Livesay, and Dr. Ilya Vakser, among others.

*Rochester, NY, USA* *Dukka B. KC*

# Acknowledgments

This work was supported in part by National Science Foundation (NSF) grants DBI-1901793 and IIS-2210356 to DBKC and a faculty start-up grant from RIT.

# Contents

# Contributors

SHANDAR AHMAD • *School of Computational and Integrative Sciences, Jawaharlal Nehru University, New Delhi, India*
T. ATHULYA • *School of Computational and Integrative Sciences, Jawaharlal Nehru University, New Delhi, India*
GIULIA BABBI • *Biocomputing Group, Department of Pharmacy and Biotechnology, University of Bologna, Bologna, Italy*
ELISA BERTOLINI • *Biocomputing Group, Department of Pharmacy and Biotechnology, University of Bologna, Bologna, Italy*
DEBSWAPNA BHATTACHARYA • *Department of Computer Science, Virginia Tech, Blacksburg, VA, USA*
FRIMPONG BOADU • *Department of Electrical Engineering and Computer Science, NextGen Precision Health Institute, University of Missouri, Columbia, MO, USA*
ALEXANDRE M. J. J. BONVIN • *Computational Structural Biology Group, Department of Chemistry, Bijvoet Centre for Biomolecular Research, Faculty of Science, Utrecht University, Utrecht, Netherlands*
LISA BRAMER • *Biological Sciences Division, Pacific Northwest National Laboratory, Richland, WA, USA*
DOINA CARAGEA • *Department of Computer Science, Kansas State University, Manhattan, KS, USA*
RITA CASADIO • *Biocomputing Group, Department of Pharmacy and Biotechnology, University of Bologna, Bologna, Italy; Biocomputing Group and the Alma Climate Institute, University of Bologna, Bologna, Italy*
ILDEFONSO CASES • *CSIC, Andalusian Center for Developmental Biology, Computational Biology and Bioinformatics Group, Seville, Spain*
MEENAL CHAUDHARI • *College of Applied Sciences and Technology, Illinois State University, Normal, IL, USA*
SIWEI CHEN • *Stanley Center for Psychiatric Research, Broad Institute of MIT and Harvard, Cambridge, MA, USA; Program in Medical and Population Genetics, Broad Institute of MIT and Harvard, Cambridge, MA, USA; Analytic and Translational Genetics Unit, Massachusetts General Hospital, Boston, MA, USA*
JIANLIN CHENG • *Department of Electrical Engineering and Computer Science, NextGen Precision Health Institute, University of Missouri, Columbia, MO, USA*
IMAN DEHZANGI • *Center for Computational and Integrative Biology, Rutgers University, Camden, NJ, USA; Department of Computer Science, Rutgers University, Camden, NJ, USA; Rutgers Cancer Institute, Rutgers University, New Brunswick, NJ, USA*
ZHENJIAO DU • *Department of Grain Science and Industry, Kansas State University, Manhattan, KS, USA*
ROSA FERNÁNDEZ • *CSIC, Institute for Evolutionary Biology, Metazoa Phylogenomics Lab, Barcelona, Spain*
LYDIA FREDDOLINO • *Gilbert S Omenn Department of Computational Medicine and Bioinformatics, University of Michigan, Ann Arbor, MI, USA; Department of Biological Chemistry, University of Michigan, Ann Arbor, MI, USA*

SWAGARIKA J. GIRI • *Department of Computer Science, Purdue University, West Lafayette, IN, USA*
MICHAEL HEINZINGER • *TUM (Technical University of Munich), School of Computation, Information and Technology (CIT), Faculty of Informatics, Chair of Bioinformatics & Computational Biology, Munich, Germany*
DUKKA B. KC • *Department of Computer Science, Golisano College of Computing and Information Sciences, Rochester Institute of Technology, Rochester, NY, USA*
DAISUKE KIHARA • *Department of Computer Science, Purdue University, West Lafayette, IN, USA; Department of Biological Sciences, Purdue University, West Lafayette, IN, USA*
NANDAN KUMAR • *Department of Grain Science and Industry, Kansas State University, Manhattan, KS, USA*
AHHYUN LEE • *Department of Electrical Engineering and Computer Science, NextGen Precision Health Institute, University of Missouri, Columbia, MO, USA*
YONGHUI LI • *Department of Grain Science and Industry, Kansas State University, Manhattan, KS, USA*
DONGFANG LIU • *Department of Computer Engineering, Rochester Institute of Technology, Rochester, NY, USA*
QUANCHENG LIU • *Gilbert S Omenn Department of Computational Medicine and Bioinformatics, University of Michigan, Ann Arbor, MI, USA*
WEI LIU • *Institute of Science and Technology for Brain-Inspired Intelligence and MOE Frontiers Center for Brain Science, Fudan University, Shanghai, China; Department of Quantitative and Computational Biology, University of Southern California, Los Angeles, California, United States*
YUFAN LIU • *Computer Science Program, Computer, Electrical and Mathematical Sciences and Engineering Division, King Abdullah University of Science and Technology (KAUST), Thuwal, Kingdom of Saudi Arabia*
MATTEO MANFREDI • *Biocomputing Group, Department of Pharmacy and Biotechnology, University of Bologna, Bologna, Italy*
PIER LUIGI MARTELLI • *Biocomputing Group, Department of Pharmacy and Biotechnology, University of Bologna, Bologna, Italy*
GEMMA MARTÍNEZ-REDONDO • *CSIC, Institute for Evolutionary Biology, Metazoa Phylogenomics Lab, Barcelona, Spain*
VIKASH K. MOHANI • *School of Computational and Integrative Sciences, Jawaharlal Nehru University, New Delhi, India*
KENTA NAKAI • *Human Genome Center, Institute of Medical Science, University of Tokyo, Tokyo, Japan*
ROBERT H. NEWMAN • *College of Science and Technology, North Carolina Agricultural and Technical State University, Greensboro, NC, USA*
HENRIK NIELSEN • *Section for Bioinformatics, Department of Health Technology, Technical University of Denmark, Lyngby, Denmark*
UDAYAN PANDEY • *Department of Computer Science, Purdue University, West Lafayette, IN, USA*
JOON HONG PARK • *Department of Computer Science, Purdue University, West Lafayette, IN, USA*
SURESH POKHAREL • *Golisano College of Computing and Information Sciences, Rochester Institute of Technology, Rochester, NY, USA; Department of Computer Science, Rochester Institute of Technology, Rochester, NY, USA*

MAHDI POURMIRZAEI • *Department of Electrical Engineering and Computer Science and Christopher S. Bond Life Sciences Center, University of Missouri, Columbia, MO, USA*
PAWEL PRATYUSH • *Golisano College of Computing and Information Sciences, Rochester Institute of Technology, Rochester, NY, USA; Department of Computer Science, Rochester Institute of Technology, Rochester, NY, USA*
FARAH QADERI • *Department of Surgical Oncology, Massachusetts General Hospital, Boston, MA, USA*
YONGFANG QIN • *Department of Electrical Engineering and Computer Science and Christopher S. Bond Life Sciences Center, University of Missouri, Columbia, MO, USA*
RAHMATULLAH ROCHE • *TSYS School of Computer Science, Columbus State University, Columbus, GA, USA*
ANA M. ROJAS • *CSIC, Andalusian Center for Developmental Biology, Computational Biology and Bioinformatics Group, Seville, Spain*
HIROTO SAIGO • *Faculty of Information Science and Electrical Engineering, Kyushu University, Kyushu, Japan*
CASTRENSE SAVOJARDO • *Biocomputing Group, Department of Pharmacy and Biotechnology, University of Bologna, Bologna, Italy*
STEFAN SCHULZE • *Thomas H. Gosnell School of Life Sciences, Rochester Institute of Technology, Rochester, NY, USA*
QING SHAO • *Chemical & Materials Engineering, University of Kentucky, Lexington, KY, USA*
ALOK SHARMA • *Institute for Integrated and Intelligent Systems, Griffith University, Brisbane, Australia; Laboratory for Medical Science Mathematics, RIKEN Center for Integrative Medical Sciences, Yokohama, Japan; College of Informatics, Korea University, Seoul, South Korea; Laboratory of Medical Science Mathematics, Department of Biological Sciences, Graduate School of Science, The University of Tokyo, Tokyo, Japan*
AUSTIN SPADARO • *Center for Computational and Integrative Biology, Rutgers University, Camden, NJ, USA*
SUMIT TARAFDER • *Department of Computer Science, Virginia Tech, Blacksburg, VA, USA*
BOXUE TIAN • *MOE Key Laboratory of Bioinformatics, State Key Laboratory of Molecular Oncology, School of Pharmaceutical Sciences, Tsinghua University, Beijing, China*
KAICHENG U • *Tri-Institutional Computational Biology & Medicine, Weill Cornell Medicine, New York, NY, USA; Department of Computational Oncology, Memorial Sloan Kettering Cancer Center, New York, NY, USA*
DANA MARY VARGHESE • *School of Computational and Integrative Sciences, Jawaharlal Nehru University, New Delhi, India*
GABRIELE VAZZANA • *Biocomputing Group, Department of Pharmacy and Biotechnology, University of Bologna, Bologna, Italy*
DUOLIN WANG • *Department of Electrical Engineering and Computer Science and Christopher S. Bond Life Sciences Center, University of Missouri, Columbia, MO, USA*
LEI WANG • *School of Software Engineering, Huazhong University of Science and Technology, Wuhan, Hubei, China*
RUHENG WANG • *Department of Biomedical Engineering, University of Texas Southwestern Medical Center, Dallas, TX, USA*
XINYU WANG • *Department of Computer Science, Virginia Tech, Blacksburg, VA, USA*
YAN WANG • *School of Life Science and Technology, Huazhong University of Science and Technology, Wuhan, Hubei, China*

ZIYE WANG • *Institute of Science and Technology for Brain-Inspired Intelligence and MOE Frontiers Center for Brain Science, Fudan University, Shanghai, China*
HONG WEI • *Department of Bioinformatics, School of Basic Medical Sciences, Tianjin Medical University, Tianjin, China*
LEYI WEI • *Centre for Artificial Intelligence driven Drug Discovery, Faculty of Applied Science, Macao Polytechnic University, Macao SAR, China; School of Software, Shandong University, Jinan, China*
CHENGHAN XIE • *School of Mathematical Sciences, Fudan University, Shanghai, China*
YI XIONG • *Department of Bioinformatics and Biostatistics, Shanghai Jiao Tong University, Shanghai, China*
DONG XU • *Department of Electrical Engineering and Computer Science and Christopher S. Bond Life Sciences Center, University of Missouri, Columbia, MO, USA*
XIAOTONG XU • *Computational Structural Biology Group, Department of Chemistry, Bijvoet Centre for Biomolecular Research, Faculty of Science, Utrecht University, Utrecht, Netherlands*
ZHIDONG XUE • *School of Software Engineering, Huazhong University of Science and Technology, Wuhan, Hubei, China*
JIANYI YANG • *MOE Frontiers Science Center for Nonlinear Expectations, Research Center for Mathematics and Interdisciplinary Sciences, Shandong University, Qingdao, China*
RONGHUI YOU • *Institute of Science and Technology for Brain-Inspired Intelligence and MOE Frontiers Center for Brain Science, Fudan University, Shanghai, China*
CHENGXIN ZHANG • *CAS Key Laboratory of Quantitative Engineering Biology, Shenzhen Institute of Synthetic Biology, Shenzhen Institutes of Advanced Technology, Chinese Academy of Sciences, Shenzhen, Guangdong, China*
SOPHIA MEIXUAN ZHANG • *College of Agriculture and Life Sciences, Cornell University, Ithaca, NY, USA; Harvard College, Harvard University, Cambridge, MA, USA*
YICHUAN ZHANG • *Department of Electrical Engineering and Computer Science and Christopher S. Bond Life Sciences Center, University of Missouri, Columbia, MO, USA*
JUNHAN ZHAO • *Department of Biomedical Informatics, Harvard Medical School, Boston, MA, USA; Department of Biostatistics, Harvard T.H. Chan School of Public Health, Boston, MA, USA*
SHANFENG ZHU • *Institute of Science and Technology for Brain-Inspired Intelligence and MOE Frontiers Center for Brain Science, Fudan University, Shanghai, China*

# Chapter 1

# A Survey of Pretrained Protein Language Models

**Suresh Pokharel, Pawel Pratyush, Meenal Chaudhari, Michael Heinzinger, Doina Caragea, Hiroto Saigo, and Dukka B. KC**

## Abstract

Inspired by the transformative success of large language models (LLMs) in natural language processing (NLP), numerous protein language models (PLMs) have recently emerged, revolutionizing the field of protein bioinformatics. PLMs have demonstrated remarkable achievements in representing proteins and designing new ones, capturing intrinsic structural and functional information trained on vast datasets of proteins, PLMs have demonstrated exceptional performance across a variety of bioinformatics tasks, including classification, function prediction, and de novo protein design. This chapter explores the evolution of PLMs, tracing their origins from NLP-based transformers and large language models (LLMs). A comprehensive summary of notable PLMs is presented, with a particular focus on encoder-only, encoder–decoder, and decoder-only architectures. Additionally, we delve into cutting-edge trends in PLM applications, such as fine-tuning methods, multimodal architectures, and the use of reduced alphabets. These innovations underscore the growing potential of PLMs to tackle complex biological problems and drive future breakthroughs in the field.

**Key words** Protein language models, Transformers in bioinformatics, Encoder-only, Encoder, Decoder, Decoder-only models

## 1 Introduction

Proteins are fundamental to life, acting as the building blocks and functional machinery of living organisms. These biomolecules are composed of one or multiple long chains of amino acids called polypeptides. Amino acids are organic molecules with a central carbon atom bonded to an amino group, a carboxyl group, a hydrogen atom, and a variable side chain. Proteins fold into three-dimensional (3D) structures that determine their roles in cellular processes that form structural components like collagen, function as enzymes, hormones, or antibodies, and regulate various biological processes [1]. Proteins make up about 42% of the human body's dry weight and are essential to nearly every biological function [2, 3]. Scientists study proteins at many levels including the

Dukka B. KC (ed.), *Large Language Models (LLMs) in Protein Bioinformatics*, Methods in Molecular Biology, vol. 2941,
https://doi.org/10.1007/978-1-0716-4623-6_1,

linear amino acid sequence (primary structure) to folded patterns like alpha helices and beta sheets (secondary structure) and complex 3D shapes (tertiary structure) to understand molecular mechanisms of life [3].

Recent studies follow an intriguing connection between proteins and natural language processing (NLP) [4–7], drawing an analogy between protein sequences and human language. Protein sequences, akin to "sentences" composed of amino acid "words," have paved the way for leveraging NLP techniques in protein bioinformatics. This analogy has facilitated groundbreaking advancements of NLP techniques to achieve remarkable achievements in protein bioinformatics, uncovering complex patterns in proteins, predicting structures, identifying functions, and even designing de novo proteins.

On one hand, recent advancements in transformer-based applications, particularly large language models (LLMs), have significantly revolutionized the NLP field [8]. Their ability to process sequential data efficiently makes them well suited for understanding complex relationships within text and for text generation. Similarly, protein language models (PLMs) are trained using large protein sequence databases, much like LLMs are trained using vast text corpora. PLMs have demonstrated remarkable achievements in representing proteins and designing new ones, capturing intrinsic structural and functional information [4, 9, 10]. This innovative application of cutting-edge LLM-driven approaches has demonstrated breakthroughs in protein bioinformatics with transformative applications in areas like drug discovery and protein engineering.

In this chapter, we start with providing a summary of transformers and describing various types of transformers architecture by examining the core principles behind LLMs and tracing the progression of various pretrained protein language models. Then, we delve into the various types of protein language models (PLMs) especially based on their architectures. Subsequently, we briefly describe the three major types of PLMS: encoder only, encoder–decoder, and decoder only architectures. We then briefly describe some applications of PLMs in protein bioinformatics and then finally summarize by outlining the recent trends in PLMs.

## 2 Transformers: A Historical Perspective and Types of Transformers

Recurrent neural networks (RNNs) and their variants, such as long short-term memory networks (LSTMs) and gated recurrent units (GRUs), were long considered the gold standard for natural language sequence modeling tasks (predicting the next element in a data sequence). These architectures excelled at processing sequential data by maintaining hidden states to encode information from previous time steps. However, these RNNs and their variants have critical limitations. First, RNNs process inputs sequentially, which hinders parallelization and makes training computationally expensive. Second, as sequence lengths increase, gradients can suffer from the problem of vanishing or exploding, hindering the process of learning. Even though LSTMs and GRUs addressed some of these challenges, they still had difficulty capturing long-range dependencies. Their sequential processing nature often caused information from earlier inputs to be forgotten in long sequences.

In 2017, researchers at Google Brain and Google Research introduced the transformer architecture [11], fundamentally changing sequence modeling tasks by moving away from traditional recurrent neural networks. The key innovation was the use of a self-attention mechanism instead of recurrent structures, allowing the model to assess the importance of each token relative to all others in a sequence. Unlike recurrent models that process data step by step, transformers handle all tokens simultaneously. They compute attention scores between every pair of elements in parallel, effectively capturing long-range dependencies without sequential processing. To incorporate sequence order information, positional encoding was added to the model. By eliminating recurrence, transformers can compute contextual relationships in a single operation through parallel processing, which significantly accelerates training and inference times. This approach also overcomes the limitations of recurrent models, such as inefficiency and challenges with long-range dependencies. The ability to process entire sequences at once enhances the scalability of transformers, especially with longer sequences, establishing them as a groundbreaking architecture for sequence modeling tasks. The transformer model outperformed existing architectures in machine translation tasks, setting new records and demonstrating the power of attention mechanisms.

The introduction of transformers sparked a surge of research and development across various domains, particularly in natural language processing (NLP). Notable early implementations include bidirectional encoder representations from transformers (BERT) [12], developed by Google in 2018, which utilized the encoder part of the transformer architecture to create deep bidirectional representations with the help of masked language modeling (MLM). During MLM, the model receives partially corrupted inputs which it tries to correct in the output. To do well in this task, the model has to learn to gather information from a non-corrupted context. This enabled BERT to excel in understanding the context of words with their surroundings, reaching state-of-the-art results across numerous NLP tasks. Central to BERT's success was pretraining on a large corpora of unlabeled data obtained from Wikipedia and BooksCourpus, followed by fine-tuning on specific tasks. This two-stage approach allowed the models to learn a general understanding of language during pretraining, which could then be adapted to downstream tasks with minimal labeled data. Another prominent architecture, OpenAI's generative pretrained transformer (GPT) [13] models, leveraged the decoder part of the transformer for language generation. GPT-2 and GPT-3, released in 2019 and 2020, respectively, showcased exceptional abilities in generating coherent and contextually relevant text, pushing the boundaries of language modeling. Another notable advancement was Transformer-XL, which addressed the fixed-length context limitation of standard transformers by incorporating recurrence mechanisms, enabling the model to capture longer-term dependencies and improving performance on tasks requiring extensive contextual understanding. Following these groundbreaking models, several modifications and improvements were proposed by researchers. Meta AI's RoBERTa refined BERT by optimizing training strategies, such as dynamic masking and longer training with larger batches, leading to improved performance. To tackle increasing computational costs, ALBERT introduced parameter reduction techniques, making models more efficient without a significant loss in accuracy. Google's T5 (text-to-text transfer transformer) [14] unified NLP tasks by framing them as a text-to-text problem, achieving state-of-the-art results across numerous benchmarks. XLNet proposed a permutation-based training objective that integrated the benefits of autoregressive and autoencoding models, surpassing BERT in several NLP tasks. These advancements underscored the adaptability of the transformer architecture, solidifying its position as a foundation for modern NLP systems.

Transformers soon surpassed their initial application in NLP, demonstrating their versatility in other domains. Vision transformers (ViT) marked a major breakthrough in image recognition by adapting transformer architectures to process images. ViT models split images into patches and treated them similarly to word tokens, achieving competitive results compared to traditional convolutional neural networks (CNNs). Multimodal transformers, such as DALL·E and CLIP, combined textual and visual data, enabling tasks like image generation from text descriptions and understanding relationships between images and text. In speech processing, transformers have been employed in speech recognition and synthesis, enhancing models' ability to understand and generate human speech with higher fidelity. These diverse applications highlighted the transformative impact of the architecture across multiple disciplines, making transformers a cornerstone of modern machine learning.

Until now, we briefly discussed the history and evolution of transformers. Now, let us examine the types of transformers. Transformers can be classified based on various aspects. One such aspect is architectural modifications, which involve changes to the original

**Table 1**
**Overview of transformer types based on pretraining method**

| Type | Context directionality | Pretraining objective | Pretraining task | Examples |
|---|---|---|---|---|
| Encoder only | Bidirectional | Masked language modeling (MLM) | Understanding/ encoding tasks | BERT, RoBERTa, ALBERT |
| Decoder only | Unidirectional (autoregressive) | Causal language modeling (CLM) | Text generation tasks | GPT series, Transformer-XL, Megatron-LM |
| Encoder–decoder | Bidirectional + autoregressive | Denoising, Seq2Seq generation | Input-to-output transformations | T5, BART |

transformer structure, such as sparse attention mechanisms, linearized attention, and hierarchical designs, aimed at improving efficiency, scalability, or adaptability. Another classification is based on pretraining methods, where transformers differ in their learning strategies. They can also be categorized by their applications, as they are tailored for diverse fields such as NLP, computer vision, and audio processing, each requiring unique adaptations. We recommend referring to the paper by Lin et al. [15] for detailed information on these classifications. In this chapter, we focus on the classification based on pretraining methods, as discussed below. Additionally, we have summarized this classification in Table 1.

### *2.1 Encoder-Only Architectures*

Encoder-only transformers utilize the encoder component of the original transformer architecture. These models are pretrained using masked language modeling (MLM), where a portion of the input tokens are masked, and the model learns to predict these masked tokens by leveraging bidirectional context. The architecture features self-attention mechanisms, allowing each token to attend to every other token in the sequence regardless of positional distance, with positional encodings incorporated to provide ordering information. By stacking multiple layers of self-attention and feed-forward networks, these models gain depth and enhance their ability to recognize complex patterns in the data. Encoder-only models are particularly suited for tasks requiring a deep understanding of the input, such as text classification, named entity recognition (NER), question answering, and semantic similarity tasks. Examples include BERT, which introduced bidirectional training for nuanced language understanding; RoBERTa, an optimized version of BERT with improved training strategies; and ALBERT, which reduces model size while maintaining high performance for efficient deployment.

### 2.2 Encoder–Decoder Architectures

Encoder–decoder transformers integrate both encoder and decoder components to convert an input sequence into a corresponding output sequence, making them well suited for sequence-to-sequence tasks, such as translation, where the output differs from the input. These models are typically pretrained using objectives like denoising autoencoding, where the input sequence is corrupted (e.g., with masking, shuffling, or deletion), and the model learns to reconstruct the original sequence or sequence-to-sequence generation, where the decoder predicts the output sequence based on the encoder's output. The encoder transforms the input sequence into continuous representations using layers of self-attention and feed-forward networks, while the decoder generates the output sequence by focusing on both the encoder's output and its own previously generated tokens. Masked self-attention ensures predictions depend only on preceding tokens, and encoder–decoder attention allows the decoder to focus on the most relevant parts of the input sequence. These models excel in tasks like machine translation, text summarization, paraphrasing, and question generation. Notable examples include the original transformer, which revolutionized neural machine translation; T5 (text-to-text transfer transformer), which reformulates all NLP tasks as text-to-text problems; and BART, which combines bidirectional encoding with autoregressive decoding for summarization and generation tasks.

### 2.3 Decoder–Only Architectures

Decoder-only transformers consist solely of the decoder component and are designed for sequence generation tasks. These models are pretrained using causal language modeling (CLM), where the model predicts the next token in a sequence based on all previous tokens, enforcing a unidirectional (left-to-right) structure. Masked self-attention ensures that the model does not have access to future tokens during training, while positional encodings provide token position information. This approach makes decoder-only transformers particularly effective for generative tasks such as language modeling, text completion, creative writing, and dialogue systems. Examples include the GPT series, known for generating human-like text; Transformer-XL, which addresses fixed-length context limitations by introducing recurrence mechanisms for long-term dependency modeling; and Megatron-LM, designed for scalable training of transformer models with billions of parameters.

In the field of protein bioinformatics, researchers have adopted transformer architectures by treating individual amino acids as "words" and entire protein sequences as "sentences" [7]. This allows transformer models to examine protein sequences in a manner analogous to processing natural language. As a result, transformer-based architectures can be pretrained on extensive protein sequence datasets, such as UniRef50 and UniRef100 [16], enabling them to learn the "grammar" of the protein language, including the structural and functional context of each amino acid with respect to the overall sequence. Language models pretrained in this way on protein sequences (and/or structures) are collectively referred to as protein language models (PLM). In the next section, we will discuss about PLMs in detail.

## 3 Protein Language Models

Large language models were developed to process human language, learning to predict word sequences and comprehend syntax and semantics using vast amounts of text data. These large language models (LLMs) like GPT and BERT [12, 13, 17, 18] revolutionized natural language processing by utilizing deep learning techniques and transformer architectures, which allowed for the modeling of long-range dependencies and the context in text sequences. Drawing inspiration from the success of human language-based large language models (LLMs) in understanding linguistic structures, protein language models (PLMs) [19] have emerged as powerful tools in bioinformatics. The evolution from human language-based models to protein language models (PLMs) [4, 20–26] capitalized on this foundation, applying similar methods to protein sequences. Proteins, like language, are composed of sequences of amino acids in place of words in sentences. Researchers realized that proteins exhibit structural and functional motifs, much like how human language has grammatical rules. By training models on a large corpus of protein sequences and protein structures, (pretrained) PLMs learn to represent this biological information in a high-dimensional space, capturing evolutionary, structural, and functional information about proteins [9, 27, 28]. PLMs have been primarily used in protein bioinformatics for two major tasks: protein representation and protein generation/design. In protein representation, pretrained protein language models are used to extract numerical vector representations (aka *embeddings*) for the sequence in hand, and in protein generation/design tasks, these pretrained PLMs are used to generate/design proteins with certain properties.

This adaptation to biological data has been transformative, as PLMs have demonstrated the ability to predict various protein functions and structures, understand sequence-function relationships, and perform multiple protein downstream tasks including protein classification [29–32] and function prediction [30, 33–37]. Their ability to generalize across diverse protein families highlights their potential to address complex problems in genomics and proteomics. Protein language models (PLMs) have been developed to learn meaningful representations of proteins, with distinct models focusing on different data types such as protein sequences, 3D structures, or a combination of both.

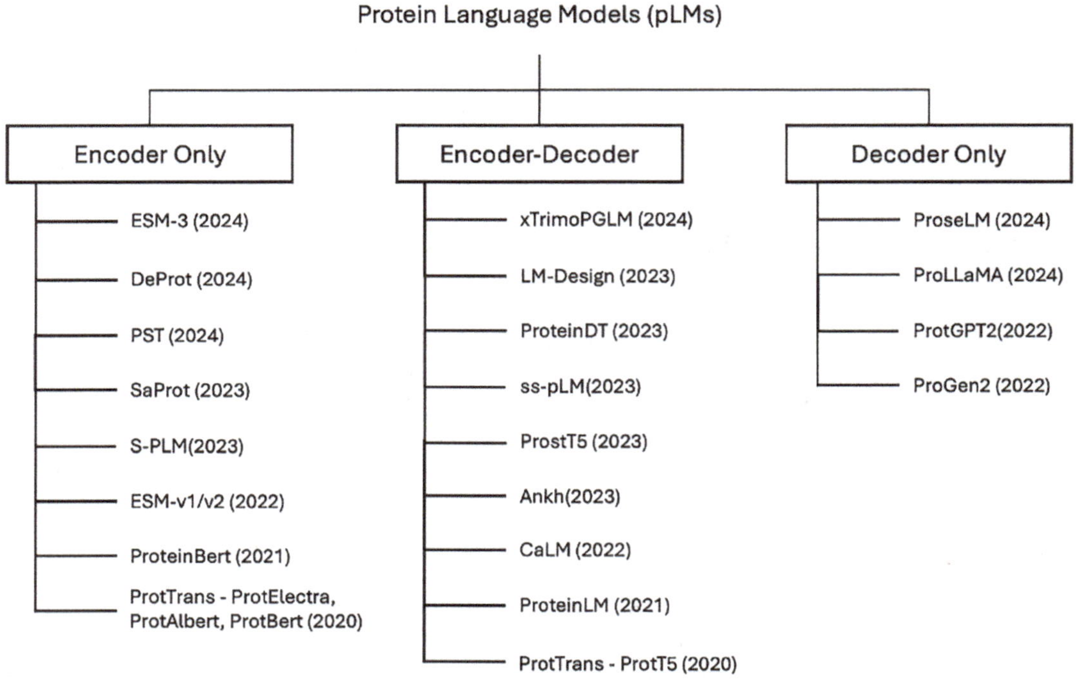


**Fig. 1** Categorization of protein language models (PLMs) based on architectural design into encoder-only, encoder–decoder, and decoder-only models

Sequence-based PLMs are trained solely on amino acid sequences, learning patterns, motifs, and evolutionary relationships within proteins. By learning rich representations from the protein sequences, PLMs provide a gateway to various applications including protein classification, structure prediction, and protein function prediction. Similarly, structure-based PLMs are trained on information derived from 3D protein structures capturing the spatial arrangements and interactions of amino acids. Furthermore, some of the PLMs are trained by integrating both AA sequence and 3D structure information to generate potentially better representations to improve performance in various tasks providing both sequence-based and structural insights. In the reminder of this section, we explore a comprehensive list of the latest PLMs (as illustrated in Fig. 1) pretrained using encoder-only, encoder–decoder, and decoder-only architecture protein sequence data, 3D structural information, and a combination of both.

### 3.1 Types of PLMS Based on Pretraining

We focus on three key transformer architecture types for the PLMs: encoder–only, encoder–decoder, and decoder-only models including their applications to the field of protein bioinformatics, as described in the following subsections. Below, we describe these three types of PLMs.

**Table 2**
**Summary of encoder-only PLMs with training datasets and architecture details**

| PLM | | Dataset | Model architecture |
|---|---|---|---|
| ProteinBert [38] | | Pretrained on 106M protein sequences, obtained from UniProtKB/UniRef90 | Bert |
| ProtTrans [23] | ProtElectra | UniRef50, UniRef100 BFD | Electra (consists of two models: generator and discriminator) |
| | ProtAlbert | UniRef100 | Albert |
| | ProtBert | UniRef100, BFD 100 | BERT |
| | ProtXLNet | UniRef100 | XLNet |
| ESM (v1/v2) [25] | | UniRef50, UniRef90, UniRef100 | Standard transformer architecture<br>Other variants: ESM-1b, ESM-MSA-1, ESM-MSA-1b, ESM-1v, ESM-IF1 |
| S-PLM [39] | | ~500K sequences from Swiss-Prot and AlphaFold2-predicted Cα contact maps | Transformer-based sequence encoder and a ResNet50 structure encoder integrated with multi-view contrastive learning |
| SaProt [26] | | ~40 million protein sequences and AlphaFold2-predicted structures | Similar to ESM-2 architecture with an expanded embedding layer |
| PST [40] | | ~500K structures AlphaFold DB (SwissProt subset) | ESM-2 architecture With GNN-based structure extractors |
| DeProt [41] | | ~18.8 million AlphaFoldDB's 3D structures | Hybrid transformer with a GNN-based structure encoder and disentangled attention to integrate sequence and structural information |
| ESM-3 [10] | | ~3 billion protein sequences, 236 million structures, 539 million annotated proteins | Bidirectional transformer integrating sequence, structure, and function through tokenized multimodal representation with geometric attention |

#### *3.1.1 Encoder–Only PLMs*

Encoder-only PLMs are designed to extract contextual representations of protein sequences, making them particularly well suited for classification tasks in various protein-related tasks. These models use the encoder component of the transformer architecture to process input sequences and capture their complex relationships and features. We describe some of the encoder-only PLMs that are used to generate information-rich protein representation useful in training models for various protein-related downstream tasks, following the summary of each model. The list of encoder-only PLMs described in the chapter is shown in Table 2.

**ProteinBert**

Built upon BERT [12] architecture, ProteinBERT [38] is a PLM trained on a large protein sequence dataset obtained from UniProtKB/UniRef90. The model is based on denoising autoencoder architecture, pretraining on two tasks of bidirectional language modeling of protein sequences and GO (Gene Ontology) annotations simultaneously, extracting both the local-level and global-level representations. The architecture comprises multiple transformer blocks with skip connections and layer normalizations. The model is evaluated on nine benchmark datasets on the tasks of protein structure, posttranslational modifications, and biophysical properties, through fine-tuning methodology. Depending on the target variable of the benchmark tasks, local or global features were used, and supervised fine-tuning has almost outperformed other state-of-the-art models in the respective protein benchmarks. Further, the ProteinBERT model is not dependent on the sequence length and thus generalizes well over different sequence lengths. Compared to other protein language models, ProteinBERT is smaller in size, having only 16 M parameters, making it less resource-extensive for inference.

**ProtTrans Models (ProtElectra, ProtAlbert, ProtBert)**

ProtTrans [23] utilizes high-performance computing to understand the language of life as defined by amino acid sequences, by training language models on billions of protein sequences leading to protein language models. The authors pretrain two auto-regressive models; Transformer-XL and XL-Net, and four auto-encoder models; BERT, Albert, Electra, and T5 on protein sequence datasets of UniRef50, UniRef100, and BFD. The ProtTrans models take as input the tokenized amino acids and pretrain them using either CLM (auto-regressive models) or MLM (auto-encoder models). The model utilizes the transformer architecture and the self-attention mechanism to obtain context-aware representations at both residue or amino acid level and protein or global level. The resulting embeddings, i.e., high-dimensional vector representations, have been shown to capture the biophysical features, protein structural class, and lineage of kingdoms through tSNE projections. Depending on the task, a CNN can be trained on residue-level (embeddings of individual amino acids) for token-level classification such as binding residue prediction. These embeddings can be concatenated and average-pooled to obtain global context-aware embeddings of fixed-length vectors, irrespective of the input sequence length. Afterward, protein-level embeddings can be used to train, e.g., feedforward neural networks for protein-level classifications such as subcellular localization prediction. ProtTrans has been used in diverse bioinformatics applications such as signal peptides classification [42], prediction of antibody structure [43], and glutarylation sites [44]. ProtBERT has been used in predicting intrinsic disorder proteins/regions [45] and protein DNA binding [31, 46].

**ESM (v1/v2)**

The ESM family of models was developed to apply natural language techniques to protein modeling. ESM 1b [47] is a transformer model, trained on masked language modeling objectives. It was trained on UniParc and UniRef databases [16]. Different datasets were developed with the objective of training, measuring generalization, and tuning toward a diverse set of sequences. The models were compared with each other on the metric of exponential cross-entropy, which measured the mean uncertainty of prediction. The ESM 1b architecture features a 34-layer transformer model with 670 million parameters, pretrained on diverse protein sequence datasets. These models were developed to question whether natural language techniques could be transferred to the protein domain. The protein sequences were considered as sentences and amino acids are characters, training in an unsupervised way to obtain latent representations of protein sequences. The learned embeddings thus obtained, are shown to encode biochemical properties, remote homology, and alignment within a protein family, and applied for the prediction of secondary structure and tertiary contacts.

ESM 2 [48] is an enhanced model, where an attention mechanism is used to boost performance. It is a self-attention transformer, trained on a UniRef database in an unsupervised setting. The model was evaluated on the metric of perplexity that measures the prediction in relation to the context. The work establishes a strong link between language modeling perplexity and the accuracy of structural prediction. With a 15B parameter model, it was the largest model to train and showed the ability to capture atomic resolution prediction. Further, it was found that the size of the model increases the ability to capture structural information. The latent representations obtained from ESM 2 were applied to develop ESMfold, which predicts secondary structure faster than AlphaFold, using only protein sequences. These variants of ESM models have been used by other researchers in various applications such as genome-wide disease variant effects [49], cleavage site prediction [50, 51], predicting intrinsically disordered prediction [52], protein-protein interaction (PPI) [53], and secondary structure prediction [30].

**S-PLM**

The S-PLM [39] was trained on a protein dataset from the Swiss-Prot library, with structural data derived from AlphaFold2-predicted 3D structures represented as Cα contact maps. The architecture integrates two encoders: a transformer-based sequence encoder adapted from ESM 2 for sequence embeddings and a

ResNet50-based structure encoder for contact map embeddings. While S-PLM incorporates both sequence and 3D structure information, it primarily emphasizes the benefits of encoding protein sequences and leverages structural data to enhance sequence-based prediction tasks. The training of S-PLM is based on a multi-view contrastive learning framework, aligning sequence and structure embeddings of the same protein. This approach employs a contrastive loss function based on normalized temperature-scaled cross-entropy (NT-Xent). The model has been applied to tasks such as protein function annotation, where the alignment of sequence and structure information improves the prediction accuracy of active protein sites and functional roles.

**SaProt**

The SaProt [26] uses a dataset of ~40 million protein sequences and their corresponding AlphaFold2-predicted 3D protein structures. This data was processed using Foldseek [54] to encode structural information into 3Di tokens, combined with residue tokens to create a novel structure-aware (SA) vocabulary. SAPROT uses the ESM-2 architecture with an expanded embedding layer to handle the SA tokens, enabling it to incorporate geometric protein structure data alongside sequence information. The model was trained using a masked language modeling (MLM) objective, focusing on predicting residue tokens while leveraging structure information as context. The training involved masking 15% of SA tokens per batch. This approach enabled SaProt to excel in tasks requiring comprehensive protein representation, including structure and function prediction. The SaProt model has been applied in secondary structure prediction, protein cleavage site prediction, and protein-protein interaction (PPI) prediction [55, 56].

**PST**

The protein structure transformer (PST) [40] was trained on a curated subset of AlphaFold's database containing ~500 K predicted protein structures. It builds on the ESM-2 language model which is used to extract embeddings for each protein sequence. Subsequently, the protein genome position and coding strand information are concatenated with ESM-2 embeddings and passed through a PST encoder that consists of multi-head attention layers. In contrast to the commonly used training objective in the masked language model (MLM), this uses triplet loss, conditioned on the relationship between proteins and genomes for improved capture of patterns in genetic diversity. The triplet loss objective ensures that the positive genome is closer to the anchor genome than the negative genome, maintaining a separation defined by a tunable margin. Finally, the PST embeddings are obtained that are further used as protein representation which is used to train downstream classification models.

**DeProt**

Disentangled protein sequence-structure model (DeProt) [41] was pretrained on ~19 million protein sequences and high-quality protein structures from AlphaFoldDB, filtered for quality using pLDDT scores, and clustered to reduce redundancy. It incorporates both amino acid information and 3D structure information using techniques called structure quantization and sequence-structure disentangled attention. Specifically, 3D protein structures are represented using a graph $G = (V, E)$, where $V$ and $E$ represent the residue-level nodes and edges, respectively. The features of the edges include the angles and distances between two residues, as well as positional encoding. Structure quantization transforms the graph representation of protein structures into discrete tokens for better integration with sequence data. The disentangled attention mechanism explicitly models the relationships between residues, structures, and their relative positions. Pretraining employed a masked language modeling objective with structure-conditioned masking, enabling the model to predict masked residues using both sequence and structural context.

**ESM-3**

Evolutionary scale modeling-3 (ESM-3) [10] is a multimodal generative language model trained on protein sequences and structures enabling the design of novel proteins. It uses a bidirectional transformer architecture with up to 98 billion parameters and integrates discrete token representations of sequence, structure, and function into a shared latent space. Trained on 3.15 billion sequences, 236 million structures, and 539 million annotated proteins (totaling 771 billion unique tokens), ESM-3 uses a generative MLM objective with variable masking rates to balance representation learning and generative capability. Its structural tokenization and geometric attention mechanisms allow efficient processing of 3D protein structures.

#### *3.1.2 Encoder–Decoder PLMs*

Encoder–decoder PLMs are designed to solve sequence-to-sequence tasks in protein biology. These models excel in various tasks, such as protein sequence generation and engineering, by combining an encoder that transforms input protein sequences into contextual representations with a decoder that generates corresponding output sequences. The encoder component of the encoder–decoder PLMs is often used for protein representation (aka as a feature extractor) and the decoder component can be used for protein generation. This architecture is used by models like ProtT5 and ProstT5, to capture the complex relationships in protein sequences/structures. The list of encoder–decoder PLMs is presented in Table 3.

**Table 3**
**Summary of encoder–decoder PLMs with training datasets and architecture details**

| PLM | Dataset | Model architecture |
|---|---|---|
| ProtT5 [Variants: T5-XL and T5-XXL] [23] | BFD, UniRef50 | T5-XL and T5-XXL model, 24-layer encoder–decoder, ~3B parameters |
| ProteinLM [57] | PFAM | BERT-based model (48 transformer layers) |
| CaLM [58] | Pretrained on ~9 million nonredundant coding sequences from whole-genome sequencing data | Transformer 12 encoder layers, 12 attention heads) |
| Ankh [24] | T5 architecture (similar to ProtT5 framework with hyperparameter tuning) | UniRef50 |
| ProstT5 [59] | 17 million sequences, AlphaFold predicted 3D structures | T5 (fine-tuned ProtT5 for both sequence-structure translations) |
| ss-PLM [60] | UniRef50 (only 1.7 Million Sequences) | Encoder–decoder transformer |
| ProteinDT [61] | SwissProtCLAP with 441K text-protein pairs from SwissProt | 3 component architecture: ProteinCLAP (contrastive learning for aligning text and protein representations), ProteinFacilitator (generates protein sequence representations from text), and a decoder (produces protein sequences from representations) |
| LM-Design [62] | CATH 4.2/4.3, AlphaFold-2 predicted structures | Enhanced PLMs using a lightweight structural adapter<br>Integrates geometric information from structure encoders like ProteinMPNN |
| xTrimoPGLM [63] | ~940 million unique protein sequences from multiple sources | General language model (GLM) Architecture |

### ProtT5 (ProtTrans)

ProtT5-XL-UniRef50 [23] is based on Google's T5 [14] architecture, specifically the T5-3B architecture, which consists of a 24-layer encoder–decoder with around 2.8 billion trainable parameters. Unlike the original T5, which uses a span denoising objective, ProtT5 is pretrained using a BERT-like MLM objective, where ~15% of amino acids are randomly masked in the input sequences given to the encoder and the decoder is asked to output the non-corrupted counterpart. This approach allows the model to predict the missing tokens, enhancing its ability to learn the biophysical and fundamental characteristics of the protein sequences. It is pretrained in a self-supervised fashion on two large datasets: the

Big Fantastic Database (BFD) and UniRef50, the latter comprising about 45 million protein sequences. The model has been extensively used for embedding extractions that have been effective in multiple downstream tasks capturing key biophysical properties in protein sequences such as PTM prediction [33, 34, 37], secondary structure prediction [30], and DNA-binding protein prediction [31, 64, 65].

**ProteinLM**

ProteinLM [57] is a large-scale PLM that captures the complex evolutionary and biological patterns from the protein sequences. The model is pretrained on the BERT architecture, specifically adapting its masked language modeling (MLM) approach to proteins with slight modifications to its loss function and model structure. Unlike BERT, ProteinLM does not employ next-sentence prediction (NSP), inferring that protein sequences lack the semantic continuity characteristic of natural language. The training data is derived from the PFAM dataset, which consists of over 32 million protein sequences clustered into evolutionarily related families. It is trained unsupervised by randomly masking 15% of amino acid tokens and training the model to predict these masked tokens, focusing solely on the MLM loss. For efficient training at this scale, ProteinLM utilizes the Megatron-LM framework, optimizing the model with techniques like model parallelism and memory optimization. ProteinLM demonstrates its capacity to capture local and global sequence dependencies, achieving superior performance in various downstream tasks, such as secondary structure prediction, remote homology detection, contact prediction, fluorescence prediction, and stability assessment.

**CaLM**

Codon adaptation language model (CaLM) [58] is built upon the ESM-2 framework that uses an architecture comprising 12 encoder layers, each with 12 attention heads, resulting in a model with 86 million parameters. Unlike other protein language models that utilize amino acid sequences, CaLM is trained on codon sequences (cDNA) derived from approximately nine million nonredundant coding sequences obtained through whole-genome sequencing. The model uses a masked language modeling (MLM) objective during pretraining with a denoising objective that enables CaLM to learn contextual representations of codons, capturing patterns of synonymous codon usage and the underlying biochemical features of codon sequences. Despite its relatively smaller parameter count compared to larger models like ESMs or ProtTrans, CaLM's focus on codon sequences tends to provide a unique advantage in capturing evolutionary signals from codon-level representation that are not present in amino acid-based models. Recently, CaLMPhosKAN

[66], a phosphorylation site prediction tool that has been developed by integrating CaLM along with ProtT5, has been proposed to enhance phosphorylation site prediction.

**Ankh**

Ankh [24] is an optimized PLM for protein sequence modeling, based on an encoder–decoder transformer architecture developed for machine translation. Unlike the recent trends of training models, which focus mostly on increasing the model size, Ankh focuses on protein-specific optimizations to improve performance efficiently. The model has two variants: Ankh_base with 450 million parameters and a 768-dimensional embedding space, and Ankh_large, with 1.15 billion parameters and a 1536-dimensional embedding space. It has an optimized architecture with a 48-layer encoder and a 24-layer decoder, optimized for various tasks of protein modeling. Based on the ProtT5 [23] framework, Ankh achieves competitive performance on most benchmarks for protein structure, function, and generation with drastically reduced computational requirements. This design prioritizes both performance and accessibility, making Ankh a versatile model for diverse downstream protein tasks such as PTM prediction [34] and protein solubility prediction [67].

**ProstT5**

ProstT5 [59] extends ProtT5, a model that was originally trained using only amino acid sequences. ProstT5 includes both protein sequence-amino acids and protein structure information, encoded as 3Di tokens [54]. The dataset for fine-tuning consists of 17 million high-quality, nonredundant protein sequences with predicted 3D structures from AlphaFoldDB [68], which were converted into 1D 3Di-token representations using Foldseek [54]. The training objective includes both the original span-based denoising objective on both sequences (amino acids and 3Di) to prevent catastrophic forgetting and a bi-directional translation task, where the model learns to translate between amino acid sequences and 3Di structure representations. This training technique is applied to capture rich features from both protein sequences and their 3D structural representations. ProstT5 excels in downstream protein tasks such as protein-binding site prediction [69], PTM prediction [70], and peptide bioactivity predictors [71].

**SS-PLM**

The Small-scale protein language model (SS-PLM) [60] is a comparatively small encoder–decoder architecture trained with self-attention mechanisms to capture long-range dependencies within protein sequences. The model is trained based on the causal language model approach with data corruption where the objective is to predict the next amino acid in a sequence. The SS-PLM was first pretrained on a subset of the UniRef50 dataset, specifically selected to focus on enzymatic sequences, reducing the dataset size to 1.7 million representative sequences from 45 million sequences. These major contributions of SS-PLM are aiming to maximize performance while reducing computational overhead, by optimizing training and architectural design considerations, making protein language modeling more accessible and efficient.

**ProteinDT**

ProteinDT [61] is a multimodal framework that bridges the gap between textual descriptions and protein design, incorporating both textual and protein sequence data. The architecture consists of three key components: ProteinCLAP, ProteinFacilitator, and a decoder. ProteinCLAP is the pretraining component that uses contrastive learning to align the representation spaces of text and protein sequences. It uses the SwissProtCLAP dataset that contains text and protein pairs which is the foundation for training the model by ensuring that both modalities (text and protein sequences) are represented in a shared latent space. Second, ProteinFacilitator generates protein sequence representations from text input, enabling the third component (i.e., decoder) to generate the corresponding protein sequences. The model is trained to maximize the alignment between textual descriptions of protein functions and the generated protein sequences, improving the framework's ability to perform text-guided protein generation.

**LM-Design**

LM-Design [62] transforms existing protein language models (PLMs) like ESMs into protein design tools by incorporating a lightweight structural adapter that introduces structural awareness into the sequence-based model. Using a framework called LM-Design, the adapter integrates geometric information from external structure encoders such as ProteinMPNN to utilize both sequential and structural knowledge effectively. It uses conditional masked language modeling (CMLM) training objectives where random amino acid residues are masked and the model learns to predict them using the surrounding sequence context and structure information. Multiple datasets including CATH 4.2/4.3 benchmarks are used to evaluate and compare the performance against other tools.

**xTrimoPGLM**

xTrimoPGLM [63] is a large protein language model built on the general language model (GLM) architecture, which uniquely combines bidirectional and autoregressive modeling capabilities to address protein understanding and generation tasks. It is trained using a dual approach incorporating the MLM and the GLM objectives that enable simultaneous optimization of token prediction in masked sequences and autoregressive sequence generation. The model was pretrained on a large dataset of ~940 million unique protein sequences containing around 200 billion residues, using ~100 billion parameter model processing over 1 trillion tokens. The overall training process was structured in two phases: an initial pretraining with the MLM objective to boost understanding capabilities, followed by a merged training phase with a 20% MLM and 80% GLM ratio to refine representation and generative abilities. Despite xTrimoPGLM having great potential to better represent proteins, its massive scale of 100 billion parameters often poses challenges for inference, particularly for small-scale downstream tasks in resource-constrained settings.

### *3.1.3 Decoder-Only PLMs*

Pure decoder-only models, like ProtGPT2 [72], are designed for autoregressive tasks, focusing on predicting sequences from prior inputs. Models built using only the decoder component predict the next amino acid in a sequence given all the previous ones. Decoder-only PLMs have been shown to excel in de novo protein sequence generation and further protein engineering and drug discovery by designing novel functional proteins. The list of decoder-only PLMs described in this chapter are presented in Table 4.

**ProGen2**

ProGen2 [73] is an autoregressive transformer model for protein design which is trained using next-token prediction language modeling learning objective. The architecture is a decoder-only transformer with left-to-right causal masking with rotary positional encoding. A mixture of Uniref90 and BFD30 databases are trained. It has small-to-large versions ranging from 1.5 M to 6.5B parameters. ProGen2 is used in various protein sequence design applications [75].

**Table 4**
**Summary of decoder-only PLMs with training datasets, and architecture details**

| PLM | Dataset | Model architecture |
|---|---|---|
| ProGen2 [73] | Uniref90, BFD30 | Decoder-only transformer with next-token prediction and left-to-right causal masking, using rotary positional encoding |
| ProtGPT2 [72] | UniRef50 | Decoder-only transformer (36 layers) |
| ProLLaMA [74] | UniRef50, InterPro (property texts from) | LLaMA2 Multitask PLP with instruction tuning |
| ProseLM [75] | CATH 4.2, SAbDab | Causal encoder for structural context + ProGen2 for structure-conditioned protein sequence design |

**ProtGPT2**

ProtGPT2 [72] is a decoder-only protein language model, closely resembling the GPT-2 large model with 36 layers. The model consists of 738 million parameters trained on the UniRef50 dataset using a self-attention mechanism optimized for generating protein sequences in an autoregressive manner. For tokenization, ProtGPT2 employs Byte Pair Encoding (BPE), creating a vocabulary of 50,256 tokens, where each token represents an average of four amino acids. This large-scale training setup is considered to enable ProtGPT2 to learn the underlying structure of protein sequences efficiently, positioning it to generate novel, diverse, and potentially functional protein sequences.

**ProseLM**

ProseLM [75] is a protein sequence design model that adapts protein language models with structural and functional context. Built on ProGen2, it integrates structural information using parameter-efficient adapters to combine embeddings from a causal encoder with each layer of the pretrained language model. The causal encoder extracts structural features and atomic context graphs to predict sequences based on protein backbones. ProseLM was trained on diverse datasets, including single-chain proteins, multi-chain complexes, and antibody structures, with frozen language model parameters updating the adapters and casual encoder.

**ProLLaMA**

Using a two-stage training approach, the ProLLaMA [74] framework introduces a protein large language model (ProLLM) for multitask protein language processing (PLP) by adapting general LLMs, such as LLaMA2 [76]. The model is trained using causal language modeling on the UniRef50 dataset and protein property texts from InterPro, with continual learning to instruct protein language while preserving natural language capabilities. ProLLaMA uses Low-Rank Adaptation (LoRA) to train only a fraction of the parameters, reducing computational costs and preventing catastrophic forgetting. The architecture integrates LoRA adapters into LLaMA2's decoder layers to effectively distinguish protein from natural language.

## 4 Application of PLMs

Protein language models (PLMs) have become essential tools in protein bioinformatics, demonstrating exceptional efficiency across various protein-related applications. These models can mainly be used for two major tasks: protein representation and protein design/engineering. Encoder-only and encoder–decoder models are particularly effective at extracting protein representations, which are crucial for classification tasks. These models have shown high accuracy in tasks such as structure prediction, disorder region identification, posttranslational modification (PTM) prediction, protein-ligand binding site identification, protein-protein interaction prediction, and cleavage site prediction. Furthermore, encoder–decoder and decoder-only PLMs are highly suited for applications in protein engineering and design [77]. Specifically, encoder–decoder models excel in sequence-to-sequence prediction tasks, such as designing protein variants or aligning sequences to specific structural templates whereas decoder-only models are highly effective in autoregressive tasks such as generation of novel protein sequences [78]. These capabilities are critical for designing de novo proteins with desired properties. Please refer to the work of Ruffolo and Madani [79] for review on designing proteins using language models. Some downstream protein bioinformatics tasks and the corresponding tools that leverage PLMs for these tasks are presented in Table 5.

Similarly, Table 6 highlights some of the exciting application of protein language models (PLMs) in protein engineering and design. These models leverage sequence and structure-based protein language models in tasks such as designing novel protein sequences, optimizing structures for specific functions, and solving inverse folding challenges.

**Table 5**
**Overview of recent applications of protein language models (PLMs) for various downstream tasks**

| Task | Applications |
|---|---|
| PTM prediction | LMSuccSite [33], LM-OGlcNAc-Site [34], PLMSNOSite [80], LMCrot [37], LMNGlyPred [81], LMPhosSite [35], CalmPhosKAN [66], DeepPTM, PTMGPT2 [82] |
| Nucleic acid protein-binding site prediction/DNA-binding sites prediction | PDNAPred [64], CLAPE [46], bindEmbed21 [83], DeepProSite[84], EquiPNAS [85], CrossBind [86] |
| DNA-binding protein prediction | PLM-DBPs [31], PreDBP-PLMs [32], ESM-DBP [87] |
| Protein-protein interaction (PPI) | ESMDNN-PPI [88], KSGPPI [89] |
| Protein function annotation/prediction | DeepGO [90], DeepFRi [91], GAT-GO [92], SPROF-GO [93] |
| Protein cleavage site prediction | DeepNeuropePred [50] |
| Disorder region prediction | SETH [94], IDP-ELM [95], DisPredict3.0 [52], IDP-LM [45] |
| Protein structure prediction | ESMFold [25], AlphaFold [68], OmegaFold [96], RoseTTA Fold [97], EMBER2 [98] HelixFold-Single [99], RGN2 [100] |

**Table 6**
**Recent applications of protein language models (PLMs) in protein engineering and design**

| Task | Application |
|---|---|
| Protein engineering/design (sequence or structure) | RFDiffusion [101], Prot-VAE [78], ProGen [102] Verkuil et al. [103] |
| Protein sequence prediction (inverse folding) | ESM-IF1 [104], ProteinMPNN [105], MIF-ST [106], ESM-Design [107], EvoDiff [108] |

## 5 Conclusion and Discussion

The last couple of years has seen tremendous advancements in the field of protein language models. The two most important uses of PLMs have been the protein understanding task aka in the downstream classification task and protein generation task—aka engineering and design of proteins/antibodies/biologics. Although there has been a tremendous improvement in the field of protein language models based on protein sequences, the same is not yet true for protein language models based on protein structures. With the availability of predicted structures for all the proteins, courtesy of AlphaFold, we are likely to see more developments in protein-structure-based language models. Moreover, the field of LLMs in NLP is developing at an alarming rate and these advances are likely to be incorporated into the PLMs. One challenge in the field has been the lack of interpretability of the embeddings generated from these models. Below we summarize some emerging trends in the field.

### 5.1 (Scaling) Small Protein Language Models

There is no doubt that PLMs have demonstrated exceptional capabilities in understanding contextual relationships, outperforming traditional approaches in downstream tasks in protein bioinformatics. More recently, following the same trend as in natural language processing (NLP), protein language models have continuously been scaled up. One thing to note is that the training of these PLMs poses a significant challenge. In this regard, there have been some studies to explore small-scale protein language models [60] which perform similarly to the larger models that are computationally more demanding. Additionally, there are also efforts [109] to make an efficient codebase, design a modern architecture, and remove data quality concerns. Amplify [109] is a newer model which is less expensive to train and deploy than other existing protein language models.

### 5.2 Multimodal Protein Language Models

As discussed above, most of the PLMs are trained on protein sequences. One new trend that has been observed in the field is the development of multimodal protein language models trained on multiple modes of information (protein sequence, structure, text, etc.). ProteinChat [110] is a multimodal protein language model that is trained on protein sequences and text. Unlike existing PLMs, this PLM takes an amino acid sequence and a (text) prompt as inputs, and generates a detailed textual prediction of protein function, unlike discrete categories as in GO. Additionally, the PST protein structure transformer is a new PLM [40] that is trained on protein structure databases. One thing to note here is that somehow the field is yet to come up with an efficient approach to tokenize structure as protein structure language models did not improve over PLMs to an extent that would amortize the computational overhead of having to predict structures first.

### 5.3 Enhancing Efficient Training of PLM

There have also been some works in the field to deal with difficulties in training a PLM in problems that lack large amounts of data. One of the recent works, Few-Shot Learning for Protein Fitness Prediction (FSFP) [111] uses a training strategy that can effectively optimize protein language models for protein fitness prediction under these data scarcity conditions by leveraging meta-transfer learning, learning to rank and parameter efficient fine-tuning. Additionally, ESM-DBP [87] uses domain-adaptive pretraining of the general PLM and DNA-binding protein repertoire to improve DNA-binding protein prediction.

### 5.4 Emergence of Domain-Specific PLM

In addition to general purpose PLM (a language model trained on a vast dataset of protein sequences, allowing it to analyze and predict various properties of proteins across a wide range of protein families, without being specifically optimized for a single protein type or function; essentially, it acts as a versatile tool for understanding and manipulating protein sequences in diverse biological contexts, similar to how a general purpose natural language model can handle various text-based tasks), one of the other trends that have been observed in the field is the development of domain-specific PLM. DR-BERT is a domain-specific PLM that is trained to predict disorder regions [112]. IgLM [113] is a generative immunoglobulin language model to generate/design antibodies. Another example is the several PLMs trained on immune-specific protein sequences (such as AbLang, Antibert, and AbMAP). Additionally, there have been studies also to assess the usefulness of domain-specific PLMs vs. general purpose PLMs [114]. At least, this study shows that it is not always true that "domain-specific models outperform general models" and also highlights the importance of selecting the appropriate layer for extracting embeddings in optimizing performance on downstream tasks.

### 5.5 PLM Trained on the Reduced Alphabet

Reduced amino acid alphabets have been of interest in the field due to their ability to simplify the protein sequence space. Ieremie et al. [115] explored the training of PLM on reduced amino acid alphabets, and as expected the PLM trained on reduced amino acid alphabets does not perform as well as amino acid alphabets due to loss of protein sequence information and due to the subsequent loss on learned representation.

### 5.6 Advanced Use of PLMs for Protein Design—AI Agents

Designing de novo proteins beyond those found in nature holds significant promise for advancements in both scientific and engineering applications. The use of emerging themes like AI agents and others has started to show their usage in the field of protein bioinformatics. ProtAgents [116] is a platform for protein design based on LLMs where multiple AI agents with distinct capabilities collaboratively address information retrieval, protein structure analysis, physics-based simulation, and results analysis.

### 5.7 Fine-Tuning and Parameter Efficient Fine-Tuning of PLMs

Until recently, most protein language model-based protein predictions just use raw embeddings from the protein language model and do not back-propagate to the language model. In NLP, fine-tuning large language models in a supervised manner to adapt to the subsequent downstream task has become the de facto standard [30]. However, the computational resources needed for fine-tuning large PLMS pose a significant challenge. Borrowing from NLP, we are observing approaches for efficient fine-tuning like parameter-efficient fine-tuning (PEFT) approaches like LoRA, DoRA, qLORA, etc. Recently, Bonnie Berger's group at MIT [117] proposed PEFT approaches for the prediction of protein-protein interaction and homo-oligomer symmetry. Similarly, Rost's group also proposed PEFT on several downstream tasks [30]. One of the main observations of these approaches is that task-specific fine-tuning is likely to improve the downstream performance and the required computational resources are always fewer than full fine-tuning.

### 5.8 Prompt-Based Fine-Tuning of PLM

We have started seeing the application of prompt-based fine-tuning of PLMs. PTMGPT2 [118] is a prompt-based fine-tuning of ProtGPT2 [72] (a GPT2 model based on protein sequences) model that fine-tunes ProtGTP2 for generating classification labels for PTMs.

**Acknowledgments** This work is supported, in part, by funds from the National Science Foundation (NSF: # 1901793 and # 2210356) and faculty start-up grant to D.B.K.

## References


1. Ezzell C (2002) Proteins rule. Sci Am 286(4): 40–47
2. Young VR, Steffee WP, Pencharz PB, Winterer JC, Scrimshaw NS (1975) Total human body protein synthesis in relation to protein requirements at various ages. Nature 253(5488):192–194. https://doi.org/10.1038/253192a0
3. Branden CI, Tooze J (2012) Introduction to Protein Structure, 2nd edn. Garland Science, New York. https://doi.org/10.1201/9781136969898
4. Heinzinger M et al (2019) Modeling aspects of the language of life through transfer-learning protein sequences. BMC Bioinformatics 20(1):723. https://doi.org/10.1186/s12859-019-3220-8
5. Ofer D, Brandes N, Linial M (2021) The language of proteins: NLP, machine learning & protein sequences. Comput Struct Biotechnol J 19:1750–1758. https://doi.org/10.1016/j.csbj.2021.03.022
6. Pakhrin SC, Pokharel S, Saigo H, KC DB (2022) Deep learning–based advances in protein posttranslational modification site and protein cleavage prediction', in computational methods for predicting post-translational modification sites. In: KC DB (ed) Methods in molecular biology. Springer US, New York, NY, pp 285–322. https://doi.org/10.1007/978-1-0716-2317-6_15
7. Pokharel S, Sidorov E, Caragea D, Dukka BKC (2022) 'NLP-based encoding techniques for prediction of post-translational modification sites and protein functions', in machine learning in bioinformatics of protein sequences. World Sci:81–127. https://doi.org/10.1142/9789811258589_0004
8. Zhao WX et al (2024) A survey of large language models. arXiv:2303.18223. https://doi.org/10.48550/arXiv.2303.18223
9. Zhang S, Fan R, Liu Y, Chen S, Liu Q, Zeng W (2023) Applications of transformer-based language models in bioinformatics: a survey. Bioinforma Adv 3(1):vbad001. https://doi.org/10.1093/bioadv/vbad001
10. Hayes T et al (2024) Simulating 500 million years of evolution with a language model. bioRxiv. https://doi.org/10.1101/2024.07.01.600583
11. Vaswani A et al (2017) Attention is all you need. In: Advances in Neural Information Processing Systems. Curran Associates, Inc. Accessed: Jul. 01, 2023. [Online]. Available: https://proceedings.neurips.cc/paper_files/paper/2017/hash/3f5ee243547dee91fbd053c1c4a845aa-Abstract.html
12. Devlin J, Chang M-W, Lee K, Toutanova K (2019) BERT: pre-training of deep bidirectional transformers for language understanding. arXiv:1810.04805. https://doi.org/10.48550/arXiv.1810.04805
13. Radford A (2018) Improving language understanding by generative pre-training. Accessed: Sep. 04, 2024. [Online]. Available: https://hayate-lab.com/wp-content/uploads/2023/05/43372bfa750340059ad87ac8e538c53b.pdf
14. Raffel C et al (2020) Exploring the limits of transfer learning with a unified text-to-text transformer. J Mach Learn Res 21(1): 5485–5551
15. Lin T, Wang Y, Liu X, Qiu X (2022) A survey of transformers. AI Open 3:111–132. https://doi.org/10.1016/j.aiopen.2022.10.001
16. Suzek BE, Huang H, McGarvey P, Mazumder R, Wu CH (2007) UniRef: comprehensive and non-redundant UniProt reference clusters. Bioinformatics 23(10): 1282–1288. https://doi.org/10.1093/bioinformatics/btm098
17. Brown TB et al (2020) Language models are few-shot learners. arXiv:2005.14165. https://doi.org/10.48550/arXiv.2005.14165
18. Grattafiori A et al (2024) The Llama 3 herd of models. arXiv:2407.21783. https://doi.org/10.48550/arXiv.2407.21783
19. Sarumi OA, Heider D (2024) Large language models and their applications in bioinformatics. Comput Struct Biotechnol J 23:3498–3505. https://doi.org/10.1016/j.csbj.2024.09.031
20. Asgari E, Mofrad MRK (2016) ProtVec: a continuous distributed representation of biological sequences. arXiv:1503.05140. https://doi.org/10.48550/arXiv.1503.05140
21. Bepler T, Berger B (2019) Learning protein sequence embeddings using information from structure. arXiv:1902.08661. https://doi.org/10.48550/arXiv.1902.08661
22. Rao R, Meier J, Sercu T, Ovchinnikov S, Rives A (2020) Transformer protein language models are unsupervised structure learners. bioRxiv. https://doi.org/10.1101/2020.12.15.422761
23. Elnaggar A et al (2021) ProtTrans: towards cracking the language of lifes code through

self-supervised deep learning and high performance computing. IEEE Trans Pattern Anal MachIntell. PP. https://doi.org/10.1109/TPAMI.2021.3095381

24. Elnaggar A et al (2023) Ankh♀: optimized protein language model unlocks general-purpose modelling. bioRxiv:2023–2001
25. Lin Z et al (2022) Language models of protein sequences at the scale of evolution enable accurate structure prediction. bioRxiv. https://doi.org/10.1101/2022.07.20.500902
26. Su J, Han C, Zhou Y, Shan J, Zhou X, Yuan F (2023) SaProt: protein language modeling with structure-aware vocabulary. bioRxiv. https://doi.org/10.1101/2023.10.01.560349
27. Unsal S, Atas H, Albayrak M, Turhan K, Acar AC, Doğan T (2022) Learning functional properties of proteins with language models. Nat Mach Intell 4(3):3. https://doi.org/10.1038/s42256-022-00457-9
28. Zhang Z et al (2024) Protein language models learn evolutionary statistics of interacting sequence motifs. bioRxiv. https://doi.org/10.1101/2024.01.30.577970
29. Alkuhlani A, Gad W, Roushdy M, Voskoglou MG, Salem AM (2022) PTG-PLM: predicting post-translational glycosylation and glycation sites using protein language models and deep learning. Axioms 11(9):9. https://doi.org/10.3390/axioms11090469
30. Schmirler R, Heinzinger M, Rost B (2023) Fine-tuning protein language models boosts predictions across diverse tasks. bioRxiv. https://doi.org/10.1101/2023.12.13.571462
31. Pokharel S, Barasa K, Pratyush P, Dukka KC (2024) PLM-DBPs: Enhanced DNA-Binding Protein Prediction in Plants Using Embeddings From Protein Language Models. bioRxiv:2024, 10.04.616755. https://doi.org/10.1101/2024.10.04.616755
32. Qi D, Song C, Liu T (2024) PreDBP-PLMs: Prediction of DNA-binding proteins based on pre-trained protein language models and convolutional neural networks. Anal Biochem 694:115603. https://doi.org/10.1016/j.ab.2024.115603
33. Pokharel S, Pratyush P, Heinzinger M, Newman RH, Kc DB (2022) Improving protein succinylation sites prediction using embeddings from protein language model. Sci Rep 12(1):16933. https://doi.org/10.1038/s41598-022-21366-2
34. Pokharel S, Pratyush P, Ismail HD, Ma J, Dukka BKC (2023) Integrating embeddings from multiple protein language models to improve protein O-GlcNAc site prediction. Int J Mol Sci 24(21). https://doi.org/10.3390/ijms242116000
35. Pakhrin SC, Pokharel S, Pratyush P, Chaudhari M, Ismail HD, Dukka BKC (2023) LMPhosSite: a deep learning-based approach for general protein phosphorylation site prediction using embeddings from the local window sequence and pretrained protein language model. J Proteome Res 22(8): 2548–2557. https://doi.org/10.1021/acs.jproteome.2c00667
36. Jahn LR, Marquet C, Heinzinger M, Rost B (2024) Protein embeddings predict binding residues in disordered regions. Sci Rep 14(1): 13566. https://doi.org/10.1038/s41598-024-64211-4
37. Pratyush P, Bahmani S, Pokharel S, Ismail HD, KC DB (2024) LMCrot: an enhanced protein crotonylation site predictor by leveraging an interpretable window-level embedding from a transformer-based protein language model. Bioinformatics 40, no. 5: btae290. https://doi.org/10.1093/bioinformatics/btae290
38. Brandes N, Ofer D, Peleg Y, Rappoport N, Linial M (2022) ProteinBERT: a universal deep-learning model of protein sequence and function. Bioinforma Oxf Engl 38(8): 2102–2110. https://doi.org/10.1093/bioinformatics/btac020
39. Wang D, Abbas UL, Shao Q, Chen J, Xu D (2023) S-PLM: structure-aware protein language model via contrastive learning between sequence and structure. bioRxiv:2023.08.06.552203. https://doi.org/10.1101/2023.08.06.552203
40. Chen D, Hartout P, Pellizzoni P, Oliver C, Borgwardt K (2024) Endowing protein language models with structural knowledge. arXiv:arXiv:2401.14819. https://doi.org/10.48550/arXiv.2401.14819
41. Li M et al (2024) DeProt: Protein language modeling with quantizied structure and disentangled attention. bioRxiv. https://doi.org/10.1101/2024.04.15.589672
42. Teufel F et al (2022) SignalP 6.0 predicts all five types of signal peptides using protein language models. Nat Biotechnol 40(7): 1023–1025. https://doi.org/10.1038/s41587-021-01156-3
43. Kenlay H, Dreyer FA, Kovaltsuk A, Miketa D, Pires D, Deane CM (2024) Large scale paired antibody language models. arXiv:2403.17889. https://doi.org/10.48550/arXiv.2403.17889

44. Indriani F, Mahmudah KR, Purnama B, Satou K (2022) ProtTrans-glutar: incorporating features from pre-trained transformer-based models for predicting glutarylation sites. Front Genet 13:885929. https://doi.org/10.3389/fgene.2022.885929
45. Pang Y, Liu B (2023) IDP-LM: Prediction of protein intrinsic disorder and disorder functions based on language models. PLOS Comput Biol 19(11):e1011657. https://doi.org/10.1371/journal.pcbi.1011657
46. Liu Y, Tian B (2024) Protein–DNA binding sites prediction based on pre-trained protein language model and contrastive learning. Brief Bioinform 25, no. 1:bbad488. https://doi.org/10.1093/bib/bbad488
47. Rives A et al (2019) Biological structure and function emerge from scaling unsupervised learning to 250 million protein sequences. Synthetic Biol Preprint. https://doi.org/10.1101/622803
48. Lin Z et al (2023) Evolutionary-scale prediction of atomic-level protein structure with a language model. Science 379(6637):1123–1130. https://doi.org/10.1126/science.ade2574
49. Brandes N, Goldman G, Wang CH, Ye CJ, Ntranos V (2023) Genome-wide prediction of disease variant effects with a deep protein language model. Nat Genet 55(9):1512–1522. https://doi.org/10.1038/s41588-023-01465-0
50. Wang L, Zeng Z, Xue Z, Wang Y (2023) DeepNeuropePred: a robust and universal tool to predict cleavage sites from neuropeptide precursors by protein language model. bioRxiv. https://doi.org/10.1101/2023.07.07.547760
51. Teufel F et al (2023) DeepPeptide predicts cleaved peptides in proteins using conditional random fields. Bioinformatics 39(10):btad616. https://doi.org/10.1093/bioinformatics/btad616
52. Kabir MWU, Hoque MT (2024) DisPredict3.0: prediction of intrinsically disordered regions/proteins using protein language model. Appl Math Comput 472:128630. https://doi.org/10.1016/j.amc.2024.128630
53. Hou Z, Yang Y, Ma Z, Wong K-C, Li X (2023) Learning the protein language of proteome-wide protein-protein binding sites via explainable ensemble deep learning. Commun Biol 6:73. https://doi.org/10.1038/s42003-023-04462-5
54. van Kempen M et al (2024) Fast and accurate protein structure search with Foldseek. Nat Biotechnol 42(2):243–246. https://doi.org/10.1038/s41587-023-01773-0
55. Li F-Z, Amini AP, Yue Y, Yang KK, Lu AX (2024) Feature reuse and scaling: understanding transfer learning with protein language models. bioRxiv. https://doi.org/10.1101/2024.02.05.578959
56. Xiong D et al (2024) A structurally informed human protein-protein interactome reveals proteome-wide perturbations caused by disease mutations. Nat Biotechnol. https://doi.org/10.1038/s41587-024-02428-4
57. Xiao Y, Qiu J, Li Z, Hsieh C-Y, Tang J (2021) Modeling protein using large-scale pretrain language model. ArXiv Prepr: ArXiv210807435
58. Outeiral C, Deane CM (2024) Codon language embeddings provide strong signals for use in protein engineering. Nat Mach Intell 6(2):170–179. https://doi.org/10.1038/s42256-024-00791-0
59. Heinzinger M, Weissenow K, Sanchez JG, Henkel A, Steinegger M, Rost B (2023) ProstT5: Bilingual language model for protein sequence and structure. bioRxiv. https://doi.org/10.1101/2023.07.23.550085
60. Serrano Y, Roda S, Guallar V, Molina A (2023) Efficient and accurate sequence generation with small-scale protein language models. bioRxiv. https://doi.org/10.1101/2023.08.04.551626
61. Liu S et al (2024) A text-guided protein design framework. arXiv:2302.04611. https://doi.org/10.48550/arXiv.2302.04611
62. Zheng Z, Deng Y, Xue D, Zhou Y, YE F, Gu Q (2023) Structure-informed language models are protein designers. arXiv:2302.01649. https://doi.org/10.48550/arXiv.2302.01649
63. Chen B et al (2024) xTrimoPGLM: unified 100B-scale pre-trained transformer for deciphering the language of protein. bioRxiv. https://doi.org/10.1101/2023.07.05.547496
64. Zhang L, Liu T (2024) PDNAPred: Interpretable prediction of protein-DNA binding sites based on pre-trained protein language models. Int J Biol Macromol 281:136147. https://doi.org/10.1016/j.ijbiomac.2024.136147
65. Zhu Y-H, Liu Z, Liu Y, Ji Z, Yu D-J (2024) ULDNA: integrating unsupervised multi-source language models with LSTM-attention network for high-accuracy protein–DNA binding site prediction. Brief Bioinform

25(2):bbae040. https://doi.org/10.1093/bib/bbae040

66. Pratyush P, Carrier C, Pokharel S, Ismail HD, Chaudhari M, Dukka BKC (2024) CaLM-PhosKAN: prediction of general phosphorylation sites in proteins via fusion of codon-aware Embeddings with amino acid-aware Embeddings and wavelet-based Kolmogorov–Arnold network. bioRxiv. https://doi.org/10.1101/2024.07.30.605530
67. Tan Y, Zheng J, Hong L, Zhou B (2024) ProtSolM: protein solubility prediction with multi-modal features. arXiv: arXiv:2406.19744. https://doi.org/10.48550/arXiv.2406.19744
68. Jumper J et al (2021) Highly accurate protein structure prediction with AlphaFold. Nature 596(7873):583–589. https://doi.org/10.1038/s41586-021-03819-2
69. Sun X, Wu Z, Su J, Li C (2024) GraphPBSP: protein binding site prediction based on graph attention network and pre-trained model ProstT5. Int J Biol Macromol 282:136933. https://doi.org/10.1016/j.ijbiomac.2024.136933
70. Wang H, Zhao L, Yu Z, Zeng X, Shi S (2025) CoNglyPred: accurate prediction of N-linked glycosylation sites using ESM-2 and structural features with graph network and co-attention. Proteomics n/a(n/a):e202400210. https://doi.org/10.1002/pmic.202400210
71. Fernández-Díaz R, Cossio-Pérez R, Agoni C, Lam HT, Lopez V, Shields DC (2024) AutoPeptideML: a study on how to build more trustworthy peptide bioactivity predictors. Bioinformatics 40(9):btae555. https://doi.org/10.1093/bioinformatics/btae555
72. Ferruz N, Schmidt S, Höcker B (2022) ProtGPT2 is a deep unsupervised language model for protein design. Nat Commun 13(1):4348. https://doi.org/10.1038/s41467-022-32007-7
73. Nijkamp E, Ruffolo J, Weinstein EN, Naik N, Madani A (2022) ProGen2: exploring the boundaries of protein language models. arXiv:2206.13517. https://doi.org/10.48550/arXiv.2206.13517
74. Lv L et al (2024) ProLLaMA: A Protein Large Language Model for Multi-Task Protein Language Processing. arXiv:2402.16445. https://doi.org/10.48550/arXiv.2402.16445
75. Ruffolo JA et al (2024) Adapting protein language models for structure-conditioned design. bioRxiv. https://doi.org/10.1101/2024.08.03.606485
76. Touvron H et al (2023) Llama 2: open foundation and fine-tuned chat models. arXiv:2307.09288. https://doi.org/10.48550/arXiv.2307.09288
77. Mardikoraem M, Wang Z, Pascual N, Woldring D (2023) Generative models for protein sequence modeling: recent advances and future directions. Brief Bioinform 24(6):bbad358. https://doi.org/10.1093/bib/bbad358
78. Winnifrith A, Outeiral C, Hie BL (2024) Generative artificial intelligence for de novo protein design. Curr Opin Struct Biol 86:102794. https://doi.org/10.1016/j.sbi.2024.102794
79. Ruffolo JA, Madani A (2024) Designing proteins with language models. Nat Biotechnol 42(2):200–202. https://doi.org/10.1038/s41587-024-02123-4
80. Pratyush P, Pokharel S, Saigo H, Kc DB (2023) PLMSNOSite: an ensemble-based approach for predicting protein S-nitrosylation sites by integrating supervised word embedding and embedding from pre-trained protein language model. BMC Bioinformatics 24(1):41. https://doi.org/10.1186/s12859-023-05164-9
81. Pakhrin SC et al (2023) LMNglyPred: prediction of human N-linked glycosylation sites using embeddings from a pre-trained protein language model. Glycobiology 33(5):411–422. https://doi.org/10.1093/glycob/cwad033
82. Shrestha P, Kandel J, Tayara H, Chong KT (2024) DL-SPhos: prediction of serine phosphorylation sites using transformer language model. Comput Biol Med 169:107925. https://doi.org/10.1016/j.compbiomed.2024.107925
83. Littmann M, Heinzinger M, Dallago C, Weissenow K, Rost B (2021) Protein embeddings and deep learning predict binding residues for various ligand classes. Sci Rep 11(1):23916. https://doi.org/10.1038/s41598-021-03431-4
84. Fang Y et al (2023) DeepProSite: structure-aware protein binding site prediction using ESMFold and pretrained language model. Bioinforma Oxf Engl 39(12):btad718. https://doi.org/10.1093/bioinformatics/btad718
85. Roche R, Moussad B, Shuvo MH, Tarafder S, Bhattacharya D (2024) EquiPNAS: improved protein–nucleic acid binding site prediction using protein-language-model-informed equivariant deep graph neural networks. Nucleic Acids Res 52(5):e27. https://doi.org/10.1093/nar/gkae039

86. Jing L et al (2023) CrossBind: collaborative cross-modal identification of protein nucleic-acid-binding residues. arXiv:2312.12094. https://doi.org/10.48550/arXiv.2312.12094
87. Zeng W, Dou Y, Pan L, Xu L, Peng S (2024) Improving prediction performance of general protein language model by domain-adaptive pretraining on DNA-binding protein. Nat Commun 15(1):7838. https://doi.org/10.1038/s41467-024-52293-7
88. Li Y, Wang C, Gu H, Feng H, Ruan Y (2024) ESMDNN-PPI: a new protein–protein interaction prediction model developed with protein language model of ESM2 and deep neural network. Meas Sci Technol 35(12):125701. https://doi.org/10.1088/1361-6501/ad761c
89. Hu J, Li Z, Rao B, Thafar MA, Arif M (2024) Improving protein-protein interaction prediction using protein language model and protein network features. Anal Biochem 693: 115550. https://doi.org/10.1016/j.ab.2024.115550
90. Kulmanov M, Khan MA, Hoehndorf R (2018) DeepGO: predicting protein functions from sequence and interactions using a deep ontology-aware classifier. Bioinformatics 34(4):660–668. https://doi.org/10.1093/bioinformatics/btx624
91. Gligorijević V et al (2021) Structure-based protein function prediction using graph convolutional networks. Nat Commun 12(1): 3168. https://doi.org/10.1038/s41467-021-23303-9
92. Lai B, Xu J (2022) Accurate protein function prediction via graph attention networks with predicted structure information. Brief Bioinform 23(1):bbab502. https://doi.org/10.1093/bib/bbab502
93. Yuan Q, Xie J, Xie J, Zhao H, Yang Y (2023) Fast and accurate protein function prediction from sequence through pretrained language model and homology-based label diffusion. Brief Bioinform 24(3):bbad117. https://doi.org/10.1093/bib/bbad117
94. Ilzhöfer D, Heinzinger M, Rost B (2022) SETH predicts nuances of residue disorder from protein embeddings. Front Bioinforma 2. https://doi.org/10.3389/fbinf.2022.1019597
95. Xu S, Onoda A (2024) Accurate and fast prediction of intrinsically disordered protein by multiple protein language models and ensemble learning. J Chem Inf Model 64(7): 2901–2911. https://doi.org/10.1021/acs.jcim.3c01202
96. Wu R et al (2022) High-resolution de novo structure prediction from primary sequence. bioRxiv. https://doi.org/10.1101/2022.07.21.500999
97. Baek M et al (2021) Accurate prediction of protein structures and interactions using a three-track neural network. Science 373(6557):871–876. https://doi.org/10.1126/science.abj8754
98. Weissenow K, Heinzinger M, Rost B (2022) Protein language-model embeddings for fast, accurate, and alignment-free protein structure prediction. Structure 30(8):1169–1177.e4. https://doi.org/10.1016/j.str.2022.05.001
99. Fang X et al (2023) A method for multiple-sequence-alignment-free protein structure prediction using a protein language model. Nat Mach Intell 5(10):1087–1096. https://doi.org/10.1038/s42256-023-00721-6
100. Chowdhury R et al (2022) Single-sequence protein structure prediction using a language model and deep learning. Nat Biotechnol 40(11):1617–1623. https://doi.org/10.1038/s41587-022-01432-w
101. Watson JL et al (2023) De novo design of protein structure and function with RFdiffusion. Nature 620(7976):1089–1100. https://doi.org/10.1038/s41586-023-06415-8
102. Madani A et al (2023) Large language models generate functional protein sequences across diverse families. Nat Biotechnol 41(8): 1099–1106. https://doi.org/10.1038/s41587-022-01618-2
103. Verkuil R et al (2022) Language models generalize beyond natural proteins. bioRxiv. https://doi.org/10.1101/2022.12.21.521521
104. Hsu C et al (2022) Learning inverse folding from millions of predicted structures. bioRxiv. https://doi.org/10.1101/2022.04.10.487779
105. Dauparas J et al (2022) Robust deep learning–based protein sequence design using ProteinMPNN. Science 378(6615): 49–56. https://doi.org/10.1126/science.add2187
106. Yang KK, Zanichelli N, Yeh H (2022) Masked inverse folding with sequence transfer for protein representation learning. bioRxiv. https://doi.org/10.1101/2022.05.25.493516
107. Jeliazkov JR, del Alamo D, Karpiak JD (2023) ESMFold Hallucinates Native-Like Protein Sequences. bioRxiv. https://doi.org/10.1101/2023.05.23.541774
108. Alamdari S et al (2023) Protein generation with evolutionary diffusion: sequence is all

you need. bioRxiv. https://doi.org/10.1101/2023.09.11.556673

109. Fournier Q, Vernon RM, van der Sloot A, Schulz B, Chandar S, Langmead CJ (2024) Protein language models: is scaling necessary? bioRxiv. https://doi.org/10.1101/2024.09.23.614603
110. Huo M et al (2024) Multi-modal large language model enables protein function prediction. bioRxiv. https://doi.org/10.1101/2024.08.19.608729
111. Zhou Z et al (2024) Enhancing efficiency of protein language models with minimal wet-lab data through few-shot learning. Nat Commun 15(1):5566. https://doi.org/10.1038/s41467-024-49798-6
112. Nambiar A, Forsyth JM, Liu S, Maslov S (2024) DR-BERT: a protein language model to annotate disordered regions. Structure 32(8):1260–1268.e3. https://doi.org/10.1016/j.str.2024.04.010
113. Shuai RW, Ruffolo JA, Gray JJ (2023) IgLM: Infilling language modeling for antibody sequence design. Cell Syst 14(11):979–989.e4. https://doi.org/10.1016/j.cels.2023.10.001
114. Deutschmann N, Pelissier A, Weber A, Gao S, Bogojeska J, Martínez MR (2024) Do domain-specific protein language models outperform general models on immunology-related tasks? ImmunoInformatics 14:100036. https://doi.org/10.1016/j.immuno.2024.100036
115. Ieremie I, Ewing RM, Niranjan M (2024) Protein language models meet reduced amino acid alphabets. Bioinformatics 40, no. 2:btae061. https://doi.org/10.1093/bioinformatics/btae061
116. Ghafarollahi A, Buehler MJ (2024) ProtAgents: protein discovery via large language model multi-agent collaborations combining physics and machine learning. Digit Discov 3(7):1389–1409. https://doi.org/10.1039/D4DD00013G
117. Sledzieski S, Kshirsagar M, Baek M, Dodhia R, Lavista Ferres J, Berger B (2024) Democratizing protein language models with parameter-efficient fine-tuning. Proc Natl Acad Sci 121(26):e2405840121. https://doi.org/10.1073/pnas.2405840121
118. Shrestha P, Kandel J, Tayara H, Chong KT (2024) Post-translational modification prediction via prompt-based fine-tuning of a GPT-2 model. Nat Commun 15(1):6699. https://doi.org/10.1038/s41467-024-51071-9

# Chapter 2

# Enhancing Structure-Aware Protein Language Models with Efficient Fine-Tuning for Various Protein Prediction Tasks

**Yichuan Zhang, Yongfang Qin, Mahdi Pourmirzaei, Qing Shao, Duolin Wang, and Dong Xu**

## Abstract

Proteins are crucial in a wide range of biological and engineering processes. Large protein language models (PLMs) can significantly advance our understanding and engineering of proteins. However, the effectiveness of PLMs in prediction and design is largely based on the representations derived from protein sequences. Without incorporating the three-dimensional (3D) structures of proteins, PLMs would overlook crucial aspects of how proteins interact with other molecules, thereby limiting their predictive accuracy. To address this issue, we present S-PLM, a 3D structure-aware PLM, that employs multi-view contrastive learning to align protein sequences with their 3D structures in a unified latent space. Previously, we utilized a contact map-based approach to encode structural information, applying the Swin-Transformer to contact maps derived from AlphaFold-predicted protein structures. This work introduces a new approach that leverages a geometric vector perceptron (GVP) model to process 3D coordinates and obtain structural embeddings. We focus on the application of structure-aware models for protein-related tasks by utilizing efficient fine-tuning methods to achieve optimal performance without significant computational costs. Our results show that S-PLM outperforms sequence-only PLMs across all protein clustering and classification tasks, achieving performance on par with state-of-the-art methods that require both sequence and structure inputs. S-PLM and its tuning tools are available at https://github.com/duolinwang/S-PLM/.

**Key words** Large protein language models (PLMs), 3D structure-aware PLM, Multi-view contrastive learning, ESM 2, Efficient fine-tuning

## 1 Introduction

Proteins are fundamental to nearly every biological process and play a crucial role in addressing challenges related to health, energy, and the environment. Consequently, the ability to rapidly and accurately obtain information about target proteins is essential. Computational prediction of protein properties based on their amino acid sequences is a key component of this effort. Protein language

Dukka B. KC (ed.), *Large Language Models (LLMs) in Protein Bioinformatics*, Methods in Molecular Biology, vol. 2941, https://doi.org/10.1007/978-1-0716-4623-6_2,

models (PLMs) have emerged as powerful tools that can uncover underlying patterns within protein sequences and predict various properties by leveraging these patterns [1]. The development and deployment of PLMs currently follow a two-stage paradigm. Initially, PLMs are trained to convert amino acid sequences into mathematical latent representations or embeddings, using techniques such as masked language modeling (MLM) or autoregressive strategies. In MLM, the model predicts masked amino acids based on their surrounding context, while autoregressive models predict subsequent amino acids based on preceding ones [2–4]. Subsequently, these pre-trained PLMs are adapted using protein property data to perform specific protein-related tasks. Notable PLMs developed under this paradigm include ProtBert [3], ESM 2 [5], and ProtGPT [6]. These models have demonstrated promising results in protein property prediction and de novo protein design, highlighting their potential for discovering and analyzing new biological insights [7–9].

Despite these advancements, a significant challenge in developing PLMs is the integration of critical biophysical information into the embeddings. Protein function is intrinsically linked to its three-dimensional (3D) structure and relying solely on amino acid sequences for training PLMs limits their predictive capabilities, especially for tasks that depend heavily on structural information. To address this limitation, some approaches have been developed to enhance sequence-based embeddings with evolutionary or functional data. For instance, methods such as AlphaFold's Evoformer [10] and MSA transformer [11] incorporate multiple sequence alignments (MSA) into PLMs. Additionally, ProteinBERT [12] integrates gene ontology (GO) [13] annotations into the MLM pretraining framework using a denoising autoencoder trained on corrupted protein sequences and GO annotations, achieving remarkable performance on various protein tasks despite its relatively small model size. These strategies enrich the information contained within sequence-based embeddings, thereby improving the performance of PLMs. However, none of these methods effectively incorporate key structural information into the embeddings.

In response to the need for integrating structural information, emerging approaches aim to develop joint embeddings that combine both sequence and structural data. Chen et al. proposed a protein structure representation method based on self-supervised learning that leverages existing pretrained PLMs [14]. Similarly, Zhang et al. explored joint protein embeddings by combining ESM 2 with three different structure encoders [15], and Hu et al. developed a joint embedding by integrating protein sequences with structural data [16]. These joint embeddings have shown superior performance in many protein property prediction tasks, underscoring the importance of including structural information in protein representations. Nevertheless, a major limitation of these joint embedding models is their requirement for both sequence and structure inputs during inference. Even with tools like AlphaFold, obtaining reliable structures for certain specific proteins remains challenging. Additionally, from a computational standpoint, these methods necessitate additional steps to acquire predicted structures, which increases both time and effort.

To overcome these challenges, we propose an alternative multimodal approach to protein representation learning that does not require simultaneous sequence and structure inputs. Instead of creating joint embeddings, we developed a sequence-based embedding that inherently incorporates structural information. We introduce S-PLM, a 3D structure-aware approach that infuses sequence-based embeddings with structural information by aligning them with structural embeddings derived from 3D protein models through multi-view contrastive learning [17]. A key advantage of S-PLM over existing joint embedding models is that, after training, it only requires amino acid sequences as input during inference, thereby eliminating the overhead associated with using predicted protein structures. This sequence-only input approach enhances computational efficiency and broadens the applicability of the model.

The potential of contrastive learning to enrich sequence-based embeddings is further supported by recent successes. ProtST [18], for example, integrates biomedical text into PLMs by aligning two modalities via a contrastive loss, resulting in enhanced protein property information and superior performance on various protein representation and classification benchmarks. Similarly, the CLEAN model [19] leverages information from EC numbers to improve PLMs' ability to predict enzyme functions based solely on protein sequences. Building on these successes, S-PLM is designed to accommodate either structure or sequence as input. This study focuses on demonstrating its advantages in encoding protein sequences and enhancing sequence representations for diverse protein prediction tasks using structural information. The sequence encoder of S-PLM is implemented based on the pretrained ESM 2 model, which is scalable and capable of gradually integrating new protein features without forgetting previously learned knowledge [20].

To further enhance the utility of S-PLM, this study builds upon our previous work by introducing a new model, S-PLM2. S-PLM1, introduced in our previous paper [17], uses contact maps and applies protein-level contrastive learning to align sequence and structure embeddings at a global level. In contrast, the newly introduced S-PLM2 incorporates 3D coordinate features and extends to include residue-level contrastive learning, capturing finer structural details. Both models employ efficient fine-tuning techniques to enable targeted modifications for downstream protein prediction tasks without requiring full parameter updates. Efficient fine-tuning strategies such as fine-tuning the top layers, adapter tuning [20], and low-rank adaptation (LoRA) [21] selectively update specific parameters while keeping others fixed. These approaches significantly reduce computational and memory requirements, mitigate issues related to data scarcity, and achieve performance comparable to or better than traditional methods. However, the application of these strategies to PLMs has been limited in existing research [22, 23]. This work explores several efficient fine-tuning strategies for both S-PLM1 and S-PLM2 across various protein prediction tasks. Through these strategies, S-PLM demonstrates competitive performance in gene ontology [13], enzymology number [24], protein secondary structure, protein fold, and enzyme reaction prediction tasks.

## 2 Materials and Methods

### 2.1 Structure-Aware Protein Language Model (S-PLM)

#### 2.1.1 Overview of the S-PLM Framework

An overview of the architectures of both S-PLM1 and S-PLM2 is shown in Fig. 1a, showcasing the integration of sequence and structure information through different encoders and the application of contrastive learning at distinct levels. S-PLM1 focuses on aligning sequence and structure representations at the protein level, while S-PLM2 extends this approach to include residue-level contrastive learning. These models aim to capture both the global and

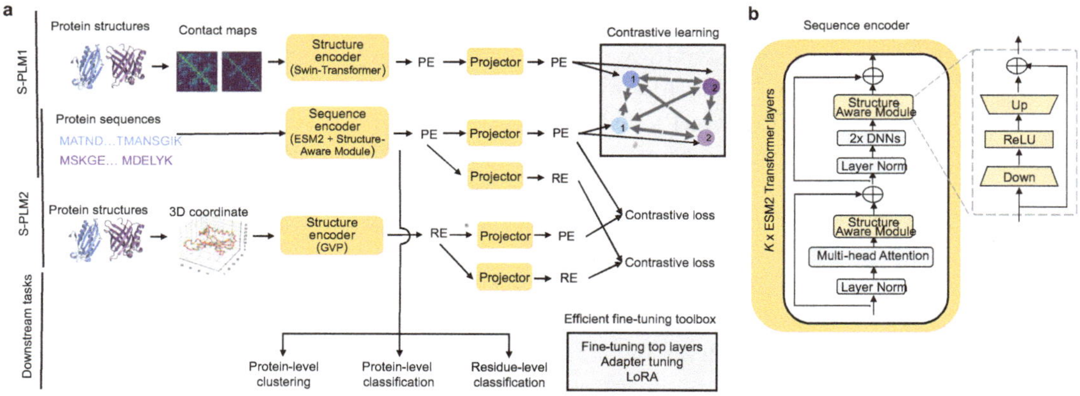


**Fig. 1** The framework of S-PLM. (**a**) The framework of S-PLM1 and S-PLM2. During pretraining, the model inputs the amino acid sequences and contact maps (S-PLM1) or 3D coordinates (S-PLM2) simultaneously. The amino acid sequences of proteins are converted into residue-level embeddings (REs) through a sequence encoder (ESM2 + structure-aware module), while the backbone Cα contact maps (S-PLM1) are converted into protein-level embeddings (PEs) through a structure encoder (Swin-Transformer). For S-PLM2, the 3D coordinates of protein structures are processed through the structure encoder (geometric vector perceptron, GVP) to generate both PEs and REs. With respective projectors, both S-PLM1 and S-PLM2 are trained through contrastive learning on the embeddings from each modality. After pretraining, the sequence encoder that generates the REs before the projector layer is used for downstream tasks, such as protein-level clustering, protein-level classification, and residue-level classification tasks. (**b**) Architecture of the sequence encoder. The sequence encoders of S-PLM1 and S-PLM2 are both fine-tuned on the top-K Transformer layers of ESM2, with each layer adding a compact structure-aware module. During pretraining, the ESM2 backbone model is frozen, and only the structure-aware modules are trainable

fine-grained structural information of proteins, thus enhancing their utility in downstream tasks such as clustering and classification.

In S-PLM1, the model takes as input both the amino acid sequences of proteins and their corresponding structural representations in the form of contact maps. The Swin-Transformer [25] serves as the structure encoder, processing contact maps into protein-level embeddings (PEs), which capture the overall structural configuration of the protein. Simultaneously, the amino acid sequences are processed by the ESM 2 model, enhanced with a structure-aware module to incorporate structural information, generating residue-level embeddings (REs). These sequence and structure embeddings are then projected into a unified latent space through independent projectors, utilizing contrastive learning inspired by the SimCLR method [26] and extending the CLIP framework [27]. This ensures that sequence and structure embeddings of the same protein are brought closer together, while maintaining distinct separation between embeddings from different proteins. Moreover, our model accounts for de-alignment within the same modality. The contrastive learning framework in S-PLM[1] allows for the effective alignment of sequence and structure at the protein level, making it suitable for tasks that require a global understanding of protein structure, such as protein-level clustering and classification.

S-PLM2, on the other hand, introduces a more detailed alignment between sequence and structure by performing contrastive learning at both the protein and residue levels. Here, the structural information is represented by 3D coordinates of the protein, which are encoded through the geometric vector perceptron (GVP) [28]. This structure encoder produces both protein-level and residue-level embeddings (PE and RE, respectively). Like S-PLM1, the sequence embeddings are generated by the ESM 2 model with the structure-aware module. The embeddings from both the sequence and structure encoders are then projected into unified latent spaces, where contrastive learning is applied not only at the protein level but also at the residue level. This residue-level contrastive learning enables the model to capture fine-grained details about the spatial relationships between residues, which is crucial for residue-level prediction tasks such as secondary structure prediction. By aligning sequence and structure at multiple levels, S-PLM2 enhances the model's ability to capture both local and global protein features, making it more effective in tasks that require detailed structural understanding.

The sequence encoder of S-PLM for both versions, depicted in Fig. 1b, is built upon the ESM 2 transformer architecture, enhanced with a structure-aware module to integrate structural information into the sequence embeddings. The structure-aware module can be implemented in several ways. One approach involves fine-tuning the pretrained weights of the ESM 2 model, allowing the structural information to be incorporated while retaining the original architecture of ESM 2. Alternatively, adapter tuning offers a more efficient solution by integrating adapter modules into the top-$K$ transformer layers of ESM 2, which are only trained during the pretraining process. These adapter modules, positioned after the multi-head attention projection and following the feed-forward layers, serve as the structure-aware components of the sequence encoder.

Adapter tuning brings several benefits. First, the adapter modules are compact, containing significantly fewer parameters than the full transformer layers of ESM 2, which reduces the training burden. Additionally, this approach enables the model to be continually trained on new protein features (e.g., protein functions) without catastrophic forgetting of previously learned information. This is possible because the ESM 2 backbone remains intact and frozen during pretraining, preserving the sequence representation capabilities of the model.

During inference, the S-PLM model is highly flexible, capable of accepting either sequence data or structural data as input. The model produces embeddings suitable for a range of downstream tasks, such as protein-level clustering, protein-level classification, and residue-level classification. In this study, we will focus on generating sequence embeddings from protein sequences by S-PLM, and the pretrained sequence encoder is utilized to produce residue-level embeddings before passing them through projectors for downstream applications. Depending on the task requirements, the entire sequence encoder can either be fully frozen or fine-tuned, ensuring the model's adaptability. To fully leverage S-PLM's potential for supervised protein prediction tasks, several efficient fine-tuning strategies, including fine-tuning top layers, adapter tuning, and LoRA, are implemented, all of which are incorporated into the efficient fine-tuning toolbox to optimize performance without excessive computational cost.

#### 2.1.2 Sequence Encoder

The sequence encoder in S-PLM is built upon the ESM 2-t33_650M_UR50D model [5], which serves as the backbone for generating contextually rich sequence embeddings. Each protein sequence is tokenized into individual amino acids and processed through 33 transformer layers, resulting in 1280-dimensional embeddings for each amino acid.

To ensure consistency in sequence representation, special tokens are added to the sequences: a BEGIN token (<cls>) is inserted at the start, an END token (<eos>) marks the end, and padding tokens (<pad>) are used to standardize sequence length for efficient batch processing. The transformer model generates an embedding for each residue, capturing not only the sequence of residues but also their interdependencies, which are essential for downstream tasks. To optimize computational resources, especially when handling large proteins, sequences exceeding 512 residues are truncated during training. Additionally, S-PLM remains flexible, allowing users to configure the model to process either full-length sequences or truncated ones based on specific task requirements. This flexibility ensures the model can handle a wide range of protein lengths and complexities, from small peptides to larger, more intricate proteins.

The sequence encoder generates residue-level embeddings, which are first projected through two multilayer perceptron (MLP) layers into a 64D representation. The protein-level embedding, on the other hand, is derived by averaging the residue embeddings after excluding the padding tokens, resulting in a global representation of the protein. This global representation is then projected through two additional MLP layers to form a 256D embedding. Both the residue-level and protein-level projections are essential for contrastive learning to ensure that the model can effectively align sequence and structural information. Further details about these projections will be introduced later in the study.

#### 2.1.3 Structure Encoder: Contact Map with Swin-Transformer

The structure encoder is a crucial component of S-PLM that enables the model to incorporate 3D structural information about proteins. Previously in S-PLM1, we used a contact map-based approach to provide a global view of protein-level residue interactions, utilizing contact maps derived from AlphaFold-predicted structures. However, in this work, we focus on a new approach of S-PLM2 that employs a GVP model, which processes 3D atomic coordinates to capture detailed local structural features at the residue level. This allows for a more precise representation of protein structures. While the contact map method is briefly introduced here for completeness, the GVP model is the primary focus of our current method.

The contact map approach offers a protein-level representation of the structural relationships between residues, simplifying the 3D structure of proteins into a 2D matrix. This makes it well suited for applying image-processing techniques like Swin-Transformer encoder. To meet the image network's requirement for three input channels, we transformed the contact map into a three-channel format. Unlike the conventional binary contact map, which assigns a value of 1 when the pairwise distance between C$\alpha$ atoms is within a defined threshold (indicating residue contact) and 0 otherwise, we converted the contact map into a continuous similarity matrix. In this transformation, $C$ represents the elements of the raw contact map. To normalize the raw distance values, any distance in $C$ exceeding a threshold $d$ (in our case, $d = 22$ Å as used in AlphaFold2) is set to $d$. The transformation formula is

$$(d - CLIP(C, d))/d$$

Here, $CLIP(C, d)$ limits values in $C$ to a maximum of $d$, ensuring that distances greater than $d$ are capped at $d$, while others remain unchanged. The formula scales the distances into a continuous range from 0 to 1, where 1 represents the closest pairwise distance and 0 the farthest. This approach generates a standardized similarity matrix, offering a more consistent representation of the protein's structural features.

The continuous similarity matrix is then processed by the Swin-Transformer (swinv2-tiny-patch4-window8–256), which extracts hierarchical features at multiple scales. This multi-scale approach enables the model to capture both local and overall architecture of the protein. The resulting protein-level embeddings capture the structural conformation and, like the sequence embeddings, are further refined and reduced to a compact 256-dimensional representation, the same shape as the sequence representation, through two projector layers.

#### 2.1.4 Structure Encoder: GVP with 3D Coordinates

In problems involving protein structure prediction, graph-based models have proven effective due to their ability to represent complex spatial relationships. The GVP model, which leverages a graph structure, excels in representing 3D protein information by incorporating key properties such as rotational invariance and equivariance. In GVP, each residue is represented as a node in a graph, with edges representing spatial relationships. The GVP model can process both scalar and vector features, where scalar features remain invariant, and vector features are equivariant under geometric transformations. These properties ensure that the model maintains geometric consistency during transformations like rotations and reflections, enabling a deeper understanding of how residue arrangements affect the protein's 3D structure and biological function.

For the node features, both scalar and vector information are utilized. Scalar information is extracted from the geometric properties of the protein structure [28, 29], such as dihedral angles, which are calculated from the directional vectors between three consecutive backbone atoms, with plane normals derived via the cross product. The angle between the normals measures the torsion between residues, and are transformed into sine and cosine representations. Vector node features include both orientation and sidechain information, derived from the backbone's 3D coordination and the positions of atoms around the backbone. Orientation vectors are calculated by normalizing the differences between consecutive backbone atoms, capturing both forward and backward directions along the protein chain. Sidechain vectors are computed by normalizing the vectors from the Cα atom to the nitrogen and carbon atoms, combining them into a bisector. A perpendicular vector is calculated via the cross product, and a linear combination of both perpendicular and bisector forms the final sidechain vector.

In addition to the node features used in the original GVP model, we incorporate optional Foldseek-derived features to enrich both scalar and vector representations [30]. For scalar features, Foldseek provides cosine similarities between backbone vectors, inter-residue distances, and features encoding the relative sequence distances between residues. For vector features, Foldseek contributes normalized direction vectors and vectors that capture the spatial orientations of residues. These combined Foldseek features further enhance the node representations within the model.

At the edge level, similar to the nodes, both scalar and vector features are incorporated. Scalar features consist of radial basis functions (RBF) to encode pairwise distances between residues [31], and rotary embeddings, which use trigonometric functions to capture relative positions in a rotation-invariant way. Vector edge features represent directional vectors between residues, describing their spatial orientation and distance, essential for maintaining the protein's structural integrity.

To construct the graph, an optional protein sequence embedding is integrated using two types of embeddings: one-hot encoding, as in the original GVP model, and Atchley factor embeddings [32], which capture key biochemical properties such as polarity, structure factor, molecular volume, amino acid composition, and electrostatic charge. However, an ablation study (Fig. 7) showed that incorporating sequence embeddings did not improve model performance.

The network's architecture processes node and edge features using a GVP-based structure encoder. Node and edge features are passed through the encoder, which captures structural information from the graph. The residue-level embeddings produced by the GVP encoder must pass through a projector before they can be used as input for contrastive learning. These embeddings are transformed through a series of MLP projection layers, reducing their dimensionality to 64, and only after this step, the contrastive loss is calculated. This approach is crucial in contrastive learning, as the non-linear projection layers adjust the learned representations to focus on features relevant to the contrastive objective. One explanation is that applying the loss to the projected embeddings helps the model avoid overfitting to specific pre-task details, although some information useful for downstream tasks might be lost in the process. This trade-off allows the model to capture more general features while retaining enough relevant information for effective learning.

Additionally, the entire protein's representation is generated by pooling the residue embeddings and then passing them through another projector layer to produce a final 256D protein-level embedding, which allows for contrastive learning at the protein level as well.

## 2.2 Multi-View Contrastive Learning

### 2.2.1 S-PLM1: Protein Level Contrastive Learning

S-PLM1 is a protein-level contrastive learning method that we introduced in a previous work. It aims to align the sequence and structure representations of a protein at the protein level. The method leverages multi-view contrastive learning to bring the sequence and structure embeddings of the same protein closer in the latent space while maintaining clear separation between embeddings from different proteins. By encouraging similar representations (i.e., sequence and structure from the same protein) to cluster together, and dissimilar representations (i.e., from different proteins) to be pushed apart, this approach ensures that the protein's overall sequence and structure information is well aligned. Understanding this approach is essential for grasping the residue-level contrastive learning introduced later, as it provides the foundation for aligning sequence and structure representations.

S-PLM1 designed a loss function inspired by the loss used in SimCLR [26]. However, instead of relying on augmented versions of the same input to create different views, it treats the sequence embedding $E^i_{\text{seq}}$ and structure embedding $E^i_{\text{str}}$ of the same protein as the positive pair. The contrastive loss is defined by the following equation, which measures the similarity between the sequence and the structure embeddings of the same protein and contrasts them against embeddings from different proteins in the mini-batch:

$$L_{E^i_{\text{seq}},E^i_{str}} = -\log \frac{\exp\left(\frac{\text{sim}\left(E^i_{\text{seq}},E^i_{\text{str}}\right)}{\tau}\right)}{\sum_{k=1}^{N}\left(1_{\{k\neq i\}}\exp\left(\frac{\text{sim}\left(E^i_{\text{seq}},E^k_{\text{str}}\right)}{\tau}\right)+1_{\{k\neq i\}}\exp\left(\frac{\text{sim}\left(E^i_{\text{seq}},E^k_{\text{seq}}\right)}{\tau}\right)+1_{\{k\neq i\}}\exp\left(\frac{\text{sim}\left(E^i_{\text{str}},E^k_{\text{str}}\right)}{\tau}\right)\right)}$$

$$\text{sim}(x,y)=\frac{x^Ty}{|x||y|}$$

$\text{sim}(x, y)$ represents the cosine similarity between embeddings $x$ and $y$, while $\tau$ is the temperature parameter controlling the sharpness of the similarity distribution. The term $k \neq i$ ensures that negative examples, corresponding to embeddings from different proteins, are effectively separated in the latent space.

By minimizing this loss, it ensures that the learned representations for a protein's sequence and structure are well aligned, capturing both views of the protein in a unified space. At the same time, the model is trained to maintain sufficient distance between the representations of different proteins, preventing confusion across protein types. This multi-view framework leverages both modalities to build a robust representation, highlighting the complementary roles that sequence and structure play in defining protein functions.

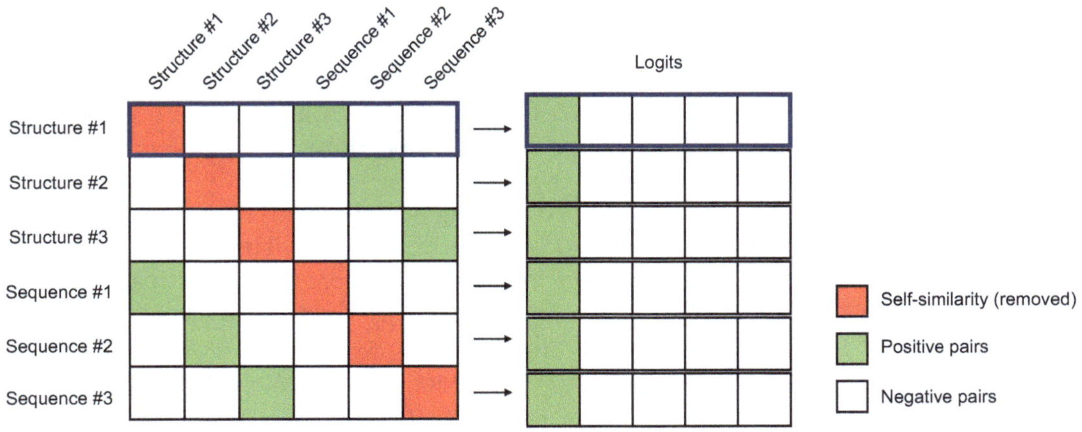


**Fig. 2** Similarity matrix including self-similarity, positive pairs, and negative pairs in a contrastive learning setting. This diagram only illustrates three residues/proteins as examples for clarity. Rows and columns represent both the structural embeddings (structure #1, structure #2, structure #3) and sequence embeddings (sequence #1, sequence #2, sequence #3). Red cells indicate self-similarity, which is excluded from the contrastive loss calculation. Green cells denote positive pairs, where the structural and sequence embeddings correspond to the same sample and are encouraged to be similar (e.g., structure #1 with sequence #1, structure #2 with sequence #2, and structure #3 with sequence #3). White cells represent the remaining negative pairs, which include both pairs from different modalities (e.g., structure #1 with sequence #2) and pairs within the same modality (e.g., structure #1 with structure #2 or sequence #1 with sequence #2)

#### *2.2.2 S-PLM2: Residue-Level Contrastive Learning*

In S-PLM2, contrastive learning is applied at the residue level to enable more detailed alignment between sequence and structure. As highlighted in a prior study [33], aligning sequence and structure embeddings at the residue level allows for the integration of spatial information from the structure, enhancing the richness of sequence-based representations. This residue-level alignment provides the model with the ability to capture fine-grained structural details, improving its effectiveness in tasks such as secondary structure prediction.

As shown in Fig. 2, for each residue, the model computes cosine similarities between the normalized sequence embeddings and structure embeddings, producing a similarity matrix. The similarity matrix is structured so that both its rows and columns follow the same order, with the sequence embeddings listed first, followed by the structure embeddings. Each element in the matrix represents the similarity between the sequence or structure embedding of one residue and the sequence or structure embedding of another residue. However, the diagonal elements of this matrix correspond to the similarity between the same residue's sequence and structure embeddings, which represents trivial self-similarity. These diagonal elements are not informative for contrastive learning and are therefore discarded.

Once the diagonal elements are removed, the remaining off-diagonal elements represent the meaningful comparisons: positive pairs, which are sequence and structure embeddings from the same residue, while all others are negative pairs. By removing the self-similarities, the model is forced to learn from these more informative comparisons. Positive pairs encourage the alignment of sequence and structure embeddings from the same residue, while negative pairs ensure that embeddings from different residues remain distinct in the latent space.

The selected positive and negative pairs are then used to construct the logits for contrastive learning. These logits are passed into a cross-entropy loss function, where the positive pair logits are expected to have high similarity, while the negative pair logits should have low similarity. The cross-entropy loss applies a softmax function to convert the logits into probabilities, where positive pairs are encouraged to have a probability close to 1 and negative pairs approach 0. This optimization ensures that the sequence and structure embeddings of the same residue are closely aligned, while maintaining separation between embeddings from different residues.

The selected positive and negative pairs are then used to construct the logits for contrastive learning. For each row in the similarity matrix, a single positive pair is combined with multiple negative pairs to form a logit vector. These logits are then passed into a cross-entropy loss function, where the positive pair logits are expected to have high similarity, while the negative pair logits should have low similarity. The cross-entropy loss applies a softmax function to convert the logits into probabilities, where positive pairs are encouraged to have a probability close to 1 and negative pairs approach 0. This optimization ensures that the sequence and structure embeddings of the same residue are closely aligned while maintaining separation between embeddings from different residues.

Formally, the contrastive loss for a positive pair $(i, j)$ is computed as

$$\text{loss}_{i,j} = -\log \frac{\exp(\text{sim}(z_i, z_j)/\tau)}{\sum_{k=1}^{2N} I_{k \neq i} \exp(\text{sim}(z_i, z_k)/\tau)}$$

where $\text{sim}(z_i, z_j)$ represents the cosine similarity between the embeddings $z_i$ and $z_j$ (the positive pair), $\tau$ is a temperature parameter that controls the softmax sharpness, and the denominator sums over all pairs, including negative pairs. This formula can be applied from either perspective (i.e., for both $(i, j)$ and $(j, i)$, ensuring that the similarity between structure and sequence embeddings is maximized in both directions. This method of computing the loss is equivalent to the S-PLM[1].

By minimizing the cross-entropy loss, S-PLM2 effectively captures residue-level relationships, aligning sequence and structure embeddings at a fine-grained level. In addition to residue-level alignment, S-PLM2 also computes global protein-level embeddings by pooling the residue embeddings. This allows for protein-level contrastive learning, complementing the detailed residue-level alignment. Together, the residue and protein-level embeddings provide a robust multi-scale representation, integrating both local structural details and global context. This dual-level alignment enhances the model's performance in tasks requiring both sequence–structure precision and broader protein-level understanding, such as binding site prediction and protein classification.

### 2.3 Training Dataset for Contrastive Learning and Training Process

We constructed the training dataset by collecting amino acid sequences from the Swiss-Prot library, and storing them in FASTA format. The corresponding 3D structures of these proteins were sourced from the AlphaFold2 database. In S-PLM1, Cα-Cα contact maps were generated based on the AlphaFold2-predicted structures for each protein. In contrast, S-PLM2 extracted the N, Cα, C, and O atomic coordinates for building a graph representation. From the Swiss-Prot library, we randomly selected 500,000 proteins for training, with an additional 41,500 proteins designated for validation. Given the Swiss-Prot library's large size, with 542,378 proteins in total, we did not remove similar sequences between the training and validation sets, as this overlap had minimal impact on performance.

During pretraining, proteins longer than 512 residues were truncated to include only the first 512 residues for both sequence and contact map representations. For optimization, we employed two separate Stochastic gradient descent (SGD) optimizers with a momentum of 0.9, one for the sequence backbone and another for the structure backbone. A cyclic learning rate schedule was used, starting at 0.001 and gradually reducing to 0 over 100 steps per cycle. The batch size was fixed at 20, and we applied a weight decay of 0.0005. No gradient accumulation was used. The temperature parameter for contrastive learning was set to 0.05. To stabilize the training process, gradient norm clipping was applied with a threshold of 1.0. Additionally, we adopted mixed precision training [34] to improve computational efficiency and optimize GPU memory usage. All experiments were conducted on a single A100 GPU, running for over 10,000 steps with a consistent batch size of 20. This setup allowed us to effectively balance training complexity with resource constraints, ensuring robust performance throughout the pretraining process.

### 2.4 Efficient Fine-Tuning on Sequence Encoder for Downstream Tasks

Given that the pretrained models S-PLM1 and S-PLM2 are large in size, full-model fine-tuning is not efficient for downstream tasks due to high computational, memory requirements, and may also lead to suboptimal performance for specific downstream tasks. To address this challenge, we employ efficient fine-tuning strategies to

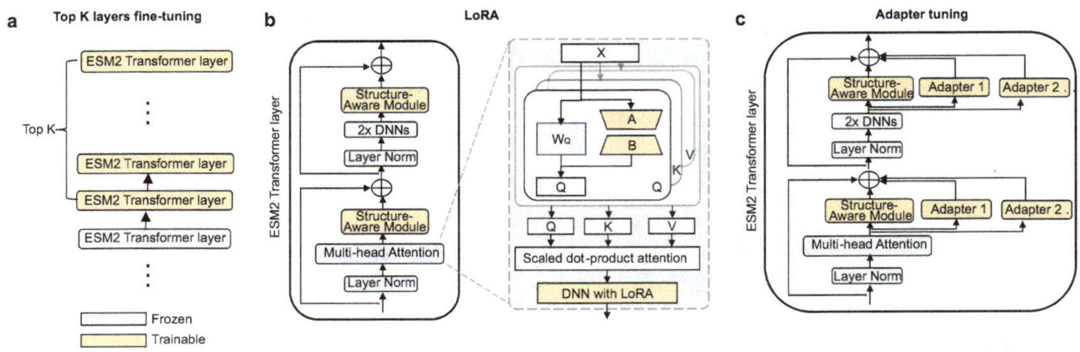


**Fig. 3** Efficient fine-tuning strategies for the sequence encoder of the S-PLM model. (**a**) Fine-tuning top layers. Only the top K transformer layers are updated, while lower layers remain frozen. (**b**) LoRA. Trainable rank decomposition matrices are inserted into the top-K self-attention layers of ESM2. (**c**) Adapter tuning. A list of parallel trainable adapters is added to the structure-adapter modules

enable targeted modification for downstream protein prediction tasks, applied to the sequence encoder of the S-PLM models, as illustrated in Fig. 3. These strategies include fine-tuning top layers, LoRA, and adapter tuning.

The fine-tuning top layers method (Fig. 3a) updates only the top $K$ transformer layers of the ESM2 backbone, which consists of 33 layers. By freezing the lower layers, which retain general knowledge from pretraining, and fine-tuning only the top layers, we not only alleviate the computational burden but also reduce the number of trainable parameters. This helps mitigate overfitting to a specific task and prevents the forgetting of pretrained knowledge. Adjusting the value of $K$ provides flexibility, enabling a balance between computational efficiency and task-specific performance.

LoRA (Fig. 3b) introduces a more parameter-efficient approach by injecting low-rank matrices into the transformer's architecture, while keeping the original pretrained weight matrices frozen. Instead of updating the entire weight matrix, LoRA represents the update as a low-rank decomposition:

$$\Delta W = BA$$

where $B \in R^{d \times r}$ and $A \in R^{r \times k}$, with $r \ll \min(d, k)$. Here, $W_0 \in R^{d \times k}$ is the original weight matrix. During training, only $A$ and $B$ are updated, allowing significant reduction in the number of trainable parameters. For an input $x$, the forward pass with LoRA can be expressed as

$$h = W_0 + \Delta Wx = W_0 + BAx$$

In practice, LoRA scales $\Delta Wx$ by $\frac{\alpha}{r}$, where $\alpha$ is a constant controlling the scaling factor. In our implementation, the low-rank decomposition matrices $B$ and $A$ are applied to the query, key, value, and output projection matrices within the self-attention layers of the top $K$ transformer layers [21] in ESM2. This approach significantly reduces the number of trainable parameters while maintaining the model's capacity for fine-tuning on new tasks. Here, $K$, $\alpha$, and $r$ are hyperparameters in our configuration.

Adapter tuning (Fig. 3c) further optimizes training by inserting adapter modules into the transformer layers of the ESM2. Unlike the original implementation [20], we only insert adapter modules into top $K$ transformer layers. Each adapter uses a bottleneck structure and a skip connection to compress and reconstruct input features with fewer parameters. Specifically, the adapter compresses the input into a lower dimensional space and then reconstructs it to match the original dimensions. In our configuration, $K$ is a hyperparameter controlling the number of transformer layers where the adapters are inserted. This method allows scalable fine-tuning for complex bioinformatics tasks while keeping the overall number of trainable parameters low.

For multiple downstream task learning, parallel adapters are employed, where each task-specific adapter is fine-tuned independently while keeping other adapters frozen. This enables efficient fine-tuning across multiple tasks. The overall transformation for each layer is described by

$$h' = h + \sum\nolimits_{i=1}^{N} a_i(h)$$

where $h$ represents the input features for a transformer layer, and $N$ is the number of parallel adapters. The task-specific transformation of each adapter $a_i(h)$ is defined as

$$a(h) = W_{\mathrm{up}} \cdot \sigma(W_{\mathrm{down}} \cdot h + b_{\mathrm{down}}) + b_{\mathrm{up}}$$

In this equation, $W_{\mathrm{up}}$ and $W_{\mathrm{down}}$ are trainable matrices for the up-projection and down-projection, respectively, while $b_{\mathrm{up}}$ and $b_{\mathrm{up}}$ are the corresponding biases. The ReLU activation function $\sigma$ is applied to the down-projection.

These efficient fine-tuning strategies, as illustrated in Fig. 3, allow us to train large models efficiently while maintaining performance on complex bioinformatics tasks, making the S-PLM model scalable for large-scale protein research.

## 3 Results

### 3.1 Unsupervised Protein Clustering

#### 3.1.1 Clustering CATH Structural Hierarchies

CATH [35] is a hierarchical classification system that categorizes protein structures based on their structural and functional features. We assessed the performance of S-PLM by determining its ability to incorporate structural information into the sequence latent space. To emphasize the structural aspects of proteins, we deliberately selected one representative sequence per CATH superfamily and

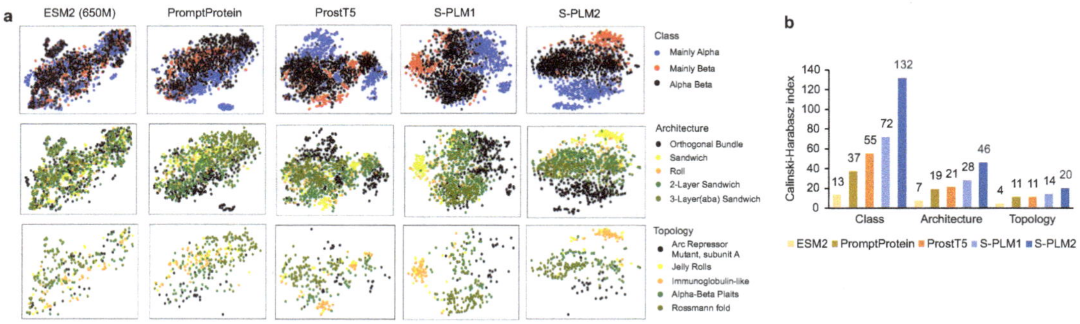


**Fig. 4** Visualization and benchmark of protein embeddings for CATH structural hierarchies—class, architecture, and topology. (**a**) T-SNE visualization of protein embeddings produced by ESM2, PromptProtein, ProstT5, S-PLM1, and S-PLM2. (**b**) Utilizing the CHI quantitatively assesses the ability of sequence embeddings generated by various models to distinguish CATH structural hierarchies

used the CATHS40 dataset, with a maximum sequence identity of 40%. The analysis focused on the class, architecture, and topology levels of the CATH hierarchy, excluding the sequence similarity-driven homologous superfamilies.

We compared the sequence representations generated by S-PLM1 and S-PLM2 with models that rely solely on sequence information, including ESM2 (ESM2_t33_650M_UR50D)—the pretrained PLM on which S-PLM is based—and two other structure-aware models, PromptProtein [36] and ProstT5 [37], which are pretrained by predicting 3D structures or 3D structure tokens from sequences. Figure 4a presents a t-distributed stochastic neighbor embedding (t-SNE) visualization of protein embeddings generated by these models across different CATH hierarchy levels, including class, architecture, and topology. The results indicate that both S-PLM1 and S-PLM2 achieve a clearer and more distinct separation of CATH structural hierarchies compared to the embeddings produced by the other models. This suggests that S-PLM more effectively incorporates structural information into the sequence latent space.

The Calinski-Harabasz index (CHI) [38] measures the ratio between inter-cluster dispersion and intra-cluster dispersion, serving as a metric for cluster validity. To quantitatively assess the ability of sequence embeddings generated by various models to distinguish CATH structural hierarchies, we employed the CHI. In this evaluation, CATH categories were used as ground-truth clusters. As illustrated in Fig. 4b, the CHI score of S-PLM1 is approximately 30% higher than that of ProsT5 across the class, architecture, and topology levels, and about 300% higher than that of ESM2. However, S-PLM2 demonstrated even more pronounced gains, with its CHI score being around 140% higher than ProsT5 in the CATH class and approximately 50% higher in both the CATH architecture and topology. Based on the CHI score,

S-PLM2 clearly outperforms S-PLM1, suggesting that S-PLM2 is more effective at clustering CATH structural hierarchies, likely due to its directly using of 3D coordinate features and residue-level contrastive learning. Given that these CATH categories are based on protein structures, this analysis strongly supports that the sequence embedding produced by both S-PLM1 and S-PLM2 demonstrate significant awareness of protein structural information, with S-PLM2 offering a more refined representation.

#### 3.1.2 Clustering Enzymes

To evaluate the effectiveness of the S-PLM1 and S-PLM2 models in clustering deaminase families, we compared their sequence embeddings with those generated by three other protein language models (PLMs): ESM2, PromptProtein, and ProstT5. The sequence data were sourced from a study by Huang et al., which employed structure-based clustering to identify deaminase functions and novel families undetectable by amino acid sequence analysis alone [39]. We generated 128-dimensional embeddings for each protein using S-PLM1 and S-PLM2, then applied t-SNE to reduce these embeddings to a 2D representation and used K-means clustering. The adjusted Rand index (ARI) was calculated by comparing the K-means clusters with known deaminase family annotations. As shown in Fig. 5a, b, S-PLM1 achieved an ARI of 0.87, S-PLM2 achieved an ARI of 0.86, outperforming ESM2 (0.63), PromptProtein (0.46), and ProstT5 (0.80).

We also examined the performance in clustering kinases groups, which facilitate phosphorylation by transferring phosphate groups to proteins. We extracted 336 kinases from GPS5.0, organized into 9 kinase groups, along with their respective kinase domain sequences. Sequence embeddings were generated for each kinase using these domain sequences. For comparison, embeddings were also generated using ESM2, PromptProtein, and ProstT5. As illustrated in the t-SNE plot (Fig. 5c) and ARI comparison (Fig. 5d), S-PLM1 achieved strong clustering performance for the kinases group, with an ARI of 0.72, while S-PLM2 demonstrated even better performance, with an ARI of 0.78, significantly higher than ESM2 (0.28), PromptProtein (0.28), and ProstT5 (0.61). This superior performance may be attributed to both S-PLM1 and S-PLM2 incorporating structural information essential for characterizing the kinases groups. In summary, S-PLM offers an effective sequence representation for enzyme clustering.

### 3.2 Efficient Fine-Tuning Strategies Enhance S-PLM'S Performance in Selected Tasks

To adapt the S-PLM1 and S-PLM2 to various supervised protein prediction tasks, we developed efficient fine-tuning strategies focused on the sequence encoder, including top-layer fine-tuning, adapter tuning, and LoRA. We benchmarked the S-PLM1 and S-PLM2 using various tuning methods across multiple tasks: gene ontology (GO) term prediction, enzyme commission (EC) number prediction, secondary structure (SS) prediction, protein fold

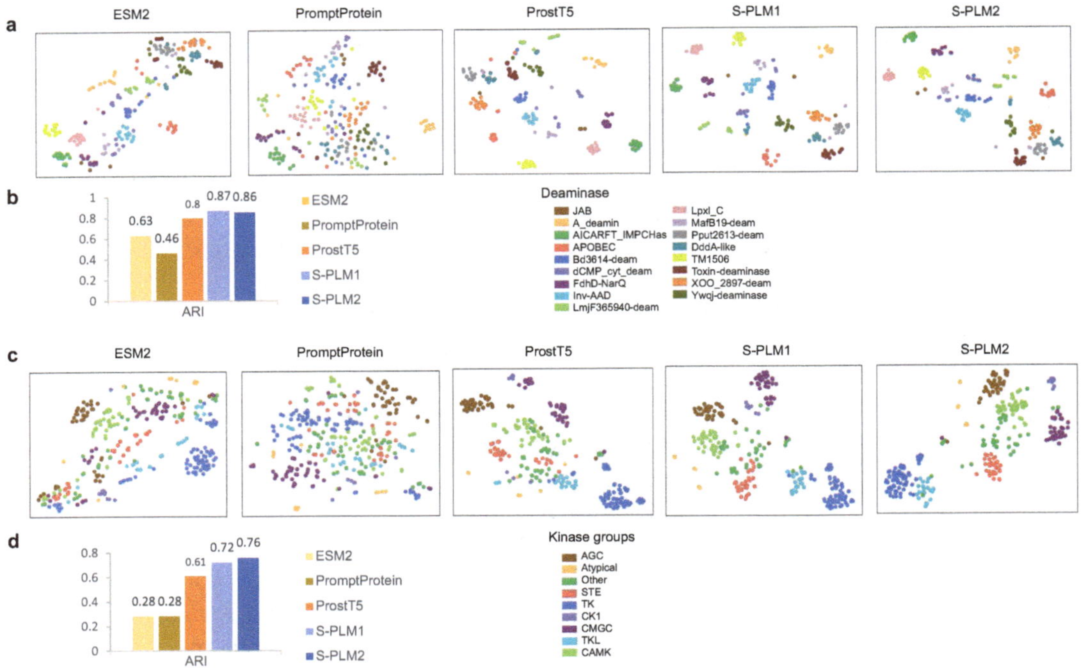


**Fig. 5** Clustering enzymes. (**a**) T-SNE visualization of protein embeddings generated from 242 deaminase proteins using models of ESM2, PromptProtein, ProstT5, S-PLM1, and S-PLM2. (**b**) Quantitative benchmarks of the ability of S-PLM's ability to cluster the deaminase proteins for sequence representation methods. (**c**) T-SNE visualization of protein embeddings from human kinase groups using pretrained PLM models of ESM2, PromptProtein, ProstT5, S-PLM1, and S-PLM2. Different kinase groups are distinguished by different colors. (**d**) Quantitative benchmarks of the ability of S-PLM1 and S-PLM2 to cluster the human kinases groups compared with other methods mentioned in (**c**). For both types of enzymes, ARI was computed by comparing K-means clustering assignments with enzyme labels

classification, and enzyme reaction (ER) classification, alongside state-of-the-art (SOTA) models. GO term and EC number predictions, as well as protein fold and ER classification, were evaluated as protein-level supervised classification tasks, while SS prediction was assessed as a residue-level task.

For fold classification, we used a dataset of 16,712 proteins classified into 1195-fold classes from the study [40]. The dataset has three test sets: "Fold," "Superfamily," and "Family". In the "Fold" set, proteins from the same superfamily are excluded from the training data. In the "Superfamily" set, proteins from the same family are excluded from training. In the "Family" set, proteins within the same family are included in the training process. For enzyme reaction classification, a dataset of 37,428 proteins categorized across 384 EC numbers was used, with training, validation, and test splits based on the methodology from the study [41]. GO term prediction and EC number prediction followed the same dataset splits as the study [42], with test sets having up to 95% sequence similarity to the training data. Finally, we adopted

**Table 1**
**Dataset statistics of downstream supervised protein prediction tasks**

| Dataset | | # Train | # Validation | # Test |
|---|---|---|---|---|
| Fold classification | Fold | 12,312 | 736 | 718 |
| | Superfamily | 12,312 | 736 | 1254 |
| | Family | 12,312 | 736 | 1272 |
| Enzyme reaction | | 29,215 | 2562 | 5651 |
| GO term | | 29,898 | 3322 | 3415 |
| EC number | | 15,550 | 1729 | 1919 |
| Secondary structure | | 8678 | 2170 | 513 |

"#" represents the number of proteins in the subset

Klausen's dataset as the training set and the CB513 dataset as the test set for secondary structure prediction, maintaining consistency with the reference. The statistics of the dataset for each task are shown in Table 1.

The training, validation, and test sets remained consistent across all tasks, with the only differences being the models and training strategies applied. Detailed results and model descriptions are provided in Table 2.

For GO and EC number predictions, seven methods using both sequence and structure inputs—GVP [15], GearNet [43], GearNet-Edge [43], CoupleNet [16], ESM-GearNet [15], PST (fine-tuned) [44], and SaProt [45]—as well as sequence-only methods ESM2 (650 M) [15] and ESM-S [46] were considered for comparison. For the SS task, sequence-only methods, TAPE (Transformer) [47], ProteinBERT [12], ESM-1b [48], and DML [49] were considered for comparison. For protein fold classification, and enzyme reaction (ER) classification, methods using both sequence and structure inputs—GVP, GearNet, GearNet-Edge, and CoupleNet were considered for comparison. Their results were taken directly from their respective publications.

Upon comparison as shown in Fig. 6, the S-PLM2 demonstrated optimum performance in GO-BP (Fmax: 0.498) and SS task (accuracy: 0.88), it also exhibits comparable performance in GO-CC (Fmax: 0.494 for CoupleNet, 0.519 for ESM-S, 0.484 for the S-PLM1 and 0.485 for the S-PLM2). At the same time, the S-PLM1 demonstrated optimum performance on GO-MF (Fmax: 0.685) and EC number prediction (Fmax: 0.890 for ESM-GearNet, 0.897 for PST, 0.888 for S-PLM1 and 0.883 for S-PLM2).

Among the methods under comparison, GVP and GearNet are designed to capture the invariant and equivariant features of protein structure, whereas GearNet-Edge is a variant of GearNet enhanced

**Table 2**
**Comparison of efficient fine-tuning methods for S-PLM1, S-PLM2, and ESM2 on protein prediction tasks with key model descriptions**

| Method | GO ($F_{max}$) BP | GO ($F_{max}$) MF | GO ($F_{max}$) CC | #Params | Sequence encoder layers | Classification layer |
|---|---|---|---|---|---|---|
| S-PLM1-fine-tune | 0.470 | 0.674 | 0.460 | 42 M, 40 M, 40 M | Fine-tune top 2 layers, freeze adapter | Mean pooling over all residues |
| S-PLM1-fine-tune | 0.475 | 0.679 | 0.435 | | | |
| ESM2-fine-tune | 0.473 | 0.677 | 0.450 | | | |
| S-PLM1-adapter | 0.495 | **0.685** | 0.484 | 55 M, 53 M, 53 M | Adapter tuning top 16 layers | |
| S-PLM2-adapter | **0.498** | 0.684 | **0.485** | | | |
| ESM2-adapter | 0.493 | 0.683 | 0.472 | | | |
| S-PLM1-LoRA | 0.480 | 0.668 | 0.472 | 3 M, 1 M, 0.7 M | LoRA on top 16 layers, $r = 2$, $\alpha = 8$ | |
| S-PLM2-LoRA | 0.476 | 0.666 | 0.473 | | | |
| ESM2-LoRA | 0.478 | 0.663 | 0.474 | | | |
| Method | EC ($F_{max}$) | | | #Params | Sequence encoder layers | Classification layer |
| S-PLM1-fine-tune | 0.885 | | | 40 M | Fine-tune top 2 layers, freeze adapter | Mean pooling over all residues |
| S-PLM2-fine-tune | 0.883 | | | | | |
| ESM2-fine-tune | 0.888 | | | | | |
| S-PLM1-adapter | 0.875 | | | 53 M | Adapter tuning top 16 layers | |
| S-PLM2-adapter | 0.872 | | | | | |
| ESM2-adapter | 0.866 | | | | | |
| S-PLM1-LoRA | **0.888** | | | 1 M | LoRA on top 33 layers, $r = 2$, $\alpha = 8$ | |
| S-PLM2-LoRA | 0.883 | | | | | |
| ESM2-LoRA | 0.864 | | | | | |
| Method | SS (accuracy%) | | | #Params | Sequence encoder layers | Classification layer |
| S-PLM1-fine-tune | 87.48 | | | 92 M | Fine-tune top 2 layers and all adapter layers | MLP (1280,100,3) on each residue |
| S-PLM2-fine-tune | **87.49** | | | | | |
| ESM2-fine-tune | 87.21 | | | | | |
| S-PLM1-adapter | 87.40 | | | 52 M | Adapter tuning top 16 layers | |
| S-PLM2-adapter | 87.33 | | | | | |
| ESM2-adapter | 86.73 | | | | | |
| S-PLM1-LoRA | 85.72 | | | 0.8 M | LoRA on top 33 layers, $r = 2$, $\alpha = 8$ | |
| S-PLM2-LoRA | 86.12 | | | | | |
| ESM2-LoRA | 87.14 | | | | | |

| Method | Protein classification (accuracy%) Fold | Superfamily | Family | #Params | Sequence encoder layers | Classification layer |
|---|---|---|---|---|---|---|
| S-PLM1-fine-tune | 37.74 | 77.59 | 98.51 | 116 M | Fine-tune top 5 layers and all adapter layers | Mean pooling over all residues |
| S-PLM2-fine-tune | **39.42** | 75.44 | 98.58 | 116 M | | |
| ESM2-fine-tune | 34.96 | 76.24 | **98.74** | 99 M | | |
| S-PLM1-adapter | 37.47 | 76,16 | 98.58 | 54 M | Adapter tuning top 16 layers | |
| S-PLM2-adapter | 38.58 | 76.71 | 98.51 | | | |
| ESM2-adapter | 37.33 | **79.59** | 98.58 | | | |
| S-PLM1-LoRA | 27.04 | 57.42 | 97.34 | 4 M | LoRA on top 33 layers, $r = 2$, $\alpha = 8$ | |
| S-PLM2-LoRA | 27.86 | 64.67 | 98.35 | | | |
| ESM2-LoRA | 27.16 | 67.22 | 97.88 | | | |
| Method | ER (accuracy%) | | | #Params | Sequence encoder layers | Classification layer |
| S-PLM1-fine-tune | 86.71 | | | 23 M | Fine-tune top1 layer and all adapter layers | Mean pooling over all residues |
| S-PLM2-fine-tune | 86.16 | | | 23 M | | |
| ESM2-fine-tune | 86.53 | | | 20 M | | |
| S-PLM1-adapter | 85.47 | | | 53 M | Adapter tuning top 16 layers | |
| S-PLM2-adapter | 85.24 | | | | | |
| ESM2-adapter | 83.63 | | | | | |
| S-PLM1-LoRA | 85.4 | | | 0.8 M | LoRA on top 33 layers, $r = 2$, $\alpha = 8$ | |
| S-PLM2-LoRA | **87.1** | | | | | |
| ESM2-LoRA | 84.5 | | | | | |

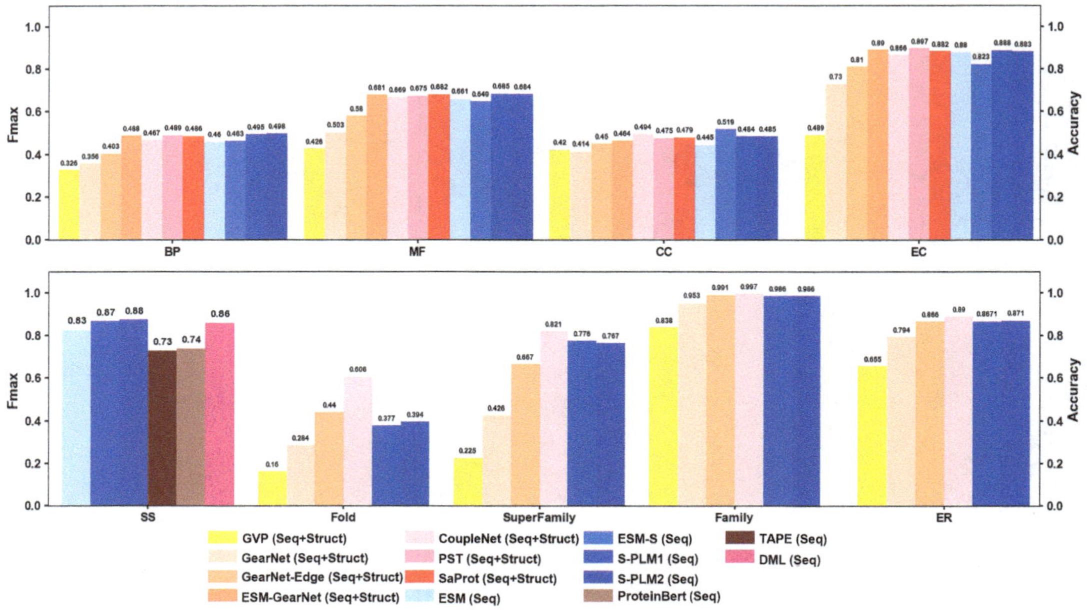


**Fig. 6** Benchmarking on GO, EC, SS prediction, protein classification, and ER classification. Seq refers to the use of protein sequences as input, while Struct refers to the use of protein structures as input. For GO and EC number predictions, seven methods using both sequence and structure inputs—GVP, GearNet, GearNet-Edge, CoupleNet, ESM-GearNet, PST (fine-tuned), and SaProt—as well as sequence-only methods ESM2 (650 M) and ESM-S were considered for comparison. For the SS task, sequence-only methods TAPE (transformer), ProteinBERT, ESM-1b, and DML were considered for comparison. For protein fold classification and enzyme reaction (ER) classification, methods using both sequence and structure inputs—GVP, GearNet, GearNet-Edge, and CoupleNet were considered for comparison

with an edge message passing layer. CoupleNet requires the integration of sequence and structure information by deeply coupling them, and ESM-GearNet is a recently proposed variant of GearNet that integrates the sequence representation from PLM with the structure representation through various fusion schemes, where the results with the best fusion scheme were reported in the comparison. SaProt is a newly introduced structure-aware PLM that explicitly utilizes a structure-aware vocabulary to integrate residue tokens with structure tokens derived from protein sequence and structure inputs. In contrast to these methods, our model relies solely on protein sequences. Remarkably, it achieves compatible results to CoupleNet, ESM-GearNet, and PST, or even superior results compared to GVP, GearNet, GearNet-Edge, and SaProt, showcasing the broader practical utility of our model.

Additionally, by employing the same efficient fine-tuning strategies on the base ESM2 encoder with an equal number of trainable parameters, we established a fair comparison among S-PLM1, S-PLM2, and the base ESM2 model. The results are presented in Table 2. In most (17/27) of the one-to-one comparisons, the

S-PLM1 encoder outperformed or matched its base ESM2 encoder. In most (16/27) one-to-one comparisons, the S-PLM2 encoder outperformed its base ESM2 encoder. Although we could not surpass ESM2 in all tasks across all training strategies, both S-PLM1 and S-PLM2 models consistently achieved the best overall performance for each task. In summary, our additional structure-aware module on ESM2, along with the new pretraining process, did not diminish ESM2's original sequence representation capability but rather led to superior performance in general.

### 3.3 Ablation Tests

In developing S-PLM, we explored various configurations of sequence encoders and pretraining strategies. This section presents the key findings from our ablation tests for both S-PLM1 and S-PLM2. Specifically, we evaluated the performance of S-PLM1 and S-PLM2 using different trainable layers across various training strategies, including adapter tuning, fine-tuning, and LoRA. The hyperparameter $K$ controlled the number of top transformer layers of the ESM2 base model into which the adapter module was inserted, as well as the number of layers that were fine-tuned or tuned using LoRA. The performance of these configurations was assessed on three protein clustering tasks, including the CATH clustering, deaminase protein clustering, and kinase group clustering. Preliminary training sessions were conducted for 10,000 steps with a consistent batch size of 20. To ensure robustness, each model variant was trained using three different random seeds. The adjusted Rand index (ARI) was computed for each task, with the average ARI score used to provide an overall evaluation.

For adapter tuning, we initially explored larger values of $K$ but discontinued further attempts after observing performance declines at higher values. In the case of fine-tuning, the number of trainable parameters increases significantly, which restricted our ability to explore higher values of $K$. As a result, we limited our exploration to values up to 6. For LoRA, larger $K$ values were possible, but $K$ was capped at 33, as this is the maximum number of layers in the base model.

The results, shown in Fig. 7, indicate that increasing $K$ for adapter tuning of S-PLM1 generally enhanced clustering performance in the CATH task, particularly in the CATH and topology categories. However, this improvement led to performance degradation in the other two tasks, which was expected. A larger number of adapter layers has a more pronounced effect on the base model, enhancing the structure-aware module's ability to learn structural features but potentially compromising the sequence representation of the base model. Fine-tuning with $K = 4$ led to modest improvements across all tasks, though comparable to the performance gains seen with higher $K$ values in adapter tuning. LoRA fine-tuning with larger $K$ values demonstrated competitive performance in the CATH clustering task, but similar to adapter tuning, it showed

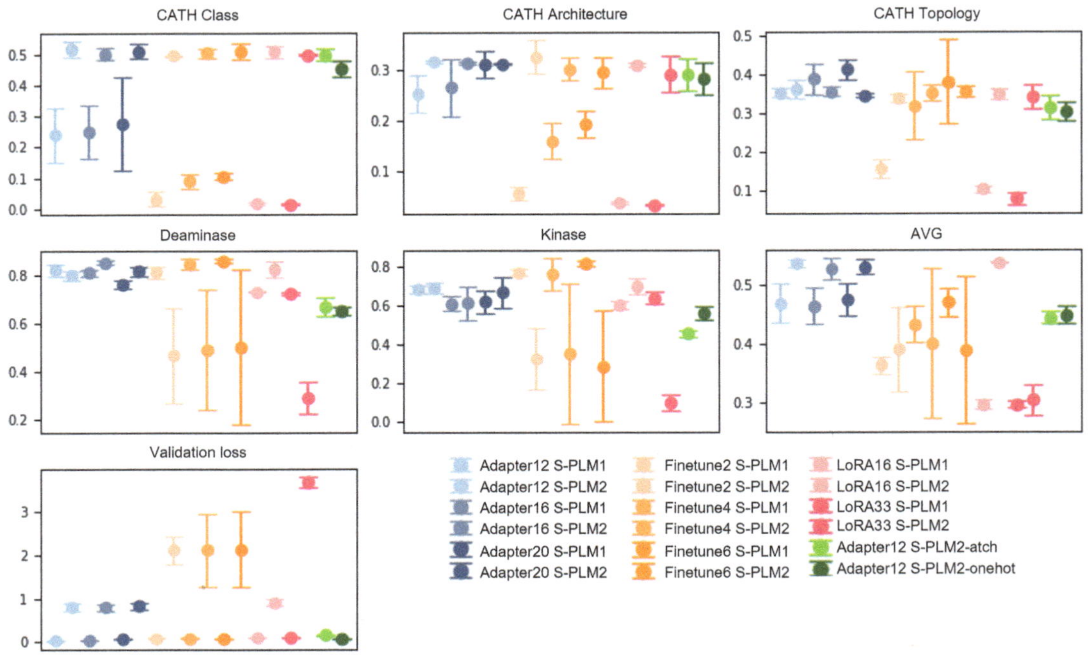


**Fig. 7** Quantitative evaluation of different configurations of S-PLM1 and S-PLM2. Their tuning methods utilize CATH clustering, which classifies based on Class, Architecture, and Topology, deaminase protein clustering, and kinase group clustering tasks. Performance is measured using adjusted Rand index (ARI) by comparing K-means clustering assignments with true labels. The numbers indicate the specific number of ESM2 transformer layers where the adapter module is inserted, or the number of layers fine-tuned or tuned using LoRA for each method

trade-offs in the other clustering tasks, especially when $K$ approached the upper limit.

Based on these findings, we ultimately selected adapter tuning with $K = 16$ for S-PLM1 and $K = 12$ for S-PLM2, achieving a balance between structural feature learning and the preservation of sequence representation performance. For S-PLM2, we introduced additional parameters to manage the usage of sequence embedding in the GVP model, specifically exploring the Atchley factor embedding and one-hot encoding. However, as demonstrated in Fig. 7, including these sequence embeddings did not improve performance and, in fact led to worse outcomes compared to not using sequence embeddings.

## 4 Discussion

In this work, we introduced two versions of S-PLM (S-PLM1 and S-PLM2), structure-aware protein language models pretrained using contrastive learning between protein sequences and their 3D structures. S-PLM1 uses contact maps and applies protein-level contrastive learning to align sequence and structure embeddings at a global level, while S-PLM2 uses 3D coordinate features and extends to include residue-level contrastive learning, capturing finer structural details. Both models employ efficient fine-tuning techniques to enable targeted modifications for downstream protein prediction tasks without requiring full parameter updates. Unlike joint-embedding methods, S-PLM independently encodes sequences and 3D structures, enabling structure-aware predictions using only sequence data.

The results of our study demonstrate S-PLM's ability to incorporate structural information into sequence representations, excelling in both protein clustering and supervised protein prediction tasks. In unsupervised clustering of CATH structural hierarchies and enzymes, both S-PLM1 and S-PLM2 showed clearer separations of protein structures compared to sequence-only models like ESM2 and structure-aware models like PromptProtein and ProstT5. Notably, S-PLM2 outperformed S-PLM1 in CATH clustering, which relies heavily on structural features. In enzyme clustering tasks, both models achieved comparable results and significantly outperformed other models, while S-PLM2 surpassed S-PLM1 in kinase clustering. These findings highlight GVP's strength in encoding intricate 3D structure features, underscoring its effectiveness as a protein structure encoder.

In supervised protein prediction tasks, S-PLM achieved competitive results without using 3D structural inputs, performing similarly to state-of-the-art models that require both sequence and structure. With efficient fine-tuning strategies like fine-tuning, adapter tuning, and LoRA, S-PLM models excelled in tasks such as GO term prediction, EC number prediction, secondary structure prediction, protein classification, and enzyme reaction classification.

We also observed that different tuning strategies performed optimally for different tasks. The variation in performance is likely due to the number of trainable parameters and the level of model adaptation provided by each strategy. Fine-tuning the top layers generally exposes more parameters, while LoRA, despite having fewer trainable parameters, impacts multiple layers, allowing it to capture complex patterns better suited for certain tasks. The nature of the task also plays a role—some tasks may benefit from fine-grained adaptation of lower layers, while others require substantial changes to higher level representations.

In conclusion, by incorporating a structure-aware module into the pretrained ESM2 model, S-PLM retains ESM2's sequence representation capabilities while effectively integrating structural information. S-PLM and its efficient fine-tuning strategies offer a powerful and flexible solution for protein analysis and prediction tasks.

## Acknowledgments

We thank Dr. Jin Chen for providing input on the development of the structure-aware protein language models. D.W. and D.X. are partially supported by the National Institutes of Health (Grant R35GM126985). Q.S. and D.X. would like to acknowledge the National Institutes of Health (Grant R01LM014510). The computation for this work was partially performed on the high-performance computing infrastructure provided by Research Computing Support Services at the University of Missouri. We also thank the University of Kentucky Center for Computational Sciences and Information Technology Services Research Computing for their support and use of the Morgan Compute Cluster and associated research computing resources. This work also used Delta-GPU at NCSA through allocation CIS230053 from the Advanced Cyberinfrastructure Coordination Ecosystem: Services & Support (ACCESS) program, which is supported by National Science Foundation grants #2138259, #2138286, #2138307, #2137603, and #2138296.

### References

1. Ofer D, Brandes N, Linial M (2021) The language of proteins: NLP, machine learning & protein sequences. Comput Struct Biotechnol J 19:1750–1758. https://doi.org/10.1016/j.csbj.2021.03.022
2. Devlin J, Chang M-W, Lee K, Toutanova K (2019) BERT: Pre-training of deep bidirectional transformers for language understanding. arXiv:1810.04805. Accessed: Sep. 27, 2024. [Online]. Available: http://arxiv.org/abs/1810.04805
3. Elnaggar A et al (2022) ProtTrans: toward understanding the language of life through self-supervised learning. IEEE Trans Pattern Anal Mach Intell 44(10):7112–7127. https://doi.org/10.1109/TPAMI.2021.3095381
4. Lin Z et al (2023) Evolutionary-scale prediction of atomic-level protein structure with a language model. Science 379(6637):1123–1130. https://doi.org/10.1126/science.ade2574
5. Rives A et al (2021) Biological structure and function emerge from scaling unsupervised learning to 250 million protein sequences. Proc Natl Acad Sci USA 118(15):e2016239118. https://doi.org/10.1073/pnas.2016239118
6. Ferruz N, Schmidt S, Höcker B (2022) ProtGPT2 is a deep unsupervised language model for protein design. Nat Commun 13(1):4348. https://doi.org/10.1038/s41467-022-32007-7
7. Vu MH et al (2023) Linguistically inspired roadmap for building biologically reliable protein language models. Nat Mach Intell 5(5):485–496. https://doi.org/10.1038/s42256-023-00637-1
8. Wang B et al (2023) Pre-trained Language Models in Biomedical Domain: A Systematic Survey. arXiv:2110.05006. Accessed: Sep. 27, 2024. [Online]. Available: http://arxiv.org/abs/2110.05006
9. Bepler T, Berger B (2021) Learning the protein language: evolution, structure, and function. Cell Systems 12(6):654–669.e3. https://doi.org/10.1016/j.cels.2021.05.017
10. Jumper J et al (2021) Highly accurate protein structure prediction with AlphaFold. Nature 596(7873):583–589. https://doi.org/10.1038/s41586-021-03819-2
11. Rao R et al (2021) MSA Transformer. https://doi.org/10.1101/2021.02.12.430858
12. Brandes N, Ofer D, Peleg Y, Rappoport N, Linial M (2022) ProteinBERT: a universal deep-learning model of protein sequence and function. Bioinformatics 38(8):2102–2110. https://doi.org/10.1093/bioinformatics/btac020
13. The Gene Ontology Consortium (2008) The Gene Ontology project. Nucleic Acids Res 36

(suppl_1):D440–D444. https://doi.org/10.1093/nar/gkm883

14. Chen (Sam) C, Zhou J, Wang F, Liu X, Dou D (2023) Structure-aware protein self-supervised learning. Bioinformatics 39(4):p. btad189. https://doi.org/10.1093/bioinformatics/btad189
15. Zhang Z et al (2023) A systematic study of joint representation learning on protein sequences and structures. arXiv:2303.06275. Accessed: Sep. 27, 2024. [Online]. Available: http://arxiv.org/abs/2303.06275
16. Hu B et al (2023) Learning complete protein representation by deep coupling of sequence and structure. https://doi.org/10.1101/2023.07.05.547769
17. Wang D et al (2023) S-PLM: structure-aware protein language model via contrastive learning between sequence and structure. Aug 07. https://doi.org/10.1101/2023.08.06.552203
18. Xu M, Yuan X, Miret S, Tang J (2023) ProtST: multi-modality learning of protein sequences and biomedical texts. arXiv:2301.12040. Accessed: Sep. 27, 2024. [Online]. Available: http://arxiv.org/abs/2301.12040
19. Yu T, Cui H, Li JC, Luo Y, Jiang G, Zhao H (2023) Enzyme function prediction using contrastive learning. Science 379(6639): 1358–1363. https://doi.org/10.1126/science.adf2465
20. Houlsby N et al (2019) Parameter-efficient transfer learning for NLP. arXiv:1902.00751. Accessed: Sep. 27, 2024. [Online]. Available: http://arxiv.org/abs/1902.00751
21. Hu EJ et al (2021) LoRA: low-rank adaptation of large language models. arXiv:2106.09685. Accessed: Sep. 27, 2024. [Online]. Available: http://arxiv.org/abs/2106.09685
22. Sledzieski S, Kshirsagar M, Baek M, Berger B, Dodhia R, Ferres JL (2023) Democratizing protein language models with parameter-efficient fine-tuning. 10. https://doi.org/10.1101/2023.11.09.566187
23. Zeng S, Wang D, Jiang L, Xu D (2024) Parameter-efficient fine-tuning on large protein language models improves signal peptide prediction. Genome Res:gr.279132.124. https://doi.org/10.1101/gr.279132.124
24. Barrett AJ (1997) Nomenclature Committee of the International Union of Biochemistry and Molecular Biology (NC-IUBMB). Enzyme nomenclature. Recommendations 1992. Supplement 4: corrections and additions. Eur J Biochem 250:1–6. https://doi.org/10.1111/j.1432-1033.1997.001_1.x
25. Liu Z et al (2021) Swin transformer: hierarchical vision transformer using shifted windows. In: 2021 IEEE/CVF international conference on computer vision (ICCV). IEEE, Montreal, QC, Canada, pp 9992–10002. https://doi.org/10.1109/ICCV48922.2021.00986
26. Chen T, Kornblith S, Norouzi M, Hinton G (2020) A simple framework for contrastive learning of visual representations. arXiv:2002.05709. Accessed: Sep. 27, 2024. [Online]. Available: http://arxiv.org/abs/2002.05709
27. Radford A et al (2021, 2024) Learning transferable visual models from natural language supervision. arXiv:2103.00020. [Online]. Available: http://arxiv.org/abs/2103.00020
28. Jing B, Eismann S, Suriana P, Townshend RJL, Dror R (2021) Learning from protein structure with geometric vector perceptrons. arXiv:2009.01411. Accessed: Sep. 27, 2024. [Online]. Available: http://arxiv.org/abs/2009.01411
29. Ingraham J, Garg V, Barzilay R, Jaakkola T. Generative models for graph-based protein design
30. Van Kempen M et al (2024) Fast and accurate protein structure search with Foldseek. Nat Biotechnol 42(2):243–246. https://doi.org/10.1038/s41587-023-01773-0
31. Defresne M, Barbe S, Schiex T (2023) Scalable coupling of deep learning with logical reasoning. In: Proceedings of the thirty-second international joint conference on artificial intelligence. International Joint Conferences on Artificial Intelligence Organization, Macau, SAR China, pp 3615–3623. https://doi.org/10.24963/ijcai.2023/402
32. Atchley WR, Zhao J, Fernandes AD, Drüke T (2005) Solving the protein sequence metric problem. Proc Natl Acad Sci USA 102(18): 6395–6400. https://doi.org/10.1073/pnas.0408677102
33. Birnbaum F, Jain S, Madry A, Keating AE. Jointly embedding protein structures and sequences through residue level alignment
34. Micikevicius P et al (2017) Mixed precision training. arXiv. https://doi.org/10.48550/ARXIV.1710.03740
35. Sillitoe I et al (2021) CATH: increased structural coverage of functional space. Nucleic Acids Res 49(D1):D266–D273. https://doi.org/10.1093/nar/gkaa1079
36. Wang Z et al (2023) Multi-level protein structure pre-training with prompt learning
37. Heinzinger M et al (2023) Bilingual language model for protein sequence and structure. 25.

https://doi.org/10.1101/2023.07.23.550085

38. Calinski T, Harabasz J (1974) A dendrite method for cluster analysis. Comm Stats - Theory Methods 3(1):1–27. https://doi.org/10.1080/03610927408827101
39. Huang J et al (2023) Discovery of deaminase functions by structure-based protein clustering. Cell 186(15):3182–3195.e14. https://doi.org/10.1016/j.cell.2023.05.041
40. Lin T-Y, Goyal P, Girshick R, He K, Dollar P (2017) Focal loss for dense object detection. In: 2017 IEEE international conference on computer vision (ICCV). IEEE, Venice, pp 2999–3007. https://doi.org/10.1109/ICCV.2017.324
41. Hermosilla P et al (2020) Intrinsic-extrinsic convolution and pooling for learning on 3D protein structures. arXiv. https://doi.org/10.48550/ARXIV.2007.06252
42. Gligorijević V et al (2021) Structure-based protein function prediction using graph convolutional networks. Nat Commun 12(1):3168. https://doi.org/10.1038/s41467-021-23303-9
43. Zhang Z et al (2022) Protein representation learning by geometric structure pretraining. arXiv. https://doi.org/10.48550/ARXIV.2203.06125
44. Chen D, Hartout P, Pellizzoni P, Oliver C, Borgwardt K (2024) Endowing protein language models with structural knowledge. arXiv. https://doi.org/10.48550/ARXIV.2401.14819
45. Su J, Han C, Zhou Y, Shan J, Zhou X, Yuan F (2023) SaProt: protein language modeling with structure-aware vocabulary. https://doi.org/10.1101/2023.10.01.560349
46. Zhang Z, Lu J, Chenthamarakshan V, Lozano A, Das P, Tang J (2024) Structure-informed protein language model. arXiv. https://doi.org/10.48550/ARXIV.2402.05856
47. Rao R et al (2019) Evaluating protein transfer learning with TAPE. Adv Neural Inf Process Syst 32:9689–9701
48. Xu M et al (2022) PEER: a comprehensive and multi-task benchmark for protein sequence understanding. arXiv. https://doi.org/10.48550/ARXIV.2206.02096
49. Yang W, Liu Y, Xiao C (2022) Deep metric learning for accurate protein secondary structure prediction. Knowl-Based Syst 242:108356. https://doi.org/10.1016/j.knosys.2022.108356

# Chapter 3

# Exploring ProtFlash: An Efficient Approach to Protein Data Analysis

**Lei Wang, Zhidong Xue, and Yan Wang**

## Abstract

This chapter introduces ProtFlash, a novel lightweight protein language model that represents a significant advancement in the field of bioinformatics. By addressing the computational challenges inherent in protein sequence analysis, ProtFlash offers an efficient and accessible solution that combines innovative architectural design with practical applicability. The model incorporates several key technological breakthroughs, including a mixed-chunk attention mechanism that integrates both global and local attention patterns, and the implementation of RoPE position encoding for handling protein sequences of arbitrary lengths. This comprehensive guide details the model's development, implementation, and practical applications in protein downstream tasks. It provides step-by-step instructions for utilizing the ProtFlash library (https://github.com/ISYSLAB-HUST/ProtFlash), including advanced techniques such as protein embedding generation and model fine-tuning using low-rank adaptation (LoRA) methodology. The chapter serves as both a theoretical foundation and practical handbook for researchers and practitioners in the field of protein sequence analysis and bioinformatics.

**Key words** ProtFlash, Protein language model, Protein embedding, LoRA, t-SNE

## 1 Introduction

Proteins are fundamental macromolecules of life and play a crucial role in maintaining human health and in the treatment of diseases. These biomolecules are involved in almost every cellular process, including enzyme catalysis, signal transduction, immune responses, and cellular structure maintenance. As such, understanding the properties and functions of proteins is paramount in biomedical research, drug design, and precision medicine [1]. The sequence information of proteins—comprising chains of amino acids—is critical for determining their three-dimensional structure and biological functions. This knowledge is essential for decoding the "biological language" of proteins, which is key to understanding

Dukka B. KC (ed.), *Large Language Models (LLMs) in Protein Bioinformatics*, Methods in Molecular Biology, vol. 2941,
https://doi.org/10.1007/978-1-0716-4623-6_3,

complex biological systems and disease mechanisms [2]. With the rapid advancement of sequencing technologies, enormous volumes of protein sequence data have been generated. Large-scale databases such as UniProt [3] and Pfam [4] have been established to store and organize this data, providing essential resources for scientists. These databases serve as valuable repositories of protein sequences, annotations, and functional information. Researchers can now tap into these datasets to develop sophisticated computational models, particularly deep learning models, for a variety of protein-related tasks.

Simultaneously, artificial intelligence (AI) technologies, particularly large language models (LLMs), have revolutionized natural language processing (NLP) and many other fields [5]. A prominent example of this is ChatGPT, a conversational model that has set new benchmarks in human–computer interaction. Inspired by the success of LLMs in NLP, protein language models have emerged as a powerful tool for representing and learning from protein sequence data. These models, which aim to capture the "semantic" meaning of proteins, are quickly becoming the standard approach for protein feature representation and have shown promise in outperforming traditional methods for protein sequence analysis [6–8]. Protein language models operate by learning distributed representations of protein sequences through deep neural networks. These models are typically trained in an unsupervised manner, meaning they can learn from large, unannotated datasets. This approach enables them to capture the complex relationships and hidden patterns embedded in protein sequences. The result is a model that can "understand" the sequence in much the same way that large language models understand human languages. As a consequence, these models have made groundbreaking strides in a variety of downstream tasks, such as predicting protein structures [9] and functional annotation [10].

However, despite the impressive progress made, several challenges remain in the development and application of protein language models. One of the most significant challenges is the inherent complexity of protein sequences. The length of proteins varies greatly, ranging from approximately 30 to over 33,000 amino acid residues. This is much larger than typical sequences in natural languages, where sentences usually contain no more than 15–30 words. The sheer length of protein sequences poses a significant challenge for computational models, particularly those based on the transformer architecture, which is widely used in protein language models [11], which has a computational complexity that scales quadratically with the input sequence length. Therefore, the high demand for computational resources has become a bottleneck for most academic communities with limited resources. A common approach is to truncate protein sequences to reduce computational complexity; however, this may compromise the semantic information of the protein sequences.

To address the computational intensity of existing protein language models, we have developed a lightweight protein language model called ProtFlash [12]. Its lightweight characteristics are mainly reflected in three aspects: linear complexity, lower memory consumption, and relatively fewer parameters. The model employs a hybrid chunked attention mechanism, incorporating both global and local attention, to meet the linear complexity requirements of attention-based models. Unlike learnable position encodings, ProtFlash integrates the latest RoPE position encoding method [13], allowing it to encode protein sequences of arbitrary lengths. Another key feature of ProtFlash is its lightweight architecture, which is reflected in its relatively small number of parameters. With only 174 million parameters, ProtFlash is substantially smaller than many of the larger protein language models available today. Despite its smaller size, ProtFlash has shown excellent performance in terms of both accuracy and computational efficiency. Specifically, the model can be trained efficiently on just eight NVIDIA A40 Graphics Processing Units (GPUs), a significant reduction in computational resource requirements compared to larger models. Additionally, ProtFlash's memory consumption increases more slowly as the protein sequence length grows. For example, for sequences ranging from 1024 to 4096 residues, the memory consumption of ProtFlash increases by approximately 3.4 times, which is a substantial improvement over existing attention-based protein models that experience much larger increases in memory usage. Given these advantages, ProtFlash is well suited for use by researchers with limited access to high-performance computing resources. By lowering the computational requirements, ProtFlash opens up new possibilities for protein sequence analysis in academic and smaller-scale research settings.

In this chapter, we will provide a comprehensive introduction to using the ProtFlash library. We will guide readers through its installation and use in Python environments, such as IPython notebooks or Python scripts. Through step-by-step examples, we will demonstrate how to apply ProtFlash for various tasks, including generating protein embeddings and fine-tuning the model using techniques like low-rank adaptation (LoRA), which helps improve model performance with fewer parameters. These examples will showcase how ProtFlash can be effectively integrated into research workflows, offering a practical tool for protein sequence analysis.

## 2 Materials

To use ProtFlash, you will need the following resources:

1. CPU: Intel Core i9 or AMD Ryzen 9 series, supporting fast processing of large datasets and complex computational tasks.
2. Memory (RAM): At least 32 GB; 64 GB or higher is recommended for handling large datasets and performing large-scale model training.
3. GPU: NVIDIA RTX 3080 or higher model, providing strong computing power and high memory bandwidth suitable for deep learning training.
4. Storage: At least 1 TB SSD; using NVMe SSD can enhance read/write speeds, and consider additional HDD for data backup.
5. Network: High-speed internet connection, suitable for users frequently downloading large datasets or synchronizing with remote servers.

Datasets:

Protein sequence data can be downloaded from public databases such as UniProt (https://www.uniprot.org/) and Pfam (http://pfam.xfam.org/).

## 3 Methods

This section is the core of this chapter and will detail how to use ProtFlash for protein sequence analysis. We will introduce it, step-by-step.

### *3.1 Installing ProtFlash*

You can customize the command based on your system configuration (e.g., CUDA version):

```
conda install pytorch torchvision torchaudio -c pytorch
```

You can use this one-liner for installation, using the latest release version:

```
# latest version

pip install git+https://github.com/isyslab-hust/ProtFlash

# stable version

pip install ProtFlash
```

### 3.2 Data Preparation

Unlike traditional bioinformatics software that directly uses the fasta format, ProtFlash parses sequences directly from dictionaries. This usually involves converting the sequence to a specific format that ProtFlash can handle. First, download protein sequence data from the UniProt database. Format the data as follows to fit the input requirements of ProtFlash:

```
data = [

   ("protein1", "MKTVRQERLKSIVRILERSKEPVSGAQLAEELSVSRQVIVQD

IAYLRSLG"),

   ("protein2", "KALTARQQEVFDLIRDHISQTGMPPTRAEIAQRLGFRSPNAA

EEHLKAL"),

]
```

### 3.3 Generating Protein Embeddings

This code snippet demonstrates how to generate protein embeddings using the ProtFlash library. It begins by importing the necessary functions, ***load_prot_flash_base*** for loading the pretrained ProtFlash model and ***batchConverter*** for preparing the input data. A list of protein identifiers and their corresponding amino acid sequences is defined. The ***batchConverter*** processes this data, returning the tokenized sequences and their lengths. After loading the ProtFlash model, embeddings for the sequences are generated without tracking gradients, which saves memory and computational resources. Finally, the code computes average representations for each protein sequence by averaging the token embeddings, excluding padding tokens, and stores these representations for further analysis. This setup facilitates various tasks in protein sequence analysis, such as classification or clustering.

```
from ProtFlash.pretrain import load_prot_flash_base
from ProtFlash.utils import batchConverter


ids, batch_token, lengths = batchConverter(data)
model = load_prot_flash_base()
with torch.no_grad():
    token_embedding = model(batch_token, lengths)
# Generate per-sequence representations via averaging
representations = []
for i, (_, seq) in enumerate(data):
    representations.append(token_embedding[i, 0: len(seq) + 1].mean(0))
```

### 3.4 t-SNE Visualization of Protein Embeddings: Uncovering Hidden Patterns

This code snippet demonstrates how to visualize protein sequence representations using t-distributed stochastic neighbor embedding (t-SNE) [14]. It begins by importing the necessary libraries: ***numpy*** [15] for array manipulation, ***matplotlib.pyplot*** for plotting, and ***sklearn.manifold.TSNE*** for dimensionality reduction. The averaged embeddings from the previous step are converted to a NumPy array. t-SNE is then applied to reduce the dimensionality of the embeddings to 2D, suitable for visualization. A scatter plot is created to display the t-SNE results, with each point representing a protein sequence, annotated with its identifier for clarity. This visualization helps to illustrate the relationships between different protein sequences in the embedding space.

```
import numpy as np
import matplotlib.pyplot as plt
from sklearn.manifold import TSNE


# Assuming 'sequence_representations' contains the averaged embeddings f
rom the previous code
# Convert the list to a NumPy array for t-SNE
embeddings_array = np.array(representations)
```

```
# Apply t-SNE to reduce dimensions to 2D for visualization
tsne = TSNE(n_components=2, random_state=42)
tsne_results = tsne.fit_transform(embeddings_array)

# Plot the t-SNE results
plt.figure(figsize=(8, 6))
plt.scatter(tsne_results[:, 0], tsne_results[:, 1], marker='o')

# Annotate points with protein identifiers
for i, (protein_id, _) in enumerate(data):
    plt.annotate(protein_id, (tsne_results[i, 0], tsne_results[i, 1]), fontsize=9)

plt.title("t-SNE Visualization of Protein Sequence Representations")
plt.xlabel("t-SNE Component 1")
plt.ylabel("t-SNE Component 2")
plt.grid()
plt.show()
```

### 3.5 Initializing ProtFlash with Custom Model Parameters

The provided code snippet demonstrates how to load custom weights into a ProtFlash model using PyTorch. It begins by importing the necessary libraries, including torch and ***FLASHTransformer*** from ProtFlash. The code then loads the model's weights and hyperparameters from a specified file using *torch.load()*, extracting the relevant hyperparameters required for model initialization. The ***FLASHTransformer*** model is instantiated with these hyperparameters, including dimensions, number of tokens, layers, and other configuration settings. Finally, the pretrained weights are loaded into the model using the ***load_state_dict()*** method, enabling the use of customized weights for specific tasks while maintaining the model's architecture.

```
import torch
from ProtFlash.model import FLASHTransformer

# Load the model data from the specified file
model_data = torch.load("your_parameters_file")  # Replace with your actual
file path

# Extract hyperparameters
hyper_parameter = model_data["hyper_parameters"]

# Initialize the FLASHTransformer model
model = FLASHTransformer(
    hyper_parameter['dim'],
    hyper_parameter['num_tokens'],
    hyper_parameter['num_layers'],
    group_size=hyper_parameter['num_tokens'],
    query_key_dim=hyper_parameter['qk_dim'],
    max_rel_dist=hyper_parameter['max_rel_dist'],
    expansion_factor=hyper_parameter['expansion_factor']
)

# Load the model state dictionary
model.load_state_dict(model_data['state_dict'])
```

### 3.6 Model Optimization

#### 3.6.1 Introduction to LoRA

Low-rank adaptation (LoRA) [16] is a parameter-efficient fine-tuning technique for neural networks. It significantly reduces the number of parameters that need to be trained during fine-tuning by applying a low-rank decomposition to the pretrained model's weights, improving training efficiency, especially in downstream tasks.

LoRA works by decomposing the weight update matrix $\Delta W$ in neural networks into the product of two smaller matrices: $\Delta W = A \times B$, where $A \in R^{d \times r}$ and $B \in R^{r \times k}$, with $r$ being the low-rank dimension (typically much smaller than $d$ and $k$). During fine-tuning, the original weight matrix $W$ is frozen, and only the low-rank matrices $A$ and $B$ are trained. This reduces the number of trainable parameters from $d \times k$ *to* $(d \times r) + (r \times k)$, making fine-tuning more efficient. The final adapted weight becomes $W + \Delta W = W + A \times B$, allowing the model to maintain pretrained knowledge while adapting to new tasks with fewer parameters.

#### 3.6.2 Installing the minLoRA Library

minLoRA (https://github.com/cccntu/minlora) is a compact yet fully functional PyTorch library for reimplementing LoRA, allowing it to be directly applied to any model that inherits from ***torch.nn.Module*** without modifying the model definition. It leverages PyTorch's ***torch.nn.utils.parametrize*** for parameterization, making it easy to extend and enabling users to add custom LoRA parameterization methods. With just around 100 lines of code, minLoRA supports the training and inference of various LoRA models, offering a simple yet powerful solution for LoRA integration. To install the minLoRA library, you can follow these steps:

```
git clone https://github.com/cccntu/minLoRA.git

cd minLoRA

pip install -e .
```

#### 3.6.3 Applying LoRA in ProtFlash

This code snippet demonstrates how to integrate the LoRA module into the ProtFlash model and optimize the LoRA parameters. The specific steps are as follows:

1. Define LoRA Configuration: Use the ***LoRAParametrization.from_linear*** function to define the LoRA weight parameterization for ***torch.nn.Linear*** layers. The rank = 3 indicates that the rank of the low-rank approximation is 3.
2. Load ProtFlash Model: Use ***load_prot_flash_base()*** to load the original ProtFlash model.
3. Add LoRA to the ProtFlash Model: Integrate the LoRA module into the ProtFlash model using the ***add_lora*** function and specify which layers use LoRA through ***lora_config***.

4. Move Model to Appropriate Device: Move the model to either GPU or CPU for computation based on the availability of a GPU.
5. Obtain LoRA Parameters for Optimization: Use the ***get_lora_params*** method to retrieve the LoRA parameters and optimize these parameters with the Adam optimizer, setting the learning rate to 3e-4.

```
from minlora import add_lora, get_lora_params, get_lora_state_dict, LoRAPa
rametrization
from functools import partial

# LoRA configuration
lora_config = {
    torch.nn.Linear: {
        "weight": partial(LoRAParametrization.from_linear, rank=3),
    },
}

# Load the original ProtFlash model
model = load_prot_flash_base()

# Add LoRA to the ProtFlash model
add_lora(model, lora_config=lora_config)
```

```
# Move model to the appropriate device

device = torch.device('cuda:0' if torch.cuda.is_available() else 'cpu')

model = model.to(device)


# Example of obtaining LoRA parameters for optimization

optimizer = torch.optim.Adam(get_lora_params(model), lr=3e-4)
```

## 4 Conclusion

This chapter serves as a comprehensive tutorial, detailing ProtFlash, a groundbreaking lightweight protein language model designed to overcome the computational limitations of traditional models. Characterized by linear complexity, low memory consumption, and a smaller number of parameters, ProtFlash is well suited for environments with limited resources. Through step-by-step guidance, this chapter demonstrates how to use ProtFlash within IPython notebooks or Python scripts, covering installation, data preparation, generation of protein embedding information, and model optimization using LoRA techniques. Additionally, we provide t-SNE visualization methods to reveal the relationships between protein sequences in a reduced-dimensional space.

In summary, this chapter offers detailed instructions for using ProtFlash and showcases its effective application in protein sequence analysis. Whether you are new to the field of bioinformatics or an expert, these guidelines will help you quickly become proficient and make the most of ProtFlash's powerful capabilities.

## Acknowledgments

This work was supported by the National Natural Science Foundation of China under Grant 62172172, the Postdoctoral Fellowship Program of CPSF under Grant Number GZC20240545, and the Scientific Research Start-up Foundation of Binzhou Medical University under Grant BY2020KYQD01.

## References


1. Carbajal-Valenzuela IA, Apolonio-Hernandez NM, Gutierrez-Chavez DV, González-Arias B, Jimenez-Hernandez A, Rico-García E, Feregrino-Pérez AA, Guevara-González RG (2022) Biological macromolecules as nutraceuticals. In: Biological macromolecules. Elsevier, pp 97–138
2. Bepler T, Berger B (2021) Learning the protein language: evolution, structure, and function. Cell Syst 12(6):654–669. e653
3. UniProt C (2023) UniProt: the universal protein knowledgebase in 2023. Nucleic Acids Res 51(D1):D523–D531. https://doi.org/10.1093/nar/gkac1052
4. Mistry J, Chuguransky S, Williams L, Qureshi M, Salazar GA, Sonnhammer ELL, Tosatto SCE, Paladin L, Raj S, Richardson LJ, Finn RD, Bateman A (2021) Pfam: the protein families database in 2021. Nucleic Acids Res 49 (D1):D412–D419. https://doi.org/10.1093/nar/gkaa913
5. Wu T, He S, Liu J, Sun S, Liu K, Han Q-L, Tang Y (2023) A brief overview of ChatGPT: the history, status quo and potential future development. IEEE/CAA J Autom Sin 10(5): 1122–1136
6. Rives A, Meier J, Sercu T, Goyal S, Lin Z, Liu J, Guo D, Ott M, Zitnick CL, Ma J (2021) Biological structure and function emerge from scaling unsupervised learning to 250 million protein sequences. Proc Natl Acad Sci 118(15):e2016239118
7. Lin Z, Akin H, Rao R, Hie B, Zhu Z, Lu W, Smetanin N, Verkuil R, Kabeli O, Shmueli Y (2023) Evolutionary-scale prediction of atomic-level protein structure with a language model. Science 379(6637):1123–1130
8. Elnaggar A, Heinzinger M, Dallago C, Rehawi G, Wang Y, Jones L, Gibbs T, Feher T, Angerer C, Steinegger M (2021) Prottrans: toward understanding the language of life through self-supervised learning. IEEE Trans Pattern Anal Mach Intell 44(10): 7112–7127
9. Wang W, Peng Z, Yang JJNCS (2022) Single-sequence protein structure prediction using supervised transformer protein language models. Nat Comput Sci 2(12):804–814
10. Yu T, Cui H, Li JC, Luo Y, Jiang G, Zhao H (2023) Enzyme function prediction using contrastive learning. Science 379(6639): 1358–1363
11. Lin T, Wang Y, Liu X, Qiu X (2022) A survey of transformers. AI Open 3:111–132
12. Wang L, Zhang H, Xu W, Xue Z, Wang Y (2023) Deciphering the protein landscape with ProtFlash, a lightweight language model. Cell Rep Phys Sci 4(10):101600
13. Su J, Ahmed M, Lu Y, Pan S, Bo W, Liu Y (2024) Roformer: enhanced transformer with rotary position embedding. Neurocomputing 568:127063
14. Van der Maaten L, Hinton G (2008) Visualizing data using t-SNE. J Mach Learn Res 9(11): 2579
15. Harris CR, Millman KJ, Van Der Walt SJ, Gommers R, Virtanen P, Cournapeau D, Wieser E, Taylor J, Berg S, Smith N (2020) Array programming with NumPy. Nature 585(7825):357–362
16. Hu EJ, Shen Y, Wallis P, Allen-Zhu Z, Li Y, Wang S, Wang L, Chen W (2021) Lora: low-rank adaptation of large language models. arXiv preprint arXiv:09685

# Chapter 4

# Ranking Protein–Protein Models with Large Language Models and Graph Neural Networks

**Xiaotong Xu and Alexandre M. J. J. Bonvin**


## Abstract

Protein–protein interactions (PPIs) are associated with various diseases, including cancer, infections, and neurodegenerative disorders. Obtaining three-dimensional structural information on these PPIs serves as a foundation to interfere with those or to guide drug design. Various strategies can be followed to model those complexes, all typically resulting in a large number of models. A challenging step in this process is the identification of good models (near-native PPI conformations) from the large pool of generated models. To address this challenge, we previously developed DeepRank-GNN-esm, a graph-based deep learning algorithm for ranking modeled PPI structures harnessing the power of protein language models. In this chapter, we detail the use of our software with examples. DeepRank-GNN-esm is freely available at https://github.com/haddocking/DeepRank-GNN-esm.


**Key words** Protein–protein scoring, Graph neural network, Protein language model, DeepRank-GNN-esm

## 1 Introduction

Protein–protein interactions (PPIs) are fundamental for driving various biological processes within cells, such as cell growth, metabolism, and regulatory pathways. Obtaining three-dimensional (3D) structural insights into these PPIs is vital for understanding both normal physiological processes and their disruption in case of disease, guiding efforts toward the development of drugs to restore a normal functioning. Next to experimental structural methods such as cryo-electron microscopy and X-ray crystallography, various computational techniques have emerged as alternatives, ranging from integrative modeling [1–3] to artificial intelligence (AI)-based methods [4–6].

Despite recent progress in predicting the structure of biomolecular complexes, accurately identifying near-native conformations (good models) from the large pool of generated models, which is

Dukka B. KC (ed.), *Large Language Models (LLMs) in Protein Bioinformatics*, Methods in Molecular Biology, vol. 2941, https://doi.org/10.1007/978-1-0716-4623-6_4,

known as the "scoring problem," remains challenging as demonstrated, for example, in the critical assessment of predicted interactions (CAPRI) [7, 8]. To address this issue, we have previously developed DeepRank-GNN [9], a deep learning architecture that enables analysis of PPI interfaces using graph neural networks. The workflow of DeepRank-GNN is as follows: Given an input PPI 3D model, it converts the interaction interface into a residue graph where nodes represent protein residues at the interface and edges denote interactions between them. All graphs are stored and handled using the HDF5 format [10] to ensure smooth I/O operations and efficient training. The generated graph is then processed through a pretrained graph neural network and the model quality is predicted in terms of $f_{nat}$, a metric between 0 and 1 reporting the fraction of native intermolecular contacts with respects to the ground truth (which would be the experimental structure of the complex if available). The network has been trained on large datasets alongside their corresponding $f_{nat}$ values, learning to rationalize between 3D structure, sequences, and interface quality. In the original DeepRank-GNN approach, five node features are used to capture various aspects of protein–protein interaction interfaces. These include residue type, polarity, buried surface area, and charge, which aims to characterize the physical properties of the interface. Additionally, we integrate evolutionary information through position-specific scoring matrix (PSSM) profiles. PSSM features have been shown to play a crucial role in improving the predictive performance of DeepRank-GNN [9]. They are, however, quite computational demanding to generate in terms of time and resources, requiring typically large sequence alignments to extract the evolutionary information.

The field of natural language processing has witnessed remarkable advancements, largely driven by the escalating scale of large language models (LLMs) [11]. This progression has extended to the realm of protein science with the development of sophisticated protein language models such as ProteinBERT [12], ProGen [13], evolutionary scale modeling-2/3 (ESM-2/EMS-3) [6, 14]. ESM-2 has been trained to predict the identity of randomly masked amino acids in a protein sequence. By solving billions of sequence puzzles, ESM-2 effectively captured evolutionary insights spanning across sequences. Consequently, sequence embeddings extracted from ESM-2 models could provide valuable features for down-stream PPI-related tasks such as protein structure prediction [15], functional annotation [16], or generating novel sequences for protein families [17]. Recognizing the great potential of such large language models, we have replaced the PSSM features with embeddings from the ESM-2 protein language model, resulting in a new version of our software called DeepRank-GNN-esm [18]. This improvement greatly reduced the computation time at no cost in-, or even better performance than the original PSSM-based model on two applications: (1) Discriminating physiological from non-physiological interfaces and (2) Ranking docked PPI models.

In this chapter, we provide comprehensive instructions for the use of DeepRank-GNN-esm, focusing on its application for scoring PPI conformations. We introduce two execution modes for the algorithm with clear guidance on their implementations. Additionally, we dive into the process of customizing the deep learning architecture to meet individual requirements, ensuring flexibility and adaptability to diverse research needs.

## 2 Materials

The following resources are required to use our deep learning tools.

### 2.1 Software

1. ESMFold [6] software package (available at https://github.com/facebookresearch/esm): Code and pretrained weights for protein language model embedding calculation.
2. DeepRank-GNN-esm [18] software package (https://github.com/haddocking/DeepRank-GNN-esm): Code and pre-trained model weights for $f_{nat}$ computation.
3. Python package PDB-tools (version 2.5.0 https://github.com/haddocking/pdb-tools): A set of useful tools for pre-processing input coordinate files from the Protein Data Bank (PDB) [19].
4. Conda (https://conda.io/projects/conda/en/latest/index.html): An open-source software environment manager.
5. Jupyter Notebook (https://jupyter.org/) (optional): A web-based interactive computing platform.

### 2.2 Hardware

We have tested the installation on CentOS 7 (with CUDA 11.6 and GeForce GTX 1080 Ti card), RockyLinux (with CUDA 12.5 and GeForce GTX 1080 Ti card), Ubuntu 20.04 LTS (with CUDA 12.4 and A100 GPU card), and openSUSE tumbleweed (with CUDA 12.3 and A4500 GPU card). A Linux computer with multiple CPUs and/or GPU is preferable.

### 2.3 Data

Structural models of protein–protein complexes in PDB format.

## 3 Methods

In this section, we outline the process for evaluating the quality of PPI conformations using the DeepRank-GNN-esm algorithm. Additionally, we offer instructions on how our deep learning architecture can be retrained and/or re-designed for new applications.

To rank PPI conformations with DeepRank-GNN-esm, the following steps should be followed:

1. Install the required software packages.
2. Prepare and preprocess the input data (the models to be scored).
3. Run $f_{nat}$ prediction using the command line interface tool from a terminal or the DeepRank-GNN-esm package at python level.
4. Analyze results.

In the following, all those steps are described in details and illustrated with two examples using different settings, highlighting the versatility and efficacy of each approach.

### 3.1 Installation of the Required Software Packages

DeepRank-GNN-esm can be used either as a command-line tool or a python package. Installation instructions for both settings are provided below.

1. Download and install Anaconda (https://docs.anaconda.com/free/anaconda/install/index.html).
2. Open a Linux terminal and clone the DeepRank-GNN-esm Github repository

```
> git clone https://github.com/haddocking/DeepRank-GNN-esm
> cd DeepRank-GNN-esm
```

3. Install the CPU or GPU version of DeepRank-GNN-esm depending on the hardware available in your system via conda:
   CPU version: Create and activate the conda environment.

```
> conda env create -f environment-cpu.yml
> conda activate deeprank-gnn-esm-cpu
```

   GPU version: Create and activate the conda environment.

```
> conda env create -f environment-gpu.yml
> conda activate deeprank-gnn-esm-gpu
```

4. Install the command line tool

```
> pip install.
```

5. In the same installation directory, clone the ESMFold github repository (*see* **Note 1**)

```
> git clone https://github.com/facebookresearch/esm.git
```

6. Install pdb-tools

```
> pip install pdb-tools
```

7. Install Jupyer Notebook (Optional)

```
> pip install notebook
```

Note that once you have installed all software, before using it in a new terminal session, you will have to activate the conda environment again with:

```
> conda activate deeprank-gnn-esm-gpu.
```

### 3.2 Preprocessing of the Input Data

Deeprank-GNN-esm is primarily tailored for interactions involving two chains. To run computation for complexes with more than two chains, refer to **Note 2**. Prior to utilizing the tool, renumbering the residue indexes for every chain in the PDB file is required (*see* **Note 2**). For this purpose, we recommend to use the provided script "pdb_renumber.py": For example, for one input PDB file (e.g., input_protein.pdb), the script will renumber all the chains in the protein starting from residue "1" and generate a new PDB file in the user-defined output directory (note that if the output directory already exists it will be overwritten). To run it type:

```
> python scripts/pdb_renumber.py input_protein.pdb 1 output_-
dir
```

### 3.3 Run DeepRank-GNN-Esm Computation

#### 3.3.1 Run Computation with Command Line Interface Tool

We use one example input to demonstrate the usage of the tool.

1. First create a separate 'tutorial' directory where we will run this example:

```
> mkdir tutorial
> cd tutorial
```

2. Download an example PDB file from the PDB (*see* **Note 3**)

```
> pdb_fetch 1B6C > 1B6C.pdb
```

3. Prepare the input structure. This python command will overwrite the original PDB file.

```
> python ../scripts/pdb_renumber.py 1B6C.pdb 1.
```

4. Run $f_{nat}$ computation:

   To assess the quality of the protein–protein interface between chain "A" and chain "B," run the following command (*see* **Note 4**). This command requires four inputs from the user: the PDB file of interest ("1B6C.pdb"), the identifiers of the two chains you wish to assess (chain_id1 "A" and chain_id2 "B"), and the number of cores you want to allocate for the prediction. In this case, we use "1" core. If you wish to score PDB ensembles, the tool can be used in exactly the same way; simply replace the single PDB input with the ensemble input file.

```
> deeprank-gnn-esm-predict 1B6C.pdb A B 1
```

For a description of the resulting output refer to Subheading 3.4.

When ranking a small number of PDB structures (<100), we recommend using the command–line interface tool. This tool consolidates all functionalities into one user-friendly command. However, the process may be relatively slow because it handles only one input PDB at a time and running the predictions directly from Python (*see* Subheading 3.3.2) might be more efficient.

#### 3.3.2 Run the Computations in Python Using the Deep Learning Architecture

DeepRank-GNN-esm can also be used as a python deep learning architecture. We will use as an example (PDBID:1ATN) to demonstrate the process. By running parallel predictions harvesting multiple CPUs for data loading, we can accelerate the process compared to the command-line tool version.

Below are the steps for using the Deeprank-GNN-esm architecture for making $f_{nat}$ predictions with example PDB files (*see* **Note 5**, for naming requirement for the PDB files) as input. For the convenience for the reader, we also provide a Jupyter notebook version of all the steps (*see* **Note 6**).

1. From within the directory where you installed the software, create a new directory for the tutorial and copy all example PDB files to the new directory:

```
> cd ../
> mkdir tutorial_python
> cp -r example/data/pdb/1ATN tutorial_python
> cd tutorial_python
```

2. Prepare all input PDBs in the folder: Inside the "tutorial_python" directory, in bash shell, run:

```
> for pdb_file in 1ATN/*.pdb; do python ../scripts/pdb_re-
number.py "$pdb_file" 1 1ATN/; done
```

3. All input PDBs have been renumbered and organized in the folder "1ATN". To compute ESM-2 embeddings for them, we begin by extracting sequences from the PDBs (*see* **Note 7**). Open a terminal, activate the deeprank-gnn-esm conda environment (see above) and then type:

```
> python ../scripts/get_fasta.py 1ATN/ A B
```

This command generates a fasta file ("1ATN.fasta") from chain "A" and chain "B" for this complex. For other chain IDs, replace "A" and "B" with specific chain identifiers. To calculate the ESM-2 embeddings from the input fasta file, as described in the supplementary material in our original publication [18], we recommend using the script provided by ESM-Fold [6] (*see* **Note 8**):

```
> mkdir embedding
> python ../esm/scripts/extract.py esm2_t33_650M_UR50D 1ATN.
fasta embedding --repr_layers 33 --include mean per_tok
```

The command generates ESM-2 embeddings for all sequences in "1ATN.fasta" with pretrained ESM-2 model "esm2_t33_650M_UR50D" and stores them in the "embedding" directory.

4. To convert the input PDBs into interface graphs that can be processed and handled by the network, in the current directory, create a python file named 'prepare_HDF5.py', copy the following python code snippet into the file:

```
from deeprank_gnn.GraphGenMP import GraphHDF5
pdb_path = "1ATN"
embedding_path = "embedding"
nproc = 20
outfile = "1ATN_residue.hdf5"
GraphHDF5(pdb_path = pdb_path,
embedding_path = embedding_path,
graph_type = "residue",
outfile = outfile,
nproc = nproc,
tmpdir="./tmpdir")
```

Run the python file in the "tutorial_python" directory with:

```
> python ./prepare_HDF5.py
```

This code will generate the input interface graphs with ESM-2 embeddings as one of their node features for all PDB files in the directory and compact them into the output HDF5 file format using 20 CPU cores.

5. To use pre-trained model to rank the input conformations in HDF5 format, create a new python file named "predict.py' in the current directory and add the following code to the file:

```
from deeprank_gnn.ginet import GINet
from deeprank_gnn.NeuralNet import NeuralNet as NN
database_test = "1ATN_residue.hdf5"
gnn = GINet
target = "fnat"
edge_feature = ["dist"]
node_features=["type", "polarity", "bsa", "charge", "embed-
ding"]
threshold = 0.3
pretrained_model="../paper_pretrained_models/scoring_of_-
docking_models/gnn_esm/treg_yfnat_b64_e20_lr0.001_foldal-
l_esm.pth.tar"
device_name = "cuda:0"
num_workers = 10
model = NN(
database_test,
gnn,
device_name = device_name,
edge_feature = edge_feature,
node_feature = node_features,
target = target,
num_workers = num_workers,
pretrained_model=pretrained_model,
threshold = threshold)
model.test(hdf5 = "GNN_esm_prediction.hdf5")
```

Run the python file in the "tutorial_python" directory with:

```
> python ./predict.py
```

As outlined in the original publication, the available node features include "type," "polarity," "charge," "bsa," and "embedding." It is important to note that since the model is trained with all these features, we highly recommend using all of them during inference time for optimal performance.

### 3.4 Analyzing the Results

#### 3.4.1 Analyzing the Output from the Command Line Interface Tool

The command line tool generates a result folder within the current working directory. The structure of this folder is as follows:

```
1B6C-gnn_esm_pred_A_B
├── 1B6C.pdb
├── all.fasta
├── 1B6C.A.pt
```

```
├── 1B6C.B.pt
├── graph.hdf5
├── GNN_esm_prediction.hdf5
└── GNN_esm_prediction.csv
```

The output folder "1B6C-gnn_esm_pred_A_B" contains the pre-processed input protein ("1B6C.pdb"), the extracted protein sequences ("all.fasta"), the calculated ESM-2 embeddings for protein chain "A" and chain "B" ("1B6C.A.pt", "1B6C.B.pt"), the generated interface graph ("graph.hdf5") and the prediction output files ("GNN_esm_prediction.hdf5" and "GNN_esm_prediction.csv") with predicted $f_{nat}$ values for the interface. In our paper [18], we use a $f_{nat}$ cutoff of 0.3 to differentiate between good and bad models, with a predicted " $f_{nat}$ " value higher than 0.3 indicating a good interface conformation and vice versa.

To access the predicted output, simply open the "GNN_esm_-prediction.csv" with a txt editor. To read node features and edge features of the generated interface graph ("graph.hdf5"), use to the following python code:

```
import h5py
f = h5py.File("1B6C-gnn_esm_pred_A_B/graph.hdf5")

#view all node features
node_features = f["1B6C"]["node_data"].keys()
print(node_features)

#view node feature embeddings
embeddings = f["1B6C"]["node_data"]["embedding"][()]
print(embeddings)

#view edge feature distance
distance = f["1B6C"]["edge_data'][‘dis’][()]
print(distance)
```

### 3.4.2 Analyzing the Output from the Python Deeprank-GNN-Esm Network Architecture

The provided code snippet above generates an output HDF5 file named "GNN_esm_prediction.hdf5". Similar to the outputs from the command line tool, you can access predicted $f_{nat}$ through python code:

```
import h5py
f = h5py.File("GNN_esm_prediction.hdf5","r+")
mol_names = f["epoch_0000"]["test"]["mol"][()]
fnats = f["epoch_0000"]["test"]["outputs"][()]
for mol_name, fnat in zip(mol_names, fnats):
print(mol_name.decode(), fnat)
```

### 3.5 Customizing the Deep Learning Network

DeepRank-GNN-esm is designed to facilitate easy adaptation for your own applications. To re-train it for customized tasks, follow these steps:

1. Construct and format training and evaluation data sets.
2. Define the training tasks. Two training modes are supported: regression ("reg") or classification ("class").
3. Define the learning parameters: Determine "batch_size," learning rate "lr" based on your system setup and hardware capacity (*see* **Note 9**).
4. Load the network.
5. Retrain the network

We use the following code example to demonstrate the process:

```
from deeprank_gnn.NeuralNet import NeuralNet
from deeprank_gnn.ginet import GINet

database_train = 'train.hdf5'
database_eval = 'eval.hdf5'

task='reg'
batch_size=64
lr = 0.001
node_feature = ['type', 'polarity', 'bsa', 'charge', 'embed-
ding']
edge_feature = ['dist']

model = NeuralNet(database_train,
GINet,
node_feature=node_feature,
edge_feature=edge_feature,
target=target,
task=task,
lr=lr,
batch_size=batch_size,
database_eval = database_eval)

model.train(nepoch=50,
validate=True,
save_model='best',
hdf5='output.hdf5')
```

Graphs, neural network layers and training target could also be customized. For advanced users, we refer you to the example here https://deeprank-gnn.readthedocs.io/en/latest/tutorial.advanced.html.

## 4 Notes

1. The environment required for ESMFold is integrated into the conda environment for DeepRank-GNN-esm. Users only need to clone the repository to access the helper scripts necessary for embeddings calculation.
2. To ensure correct matching between the calculated ESM-2 embeddings (shaped [AA_len, 1280]) and their corresponding residue nodes in the graph, we employ residue renumbering prior to computations. This approach will ensure the residue numbering in all chains is continuous and starts from residue "1." As a result, the *ith* element in the embeddings corresponds to the *ith* residue in the sequence. It is worth noting that gaps in the structure do not affect this process, as long as the sequence order in the PDB file matches the sequence provided to ESM-2.
3. To use our scoring function with files in the default CIF format from the Protein Data Bank, you need to first convert them to the PDB format. We recommend using the "pdb_fromcif" command from PDB tools for this conversion.
4. Deeprank-GNN-esm is primarily tailored for interactions involving two chains. While our method may not directly evaluate the overall quality of complexes with more than two chains, it can still be effectively applied to assess interactions between specific pairs of chains within such complexes. In such cases, providing the two relevant chain IDs will enable our method to function probably.
5. All input PDB file should be named according to the following format: PDBID (four-letter code) followed by any combination of characters of any length and ending with the ".pdb" extension.
6. A Jupyter notebook version of the tutorial ("tutorial.ipynb") is distributed with the package. To run the tutorial, first install the conda environment and Jupyter Notebook. Then initiate Jupyter Notebook with command "jupyter notebook" and run the tutorial inside the DeepRank-GNN-esm installation folder.
7. We assume that all input PDB conformations share the same fasta sequence, which is typically the case in scoring scenarios in which the models are derived from one specific docking software. All modeled conformations for one complex should be stored in one directory. In such scenarios, it suffices to calculate the fasta sequence for only one PDB. However, if the input PDB conformations possess different sequences, it is necessary to generate fasta sequences for each of them.

8. Refer to the ESM-Fold repository or the Supplementary Material of our original DeepRank-GNN-esm publication for detailed explanations of each input parameter.
9. As an example of learning parameters, for a GPU GeForce GTX 1080 Ti card (11G memory), we use "batch _size"=64 and "lr"=0.001.

## Acknowledgments

This work was supported by the European Union Horizon 2020, project BioExcel [823830]. X. Xu acknowledges financial support from the China Scholarship Council, Grant 202208310024.

### References

1. Koukos PI, Bonvin AMJJ (2020) Integrative modelling of biomolecular complexes. J Mol Biol 432:2861–2881
2. van Noort CW, Honorato RV, Bonvin AMJJ (2021) Information-driven modeling of biomolecular complexes. Curr Opin Struct Biol 70:70–77
3. Braberg H, Echeverria I, Kaake RM, Sali A, Krogan NJ (2022) From systems to structure — using genetic data to model protein structures. Nat Rev Genet 23:342–354
4. Jumper J, Evans R, Pritzel A, Green T, Figurnov M, Ronneberger O, Tunyasuvunakool K, Bates R, Žídek A, Potapenko A, Bridgland A, Meyer C, Kohl SAA, Ballard AJ, Cowie A, Romera-Paredes B, Nikolov S, Jain R, Adler J, Back T, Petersen S, Reiman D, Clancy E, Zielinski M, Steinegger M, Pacholska M, Berghammer T, Bodenstein S, Silver D, Vinyals O, Senior AW, Kavukcuoglu K, Kohli P, Hassabis D (2021) Highly accurate protein structure prediction with AlphaFold. Nature 596:583–589
5. Evans R, O'Neill M, Pritzel A, Antropova N, Senior A, Green T, Žídek A, Bates R, Blackwell S, Yim J, Ronneberger O, Bodenstein S, Zielinski M, Bridgland A, Potapenko A, Cowie A, Tunyasuvunakool K, Jain R, Clancy E, Kohli P, Jumper J, Hassabis D (2022) Protein complex prediction with AlphaFold-Multimer 2021.10.04.463034
6. Lin Z, Akin H, Rao R, Hie B, Zhu Z, Lu W, Smetanin N, Verkuil R, Kabeli O, Shmueli Y, dos Santos Costa A, Fazel-Zarandi M, Sercu T, Candido S, Rives A (2023) Evolutionary-scale prediction of atomic-level protein structure with a language model. Science 379:1123–1130
7. Collins KW, Copeland MM, Brysbaert G, Wodak SJ, Bonvin AMJJ, Kundrotas PJ, Vakser IA, Lensink MF (2024) CAPRI-Q: the CAPRI resource evaluating the quality of predicted structures of protein complexes. J Mol Biol 436:168540
8. Schweke H, Xu Q, Tauriello G, Pantolini L, Schwede T, Cazals F, Lhéritier A, Fernandez-Recio J, Rodríguez-Lumbreras LA, Schueler-Furman O, Varga JK, Jiménez-García B, Réau MF, Bonvin AMJJ, Savojardo C, Martelli P-L, Casadio R, Tubiana J, Wolfson HJ, Oliva R, Barradas-Bautista D, Ricciardelli T, Cavallo L, Venclovas Č, Olechnovič K, Guerois R, Andreani J, Martin J, Wang X, Terashi G, Sarkar D, Christoffer C, Aderinwale T, Verburgt J, Kihara D, Marchand A, Correia BE, Duan R, Qiu L, Xu X, Zhang S, Zou X, Dey S, Dunbrack RL, Levy ED, Wodak SJ (2023) Discriminating physiological from non-physiological interfaces in structures of protein complexes: a community-wide study. Proteomics 23:2200323
9. Réau M, Renaud N, Xue LC, Bonvin AMJJ (2023) DeepRank-GNN: a graph neural network framework to learn patterns in protein–protein interfaces. Bioinformatics 39:btac759
10. HDF5 for Python — h5py 3.11.0 documentation. https://docs.h5py.org/en/stable/. Accessed 14 May 2024
11. Wei J, Tay Y, Bommasani R, Raffel C, Zoph B, Borgeaud S, Yogatama D, Bosma M, Zhou D, Metzler D, Chi EH, Hashimoto T, Vinyals O, Liang P, Dean J, Fedus W (2022) Emergent abilities of large language models. arXiv:2206.07682
12. Brandes N, Ofer D, Peleg Y, Rappoport N, Linial M (2022) ProteinBERT: a universal

deep-learning model of protein sequence and function. Bioinformatics 38:2102–2110

13. Madani A, Krause B, Greene ER, Subramanian S, Mohr BP, Holton JM, Olmos JL, Xiong C, Sun ZZ, Socher R, Fraser JS, Naik N (2023) Large language models generate functional protein sequences across diverse families. Nat Biotechnol 41:1099–1106
14. Hayes T, Rao R, Akin H, Sofroniew NJ, Oktay D, Lin Z, Verkuil R, Tran VQ, Deaton J, Wiggert M, Badkundri R, Shafkat I, Gong J, Derry A, Molina RS, Thomas N, Khan Y, Mishra C, Kim C, Bartie LJ, Nemeth M, Hsu PD, Sercu T, Candido S, Rives A (2024) Simulating 500 million years of evolution with a language model. bioRxiv. https://doi.org/10.1101/2024.07.01.600583
15. Lin Z, Akin H, Rao R, Hie B, Zhu Z, Lu W, Smetanin N, Verkuil R, Kabeli O, Shmueli Y, dos Santos CA, Fazel-Zarandi M, Sercu T, Candido S, Rives A (2023) Evolutionary-scale prediction of atomic-level protein structure with a language model. Science 379:1123–1130
16. Unsal S, Atas H, Albayrak M, Turhan K, Acar AC, Doğan T (2022) Learning functional properties of proteins with language models. Nat Mach Intell 4:227–245
17. Nijkamp E, Ruffolo JA, Weinstein EN, Naik N, Madani A (2023) ProGen2: exploring the boundaries of protein language models. Cell Syst 14:968–978.e3
18. Xu X, Bonvin AMJJ (2024) DeepRank-GNN-esm: a graph neural network for scoring protein–protein models using protein language model. Bioinformatics. Advances 4:vbad191
19. Rodrigues JPGLM, Teixeira JMC, Trellet M, Bonvin AMJJ (2018) pdb-tools: a Swiss army knife for molecular structures. F1000Research 2018 7:1961

# Chapter 5

# Translating a GO Term List to Human Readable Function Description Using GO2Sum

**Swagarika J. Giri, Udayan Pandey, Joon Hong Park, and Daisuke Kihara**

## Abstract

Understanding the functions of proteins is one of the most important challenges in modern biology. Typically, protein function prediction methods generate a list of gene ontology (GO) terms, sometimes consisting of 50–100 functional terms. While GO serves the purpose of standardizing terms, interpreting a long list of GO terms is difficult for biologists. To address this challenge, we developed Gene Ontology terms Summarizer (GO2Sum), a language model–based summarizer that takes a list of GO terms as input and converts them into a concise, free-text summary describing the protein's function, subunit structure, and pathway information. GO2Sum was fine-tuned on GO term assignments and free-text function descriptions from UniProt entries. We built a Web server of GO2Sum, which offers an easy use of GO2Sum for biology users.

**Key words** GO2Sum, Protein function, Gene ontology, Summarization, Large language model, LLM, Function prediction

## 1 Introduction

Elucidating the biological function of proteins is a central challenge in modern biology. While protein functions are ultimately determined through wet lab experiments, computational methods play a crucial role in hypothesizing potential functions beforehand [1]. Usually, protein function prediction outputs possible functions that a protein may perform in terms of a list of gene ontology (GO) terms. The gene ontology (GO) [2] is a standardized vocabulary, which facilitate description of protein function by computer programs. GO categorizes protein functions into three sub-ontologies: biological process (BP), referring to biological programs accomplished by multiple molecular activities; cellular component (CC), specifying the location in the cell where the gene product is active; and molecular function (MF), focusing on

Dukka B. KC (ed.), *Large Language Models (LLMs) in Protein Bioinformatics*, Methods in Molecular Biology, vol. 2941,
https://doi.org/10.1007/978-1-0716-4623-6_5,

the biochemical activity of the protein. Despite its convenience for computational analysis, understanding a list of GO terms is challenging for researchers.

This chapter explains Gene Ontology terms Summarizer (GO2Sum) [3], a novel tool that converts a list of GO terms into human-readable function descriptions in free text. GO2Sum uses a large language model (LLM), which has been extensively studied in the natural language processing (NLP) area in recent years. GO2Sum uses text summarization technique [3] with LLM. GO2Sum processes GO term lists, which are often generated by protein function prediction methods [4–6] or obtained from databases like UniProt [7] and InterPro [8]. It then converts these lists into three functional descriptions, Function, Subunit Structure, and Pathway, which are presented in UniProt entries. The Function description provides the overall role of the protein, Subunit Structure details molecular interactions with other proteins, and a Pathway paragraph explains the metabolic pathways associated with the protein. In this chapter, we first explain the workflow of GO2Sum and then discuss evaluation results for GO2Sum with examples of paragraphs generated by GO2Sum. Finally, we will introduce the GO2Sum Web server, which provides an easy use of GO2Sum to researchers.

## 2 Workflow of GO2Sum

The GO2Sum workflow is summarized in Fig. 1. GO2Sum takes a list of GO terms as input. These GO terms are annotations of the protein, generated either from protein function databases like UniProt [7] and InterPro [8], or through experimental methods or computational protein function prediction tools [1, 4–6, 9–13]. Each GO term is associated with a text description in the Gene Ontology Consortium (GOC) [2] database. For instance, GO: 0034498 is associated with the description "The directed movement of substances from early endosomes to the Golgi." These descriptions are concatenated into a single document, which serves as input for the summarization module. In GO2Sum, text-to-text transfer transformer (T5) model [14] was used to summarize functional descriptions from a list of GO terms. T5 uses an encoder–decoder deep learning architecture, where the encoder is assigned to extract a rich latent representation of the text, irrespective of the task, and the decoder aims at generating task-specific output based on the encoder embeddings. T5 was originally trained on the Colossal Clean Crawled Corpus (C4) [15], which combines data from Wikipedia and Web scraping and achieved state-of-the-art results across various text-to-text tasks, including translation, summarization, question answering, and classification.

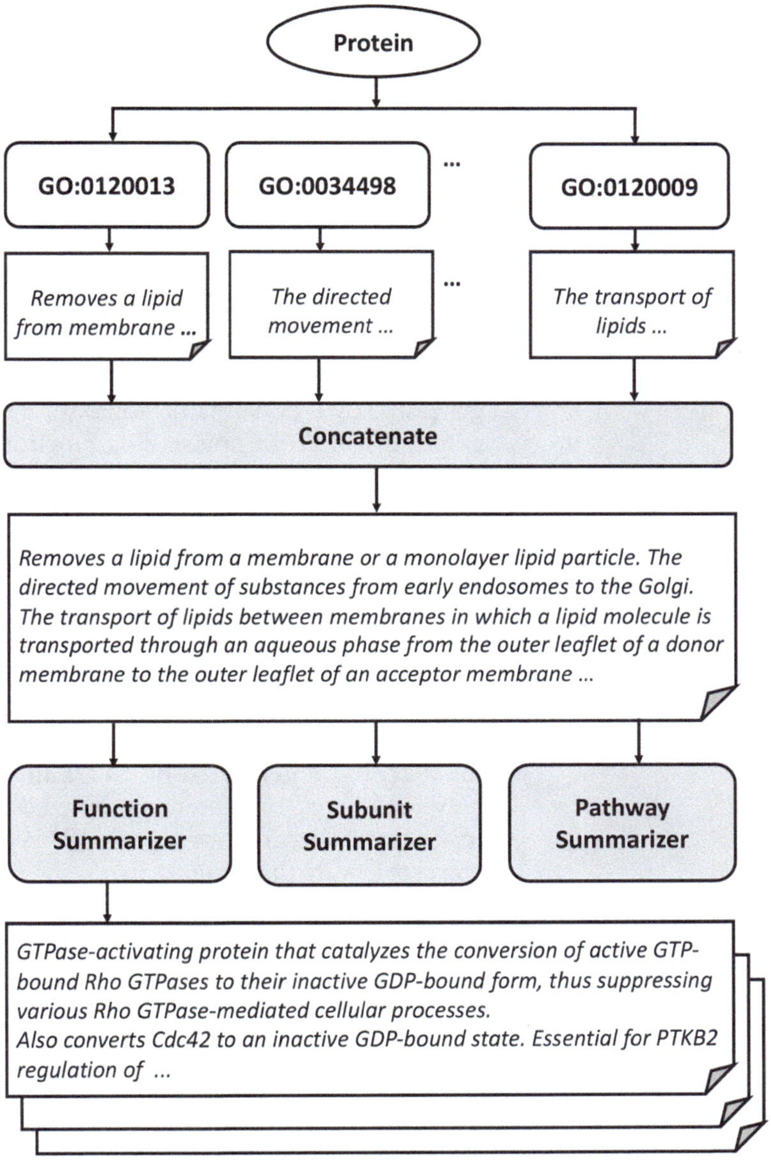


**Fig. 1** The GO2Sum workflow. UniProt entry A1A4S6, "Rho GTPase-activating protein 10" is used in this example

In GO2Sum, we downloaded the entire SwissProt portion from UniProt (Release: February 11, 2022). After applying filtration to remove redundancy, the dataset was reduced to 109,658 proteins from an initial 542,953. The original paper of GO2Sum provides a detailed discussion on preprocessing. For each protein, we gathered three different paragraphs, Function, Subunit Structure, and Pathway, from the UniProt database. As an entry often does not have all three functional descriptions of Function, Subunit Structure, and Pathway, we constructed three separate datasets for each. The dataset for Function, Subunit Structure, and Pathway had 97,600, 62,340, and 14,600 proteins, respectively. We fine-tuned three separate T5 models on GO assignments, each designed to produce paragraphs for the Function, Subunit Structure, and Pathway sections in UniProt, respectively.

## 3 Evaluation of GO2Sum

In this section, we briefly summarize the benchmark study we did to evaluate the performance of GO2Sum. More results are available in the original paper of GO2Sum [3]. We evaluated paragraphs generated by GO2Sum in comparison with the ground-truth paragraphs in UniProt using scoring functions used in text comparison in NLP. Scores used are three embedding-based metrics, BERT [16], MiniLM [17], and BioSentVec [18], and additional three sentence-mover-based metrics, SMS, WMS, and S + WMS [19]. The embedding-based metrics represent paragraphs as embeddings, numerical vectors, from the deep learning models for NLP. These vectors capture the semantic meaning of the text by encoding it into a high dimensional space, where similar texts locate close to each other. We used the Cosine similarity between an embedding of the ground-truth UniProt description and that of the description generated by GO2Sum to quantify the agreement between them. In sentence-mover-based metrics, sentences are treated as collections of words. We represented words and sentences of both the ground truth and the descriptions generated by GO2Sum in the embedding space. Next, we solve a linear optimization problem to minimize the cost of transforming the ground truth into the descriptions generated by GO2Sum, and the cost is considered as the distance between the two descriptions.

We compared the paragraphs generated by GO2Sum with those produced by the vanilla (original) T5 model and SciFive [20], which is specifically pretrained LLM on extensive biomedical literature. In Fig. 2a–c, we compared the evaluation scores of GO2Sum with SciFive using the average of the three embedding-based scores, BERT, MiniLM, and BioSentVec. As shown, GO2Sum demonstrated superior performance, achieving higher scores than SciFive in 97.60% of cases for Function, 98.89% for Subunit Structure, and 99.79% for Pathway descriptions, respectively (Fig. 2a–c). In Fig. 2d, we used the three mover-distance-based similarity scores to compare GO2Sum and the vanilla T5. The results show that the summary by GO2Sum was more similar to the ground truth in UniProt than vanilla T5 for almost all UniProt entries, 95.3%, 97.3%, and 97.0%, of the cases evaluated by WMS, SMS, and S + WMS, respectively.

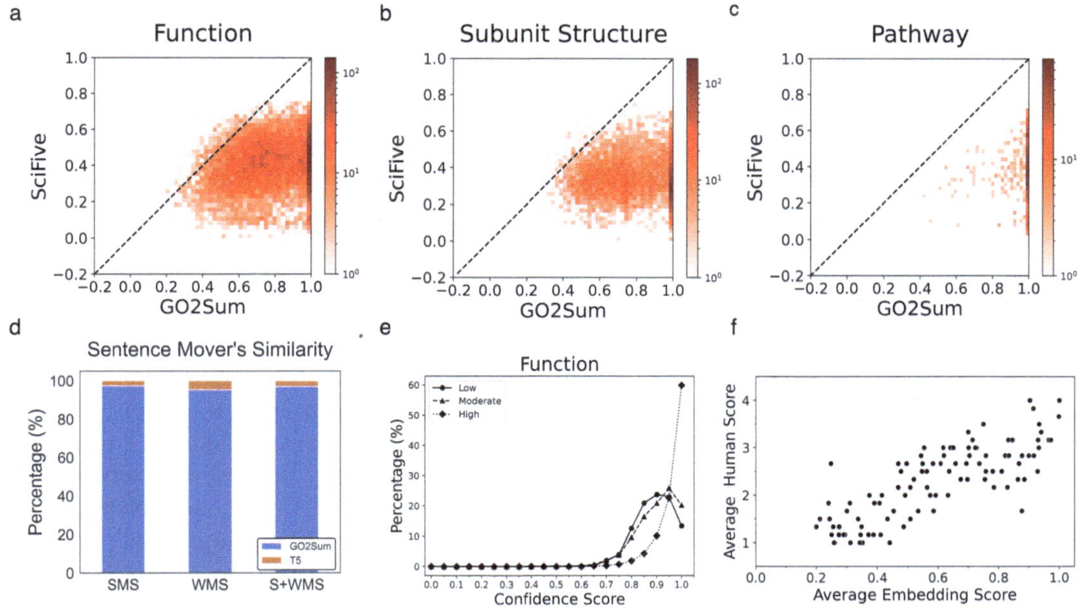


**Fig. 2** Evaluation results for GO2Sum. (**a**–**c**) Comparison of GO2Sum and SciFive using average scores of three embedding scores, BERT, MiniLM, and BioSentVec. (**a**) Benchmarked on the UniProt Function description. (**b**) Benchmarked on UniProt Subunit Structure description. (**c**) Benchmarked on UniProt Pathway description. (**d**) Comparison between GO2Sum and vanilla (original) T5 using three Sentence Mover's similarity-based metrics. (**e**) The distribution of the confidence score for entries of Function. Entries were classified into three classes based on average embedding score: high, [0.8–1.0]; moderate, [0.5,0,8); and low, [0,0.5). (**f**) The correlation between evaluation by human biology experts and the evaluation score we used for the benchmark. This comparison was performed on 100 entries

Figure 2e shows the confidence score of generated description by GO2Sum. GO2Sum employs beam search, which generates many different paragraphs with a probability value assigned. We used this probability value as a confidence score, which can indicate the quality of the generated paragraph. In Fig. 2e, the distribution of confidence scores for generated paragraphs for UniProt Function section is shown. Generated paragraphs were categorized into three accuracy classes based on the average embedding scores: high (0.8 to 1.0), moderate (0.5 to 0.8), and low (below 0.5). The plot shows that confidence scores for highly accurate summaries tend to be high: 82.6% of paragraphs in the high-scoring class have confidence scores above 0.9, while this proportion drops to 45.0% for the moderate class and 36.4% for the low-scoring class, indicating that the confidence score corresponds well to their accuracy.

In Fig. 2f, we evaluated the alignment between our average embedding score we used for evaluation with assessments by human experts to confirm that the automatic evaluation we performed is reasonable. We sampled 100 paragraphs generated by GO2Sum and asked six human evaluators to give a score from 1 (the worst) to 4 (the best) by comparing them with the ground truth paragraphs in UniProt. The y-axis shows the average score from the six evaluators. The correlation between the human evaluators' score and the average embedding score we used 0.804, indicating that overall the average embedding score agrees well with human evaluation. Particularly, high-scoring summaries generated by GO2Sum are deemed highly reliable, as confirmed by the human evaluators.

## 4 GO Summary for Function

In this section, we show examples of descriptions generated by GO2Sum for UniProt Function paragraph. We used UniProt entries taken from its latest release (April 2024). GO terms are taken from the top page of entries (but not from a complete list of GO terms on QuickGO that is linked from each entry page). In Table 1, the first protein, Q9LNA5 (ACT domain-containing protein ACR8), is associated with two GO term

**Table 1**
**Examples of Function description generated by GO2Sum**

| Protein | GO IDS | GO2Sum generated summary | UniProt description |
|---|---|---|---|
| Q9LNA5 | GO:0016597;GO:0009737 | May bind amino acids | May bind amino acids |
| Q9ZUL7 | GO:0005576;GO:0005739; GO:0006952; GO:0009506; GO:0031640; GO:0050832 | Confers broad-spectrum resistance to pathogens | Confers broad-spectrum resistance to pathogens |
| Q51472 | GO:0000976;GO:0001216; GO:0003677; GO:0005829; GO:0006310; GO:0006417; GO:0030527; GO:0032993; GO:0045893 | This protein is one of the two subunits of integration host factor, a specific DNA-binding protein that functions in genetic recombination as well as in transcriptional and translational control | This protein is one of the two subunits of integration host factor, a specific DNA-binding protein that functions in genetic recombination as well as in transcriptional and translational control |
| Q41062 | GO:0005524;GO:0005829; GO:0006605; GO:0006886; GO:0008320; GO:0009535; GO:0009570; GO:0009941; GO:0016464; GO:0017038; GO:0071806 | Has a central role in coupling the hydrolysis of ATP to the transfer of proteins across the thylakoid | Has a central role in coupling the hydrolysis of ATP to the transfer of proteins across the thylakoid membrane. Facilitates the transport of precursor proteins from the chloroplast stroma to thylakoid lumen |

annotations, GO:0016597 and GO:0009737, in the UniProt entry. The term GO:0016597 describes its function as "binding to an amino acid or organic acids containing one or more amino substituents" and the second one, GO:0009737 is for "response to abscisic acid." UniProt has a brief function description for this entry, "may bind amino acids," which perfectly agrees with the output from GO2Sum.

The second protein, Q9ZUL7 (Defensin-like protein 1), is associated with GO terms, including GO:0006952 (defense response), GO:0050832 (defense response to fungus), and GO:0031640 (killing of cells of another organism), which are linked to pathogen resistance. The remaining three GO terms (GO:0005576, GO:0005739, GO:0005739) are related to the cellular location of the protein in the extracellular region, mitochondria, and plasmodesma, respectively. Again, for this example, GO2Sum outputs the identical function description as UniProt Function, "Confers broad-spectrum resistance to pathogens."

The third example, Q51472 (integration host factor subunit alpha) is annotated with nine different terms, including GO:0032993 (protein–DNA complex), GO:0003677 (DNA binding), GO:0001216 (DNA-binding transcription activator activity), GO:0045893 (positive regulation of DNA-templated transcription), GO:0000976 (transcription cis-regulatory region binding), GO:0006310 (DNA recombination), GO:0006417 (regulation of translation), GO:0005829 (cytosol), and GO:0030527 (structural constituent of chromatin). Function summarizer correctly choose the seven GO terms related to Function
(GO:0032993; GO:0003677; GO:0001216; GO:0045893; GO:0000976; GO:0006417; and GO:0006310) and was able to produce identical annotations as UniProt.

The fourth example, Q41062 (protein translocase subunit SecA, chloroplastic), is annotated with 11 different terms: GO:0005524 (ATP binding), GO:0005829 (cytosol), GO:0006605 (protein targeting), GO:0006886 (intracellular protein transport), GO:0008320 (protein transmembrane transporter activity), GO:0009535 (chloroplast thylakoid membrane), GO:0009570 (chloroplast stroma), GO:0009941 (chloroplast stroma), GO:0016464 (chloroplast protein-transporting ATPase activity), GO:0017038 (protein import), and GO:0071806 (protein transmembrane transport). The function description of this protein has two sentences: coupling ATP hydrolysis to the transfer of proteins across the thylakoid membrane and facilitating the transport of precursor proteins from the chloroplast stroma to the thylakoid lumen. GO2Sum successfully reproduced the first sentence, but did not use the GO term of chloroplast storma and reproduced the second sentence.

## 5 GO Summary for Subunit Structure

We showed examples of Subunit Structure paragraphs generated by GO2Sum in Table 2. Compared to Function description, Subunit Structure descriptions in UniProt are usually short, typically consisting of a few words. The first example, P41912 (Ras modification protein ERF4), functions as part of the ERF2–SHR5 complex and is a palmitoyltransferase specific for Ras proteins. GO2Sum correctly generated its Subunit Structure as "Interacts with ERF2," reproducing the description in UniProt. All the four GO terms GO:0006612 (protein targeting to membrane), GO:0005789 (endoplasmic reticulum (ER) membrane), GO:0031211 (ER palmitoyltransferase complex), and GO:0005783 (ER) contributed to this accurate output by GO2Sum. While these GO terms do not contain the name ERF2 as its interacting partner, they provide functional and locational clues. GO:0031211 (ER palmitoyltransferase complex) GO term seems to be a key to produce the correct output, as ERF2 is known to be a subunit of this complex.

The next example is Q752A3, topoisomerase 1–associated factor 1 (TOF1). It plays a crucial role in forming a fork protection complex (FPC) with CSM3. This complex is essential for chromosome segregation during meiosis and for the repair of DNA

**Table 2**
**Examples of Subunit Structure description generated by GO2Sum**

| Protein | GO IDS | GO2Sum generated summary | UniProt description |
|---|---|---|---|
| P41912 | GO:0005783;GO:0005789; GO:0006612;GO:0031211 | Interacts with ERF2 | Interacts with ERF2 |
| Q752A3 | GO:0000076;GO:0003677; GO:0005829;GO:0006281; GO:0007064;GO:0031298; GO:0043111;GO:0043570; GO:0051321 | Component of the fork protection complex (FPC) consisting of TOF1 and CSM3 | Component of the fork protection complex (FPC) consisting of TOF1 and CSM3 |
| P63031 | GO:0005743;GO:0006086; GO:0006850;GO:0050833; GO:1990830 | The functional 150 kDa pyruvate import complex is a heteromer of MPC1 and MPC2 | Homodimer. Forms heterodimer with MPC2. The heterodimer is the more stable and dominant form |
| Q8BM54 | GO:0004842; GO:0005737; GO:0005856; GO:0005886; GO:0006511; GO:0007399; GO:0008092; GO:0010977; GO:0010989; GO:0016567; GO:0031648; GO:0032803; GO:0042632; GO:0045732; GO:0046872; GO:0061630; GO:0071404 | Homodimer. Interacts with the E2 ubiquitin-conjugating enzyme, UBE2D1 | Homodimer. Interacts with the E2 ubiquitin-conjugating enzyme, UBE2D1 (via RING-type zinc finger). Interacts with myosin regulatory light chain (MRLC) and TMEM4 |

damage. This protein is annotated with nine GO terms, and among them two GO terms, GO:0031298 (replication fork protection complex) and GO:0043111 (replication fork arrest) are related to Subunit Structure. GO2Sum generated a summary, which did not include the abbreviation "(FPC)" in UniProt but besides this, it has a perfect match.

The next protein, P63031, mitochondrial pyruvate carrier 1 (MPC1), is responsible for transporting pyruvate into the mitochondria. This protein is annotated with five GO terms, two of which, GO:0006850 (mitochondrial pyruvate transmembrane transport) and GO:0050833 (pyruvate transmembrane transporter activity), are closely related to the subunit structure predicted by GO2Sum. Both GO2Sum and UniProt agree on the formation of the heterodimer (MPC1 and MPC2), but UniProt provides additional detail about the homodimeric potential of MPC1, as well as a clearer statement on the stability and dominance of the heterodimer over the homodimer. The size of the complex, 150 kDa is not information included in these five GO terms, but it probably came from the same protein from a different organism (P53157), which has a similar set of GO terms in its annotation and has the identical the subunit description as this entry.

The fourth protein, Q8BM54 (E3 ubiquitin–protein ligase MYLIP), has 17 GO term annotations, four of which are related to ubiquitin. These include GO:0016567 (protein ubiquitination), GO:0061630 (ubiquitin–protein ligase activity), GO:0004842 (ubiquitin–protein transferase activity), and GO:0016567 (protein ubiquitination). These annotations align with the Subunit Structure paragraph generated by GO2Sum and highlight the protein's interaction with the E2 ubiquitin-conjugating enzyme, UBE2D1. In this case, GO2Sum was successful in capturing a part of interaction information, missing the specific interaction between the myosin regulatory light chain (MRLC) and TMEM4. Interaction with myosin could be inferred from other GO terms pointing the localization at cytoskeleton. But according to UniProt, this Subunit Structure description is annotated by similarity, thus not all the information are embedded in GO term annotations.

## 6 GO Summary for Pathway

Table 3 lists three examples for pathway paragraphs from GO2Sum. Protein Q8LPK5 (cellulose synthase A catalytic subunit 8) which is essential for glycan metabolism and cellulose biosynthesis. The GO2Sum output "glycan metabolism; plant cellulose biosynthesis," seems determined mainly from two terms GO:0016760 (cellulose synthase (UDP-forming) activity) and GO:0030244 (cellulose biosynthetic process). The rest of GO terms are primarily related to the protein's role in stress response and structural integrity.

**Table 3**
**Examples of Pathway description generated by GO2Sum**

| Protein | GO IDS | GO2Sum generated summary | UniProt description |
|---|---|---|---|
| Q8LPK5 | GO:0005886;GO:0006970; GO:0009414;GO:0009834; GO:0016760;GO:0030244; GO:0042742;GO:0046872; GO:0050832;GO:0071555 | Glycan metabolism; plant cellulose biosynthesis | Glycan metabolism; plant cellulose biosynthesis |
| P21177 | GO:0003857;GO:0004165; GO:0004300;GO:0006635; GO:0008692;GO:0016509; GO:0036125;GO:0070403 | Lipid metabolism; fatty acid beta-oxidation | Lipid metabolism; fatty acid beta-oxidation |
| Q8YA44 | GO:0000287;GO:0004789; GO:0005737;GO:0009228; GO:0009229 | Cofactor biosynthesis; thiamine diphosphate biosynthesis | Cofactor biosynthesis; thiamine diphosphate biosynthesis; thiamine phosphate from 4-amino-2-methyl-5-diphosphomethylpyrimidine and 4-methyl-5-(2-phosphoethyl)-thiazole: step 1/1 |

The next protein, P21177 (fatty acid oxidation complex subunit alpha), is integral to lipid metabolism, particularly in fatty acid beta-oxidation. The associated GO terms, GO:0006635 (fatty acid beta-oxidation) and GO:0036125 (fatty acid beta-oxidation multienzyme complex), highlight its participation in the multienzyme complex that facilitates the breakdown of fatty acids. A supporting publication further validates the pathway summary generated by GO2Sum [21].

For the last example in this table, Q8YA44 (thiamine–phosphate synthase), GO2Sum successfully captures the pathway information, "cofactor biosynthesis; thiamine diphosphate biosynthesis," but did not provide more detailed chemical reaction that was included in UniProt. The specific reaction information is included in GO:0004789, but GO2Sum did not take information from it.

## 7 GO2Sum Code and Web Server

The source code of GO2Sum can be downloaded from Github at https://github.com/kiharalab/GO2Sum under the terms of GNU Lesser General Public License Ver. 3.0. We also developed a Web server of GO2Sum that provides an easy access. The GO2Sum server is freely accessed at https://kiharalab.org/go2sum_server/. It is a Web application and does not require any platform other than Web browser.

This server takes a list of GO terms for protein(s) and outputs a summary paragraph in Function, Pathway, and Subunit categories of the protein(s). Figure 3 shows the front page of the GO2Sum server. The input file should have two columns: a protein name and

**Fig. 3** Screenshot of the GO2Sum server. The server can take a list of GO terms or take protein and their GO term annotations for multiple proteins. For a single protein, GO term annotations can be selected from the dropdown menu, manually entered into the input space, or uploaded from a file. At the end of the page, there is a checkbox for selecting summary categories to generate, either Function, Subunit Structure, or Pathway. Multiple categories can be selected. Users can receive the result by email if the email address is provided

a list of GO term IDs for each row. The column headers should be "Protein" and "GO_IDS." A space is used to separate the columns, and a semicolon is used to separate multiple GO terms in the list. A sample input file can be downloaded to help users create input files for multiple proteins.

Users can choose GO terms from a drop-down menu instead of uploading a file. When a GO term ID is typed, GO2Sum automatically recognizes and displays matching options in a pull-down list, allowing for easy selection. For instance, typing "5737" will display "GO:0005737" as a selectable option. At the end of the page, there are three checkboxes, one for each summary type, which the server will output: Function, Subunit Structure, and Pathway. Users can select one or more checkboxes to generate summaries. The GO2Sum server also offers an option to receive descriptions via email, which is particularly useful for large datasets.

An example of output is shown in Fig. 4. We present the functional output for two proteins, O00237 and O00462. The input windows of Option 1 shows GO terms input for these two proteins. For example, protein O00237 (E3 ubiquitin–protein ligase RNF103) is annotated with eight GO terms: GO:0004842 (ubiquitin–protein transferase activity); GO:0005783 (endoplasmic reticulum); GO:0005789 (endoplasmic reticulum membrane); GO:0007417 (central nervous system development); GO:0016567 (protein ubiquitination); GO:0036503 (ERAD pathway); GO:0046872 (metal ion binding); and GO:0061630 (ubiquitin–protein ligase activity). The output is displayed in a table format with three columns at the bottom of the page: the first column contains the protein name, the second describes the summary type, and the third provides the GO2Sum summary. GO2Sum correctly predicts the function of the protein as "Acts as an E2-dependent E3 ubiquitin–protein ligase, probably involved in the ER-associated protein degradation pathway." The pathway is predicted as "protein modification; protein ubiquitination," and the subunit structure as "Interacts with DERL1 and VCP."

## 8 Conclusion

In this chapter, we discussed GO2Sum, an LLM-based summarizer that takes a list of GO terms as input and generates a concise, free-text summary detailing the protein's functional descriptions, including its function, subunit structure, and pathway information. The summaries generated by GO2Sum demonstrate a substantially higher level of agreement with UniProt data when compared to the results obtained using the standard T5 and SciFive.

For many years, protein function prediction has relied on GO term annotations to describe protein functions. However, recent

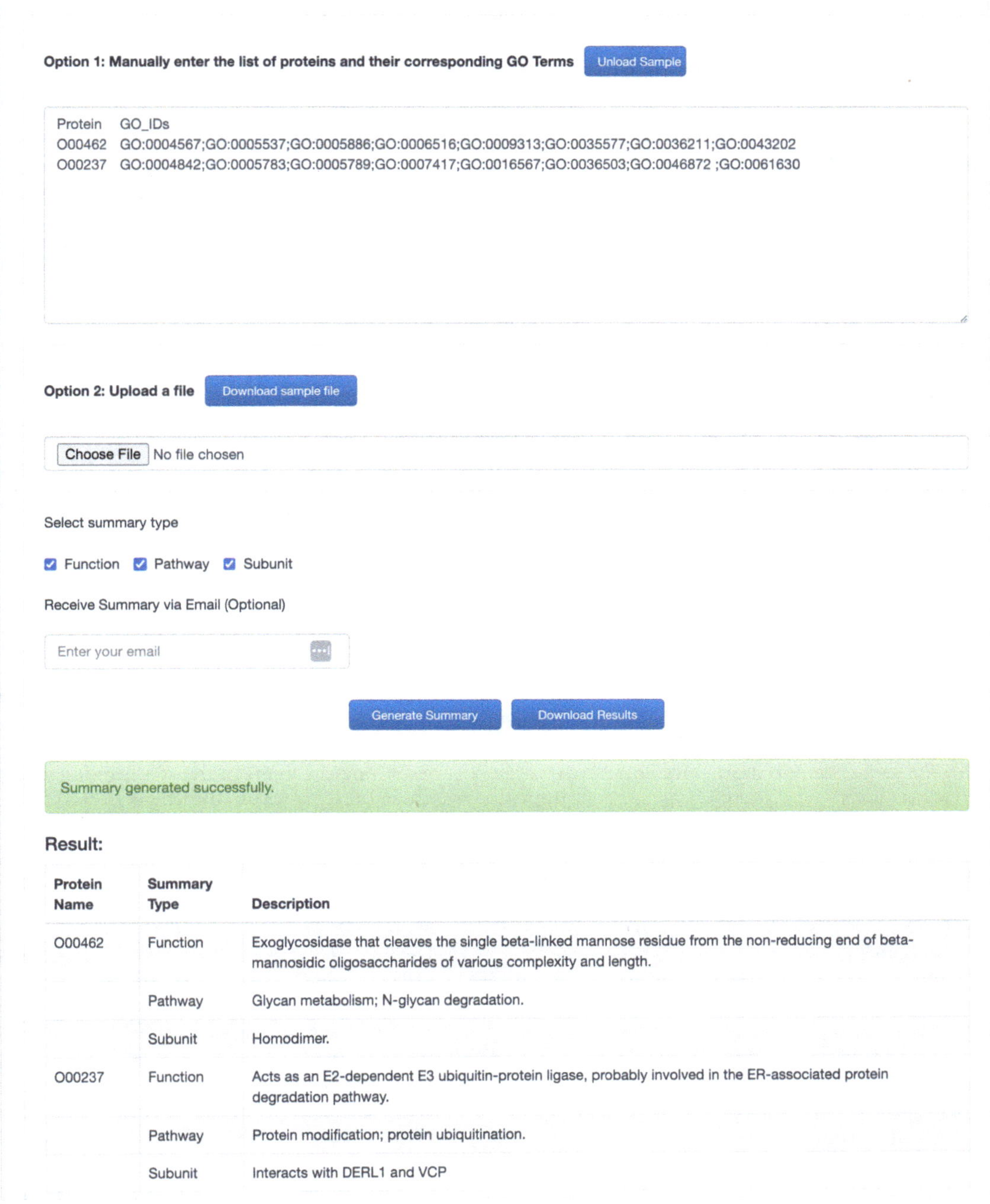


**Fig. 4** Screenshot of the summary generation for two proteins, 000237 and 000462. The upper part of the page shows the input provided for the two proteins. The screenshot below displays the generated functional summaries, shown in a table format, for three functional categories selected by the user: Function, Pathway, and Subunit Structure

advancements in NLP, particularly the development of LLM technologies, have enabled us to translate these predictions into concise, human-readable text that describes these functions. We believe that GO2Sum significantly enhances the interpretability of GO term predictions, especially in supporting biologists and medical scientists in their research endeavors.

## Acknowledgments

This work was partly supported by the National Science Foundation (DBI2003635, DBI2146026, IIS2211598, DMS2151678, CMMI1825941, and MCB1925643) and by the National Institutes of Health (R01GM133840, 3R01GM133840-02S1).

### References


1. Kagaya Y, Flannery ST, Jain A, Kihara D (2022) ContactPFP: protein function prediction using predicted contact information. Front Bioinf 2:896295
2. Ashburner M, Ball CA, Blake JA et al (2000) Gene ontology: tool for the unification of biology. Nat Genet 25:25–29
3. Giri SJ, Ibtehaz N, Kihara D. GO2Sum: generating human readable functional summary of proteins from GO terms. npj Syst Biol Appl 10: 29
4. Kulmanov M, Hoehndorf R (2020) DeepGOPlus: improved protein function prediction from sequence. Bioinformatics 36:422–429
5. Hawkins T, Luban S, Kihara D (2006) Enhanced automated function prediction using distantly related sequences and contextual association by PFP. Protein Sci 15:1550–1556
6. Ibtehaz N, Kagaya Y, Kihara D (2023) Domain-PFP allows protein function prediction using function-aware domain embedding representations. Commun Biol 6:1103
7. (2019) UniProt: a worldwide hub of protein knowledge. Nucleic Acids Res 47:D506–D515
8. Hunter S, Apweiler R, Attwood TK et al (2009) InterPro: the integrative protein signature database. Nucleic Acids Res 37:D211–D215
9. Jain A, Kihara D (2019) Phylo-PFP: improved automated protein function prediction using phylogenetic distance of distantly related sequences. Bioinformatics 35:753–759
10. Hawkins T, Chitale M, Luban S, Kihara D (2009) PFP: automated prediction of gene ontology functional annotations with confidence scores using protein sequence data. Bioinformatics 74:566–582
11. Chitale M, Hawkins T, Park C, Kihara D (2009) ESG: extended similarity group method for automated protein function prediction. Bioinformatics 25:1739–1745
12. You R, Zhang Z, Xiong Y et al (2018) GOLabeler: improving sequence-based large-scale protein function prediction by learning to rank. Bioinformatics 34:2465–2473
13. Wang S, You R, Liu Y et al (2023) NetGO 3.0: protein language model improves large-scale functional annotations. Genomics Proteomics Bioinformatics 21:349–358
14. Raffel C, Shazeer N, Roberts A et al (2019) Exploring the limits of transfer learning with a unified text-to-text transformer. J Mach Learn Res 21:1–67
15. Dodge J, Sap M, Marasović A, et al (2021) Documenting large Webtext corpora: a case study on the colossal clean crawled corpus. arXiv preprint arXiv:2104.08758
16. Devlin J, Chang M-W, Lee K, Toutanova K (2018) BERT: pre-training of deep bidirectional transformers for language understanding. arXiv preprint arXiv:1810.04805
17. Wang W, Wei F, Dong L, et al (2020) MiniLM: deep self-attention distillation for task-agnostic compression of pre-trained transformers. arXiv preprint arXiv:1810.04805
18. Chen Q, Peng Y, Lu Z (2019) BioSentVec: creating sentence embeddings for biomedical texts. In: 2019 IEEE international conference on healthcare informatics (ICHI). IEEE, pp 1–5
19. Clark E, Celikyilmaz A, Smith NA (2019) Sentence mover's similarity: automatic evaluation for multi-sentence texts. In: Proceedings of the 57th annual meeting of the Association for Computational Linguistics. Association for

Computational Linguistics, Stroudsburg pp 2748–2760
20. Phan LN, Anibal JT, Tran H, et al (2021) SciFive: a text-to-text transformer model for biomedical literature. arXiv preprint arXiv:2106.03598
21. Kamijo T, Aoyama T, Komiyama A, Hashimoto T (1994) Structural analysis of cDNAs for subunits of human mitochondrial fatty acid β-oxidation trifunctional protein. Biochem Biophys Res Commun 199:818–825

# Chapter 6

# TransFun: A Tool of Integrating Large Language Models, Transformers, and Equivariant Graph Neural Networks to Predict Protein Function

**Frimpong Boadu, Ahhyun Lee, and Jianlin Cheng**

## Abstract

Experimentally determining the functions of proteins is a complex and time-consuming process. This challenge contributes to a gap, where many proteins have known sequences, predicted structures, and other crucial information, yet lack functional annotations. This gap underscores the critical importance of automated function prediction (AFP) methods, which aim to develop computational techniques dedicated to predicting protein functions. Most AFP methods leverage the wealth of diverse protein information available, such as sequences, structures, protein-protein interactions, and domain characteristics. These methods often utilize individual features or integrate multiple features to enhance the accuracy of function prediction. In this chapter, we focus on TransFun, a structure-based protein function prediction technique. TransFun leverages the embeddings provided by the ESM-1b pretrained protein language models to distill intricate sequence features and combines them with AlphaFold's predicted structures to predict protein functions. **Availability:** https://github.com/jianlin-cheng/TransFun

**Key words** Protein function prediction, Gene ontology, Transformer, Graph neural networks, Equivariant graph neural networks, GO terms, Deep learning

## 1 Introduction

In this section, we delve into the automated function prediction (AFP) [3, 11, 19] of proteins problem, a crucial endeavor aimed at elucidating the functions of proteins. Understanding protein functions is pivotal for comprehending biological systems and manipulating biological processes, thus holding significant implications for biomedical research and biotechnology development. We discuss the problem in detail and introduce TransFun [4], a method for protein function prediction that operates on both sequence and structural information. Leveraging the ESM-1B [12] pretrained protein language model, TransFun extracts representative embeddings from protein sequences. These embeddings are then

Dukka B. KC (ed.), *Large Language Models (LLMs) in Protein Bioinformatics*, Methods in Molecular Biology, vol. 2941,
https://doi.org/10.1007/978-1-0716-4623-6_6,

combined with predicted protein structures to make predictions on protein functions.

To streamline the organization and understanding of protein functions, they are structured into ontologies, which categorize functions into terms with defined relationships, capturing the intricate connections among various functional terms. This structured approach not only provides a clear understanding of functional terms but also facilitates knowledge exchange, enabling efficient analysis and interpretation of protein functions.

In the realm of protein function prediction, many methods focus on the biomedical ontologies developed by the Gene Ontology (GO) Consortium [1, 2, 7, 15]. These ontologies consist of three main groups: Molecular Function Ontology (MFO), Biological Process Ontology (BPO), and Cellular Component Ontology (CCO), each represented as a directed acyclic graph.

The function prediction problem can be succinctly defined as follows: Given a set of proteins $D_l = \{(p_1, O_1), (p_2, O_2), \ldots, (p_n, O_n)\}$, where $p_i$ denotes the $i$th protein and $O_i$ represents its true functional annotation labels (specifically, a set of Gene Ontology terms), the objective is to accurately predict $O_i$ for any given protein $p_i$. Since annotations are organized hierarchically, with a general root term at the apex, when a GO term $x$ is linked to protein $p_i$, all ancestor terms of $x$ in the GO graph are also attributed to $p_i$. Hence, we aim to predict the subgraph $g$ within the GO graph $G$, comprising all GO terms associated with the protein, as shown in Fig. 1 [6].

As previously mentioned, proteins possess essential features or properties—such as sequence, structure, domain, and interaction—that distinguish one protein from another. Experiments have demonstrated that protein function is closely related to these properties, and inference about protein function can be derived from them.

In TransFun [4], we leverage two fundamental features—sequence and structure information—to predict protein functions. Utilizing representative embeddings derived from the ESM-1B [12] pretrained protein language model, TransFun generates feature representations from protein sequences. These embeddings are then seamlessly integrated with AlphaFold's [10, 18] predicted protein structures, effectively combining the insights from both sequence and 3D structural data to predict protein functions accurately. This combined approach not only enhances the predictive power of TransFun but also underscores the importance of leveraging multiple sources of information for comprehensive protein function prediction. To incorporate the intricate 3D features, we employ the equivariant neural network (EGNN) [13] to accurately model the protein's three-dimensional structures. In the next section, we discuss the data collection process, delve into the TransFun

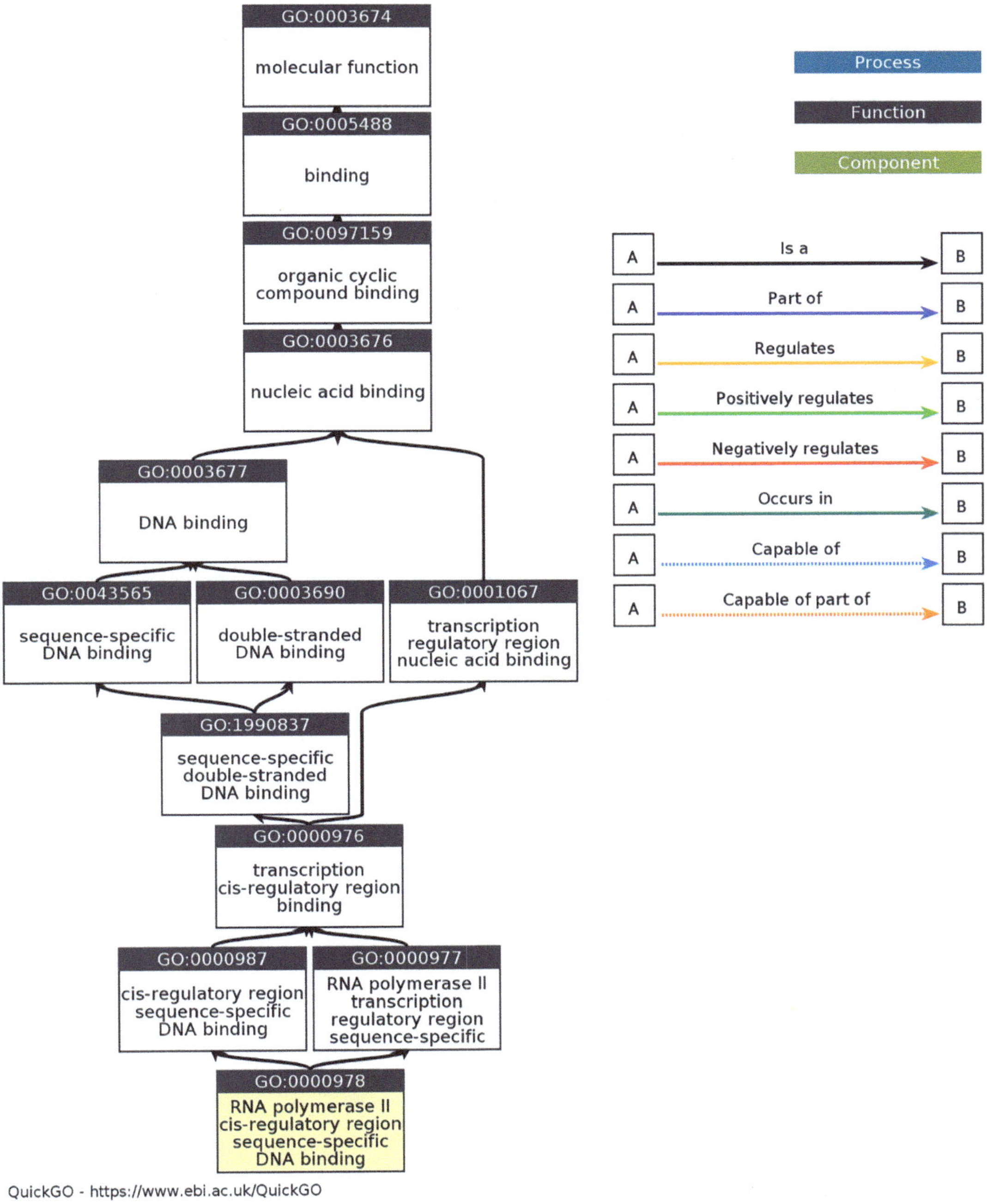


**Fig. 1** The image illustrates how Gene Ontology (GO) terms are structured hierarchically. For example, if a protein is annotated with the specific GO term "GO:0000978," it automatically includes broader terms like "GO:0000977" and "GO:0003676." However, a higher-level annotation like "GO:0003676" does not guarantee the specific function "GO:0000978." This hierarchy is key in predicting related protein functions. The image source is QuickGO, provided by the European Bioinformatics Institute (EBI)

methodology—including the data preprocessing steps that involve leveraging the ESM-1B language model and EGNN network—and elucidate the integration of these preprocessed features. Furthermore, we explore how to use the tool to predict protein functions.

### *1.1 Dataset*

In our study, we curated protein sequences with functional annotations from the UniProt/Swiss-Prot database [16], with release date of 23 February 2022. Additionally, we obtained predicted protein structures from the AlphaFold Protein Structure Database (AlphaFoldDB) [18], with release date January 2022. To ensure data consistency, we cross-referenced the sequences extracted from the predicted structures with their corresponding entries in UniProt. We then filtered the dataset to include only proteins with sequence lengths ranging from 100 to 1022 residues for our training and validation sets. The upper limit of 1022 residues corresponds to the maximum sequence length accepted by the pretrained ESM model used for generating sequence embeddings. Functional annotations for the proteins were retrieved from UniProt, while ontology graph data were sourced from the Open Biological and Biomedical Ontology (OBO) Foundry data repository [7]. Protein functions were collected from the Gene Ontology (GO), which organizes relationships between GO terms in directed acyclic graphs (DAGs). Adhering to CAFA standards [20], we exclusively considered terms curated under specific evidence codes (EXP, IDA, IPI, IMP, IGI, IEP, TAS, IC, HTP, HDA, HMP, HGI, HEP). To mitigate the inclusion of rare GO terms, we retained only those terms associated with at least 60 proteins for both training and testing purposes. Additionally, to investigate how sequence identity influences the accuracy of protein function prediction, we used MMseqs2 [14] to cluster the proteins in our curated dataset at sequence identity thresholds of 30%, 50%, 90%, and 95%. This resulted in the following data statistics (Table 1):

**Table 1**
**The statistics of the curated protein function prediction dataset**

| Ontology | No. of Proteins | No. of GO terms | Sequence identity threshold | | | |
|---|---|---|---|---|---|---|
| | | | 0.3 | 0.5 | 0.9 | 0.95 |
| MF | 35,507 | 600 | 14,667 | 19,512 | 26,876 | 28,067 |
| CC | 50,340 | 547 | 20,679 | 26,808 | 36,721 | 38,509 |
| BP | 50,320 | 3,774 | 20,180 | 26,647 | 37,536 | 39,348 |

[a]The first three columns are the GO category, the total number of proteins in each category, and the number of GO terms in each category. The remaining four columns list the number of protein clusters at each sequence identity threshold (0.3, 0.5, 0.9, 0.95)

### 1.2 Data Preprocessing

Each protein is represented as a graph, with the extraction process from the predicted structure following a systematic approach. For every protein, we construct a graph encapsulated by an adjacency matrix $n \times n$, where $n$ denotes the number of residues within the protein. This graph, denoted as $G = (V, E, X)$, comprises a vertex set $V = \{v_1, v_2, ..., v_n\}$, an edge set $E$, and a node feature matrix $X \in \mathbb{R}^{m \times |V|}$, where $m$ represents the dimension of the feature and $|V|$ denotes the number of vertices (or residues).

The establishment of edges between residues is computed using the K-nearest neighbor (KNN) method. Specifically, for the KNN approach, a node $v$ is included in the neighborhood $N_k$ if it is among the $K$ nearest neighbors of a given node $u$. Here, $K$ is determined as $\sqrt[3]{n}$, where $n$ signifies the number of residues. Alternatively, we explored two additional methods for graph generation: setting $K$ to $\sqrt{n}$ and employing a distance threshold approach. In the latter method, edges are established based on the Euclidean distance between carbon alpha atoms, with five distinct distance thresholds examined: 4 Å, 6 Å, 8 Å, 10 Å, and 12 Å.

To extract sequence features, we utilize the ESM-1B pretrained protein language model. This model operates at the residue level, generating embeddings for each residue in the protein sequence. These embeddings form a tensor of dimensions $l \times d$, where $l$ represents the sequence length and $d$ denotes the embedding dimension (1280). To generate embeddings for a given sequence, we adopt a mean aggregation approach over the residue-level embeddings. Furthermore, the ESM-1B transformer model produces embeddings at multiple layers, and for our purposes, we specifically use embeddings from the last (33rd) layer [12].

### 1.3 Method

The architecture of TransFun [4] is designed with blocks of rotation- and translation-equivariant graph neural networks, leveraging the equivariant graph neural network (EGNN) [13] as the foundation. Our design comprises four EGNN blocks, each interspersed with a rectified linear unit (ReLU) activation function and a batch normalization layer.

The equivariant graph neural network (EGNN) [13] embodies key principles of equivariance, including translation, rotation, reflection equivariance (E(n)), and permutation equivariance. Permutation equivariance ensures that the architecture respects input symmetries under permutations, maintaining invariance to changes in element order. Similarly, translation equivariance preserves structural consistency by ensuring that network output remains unaffected by translations of the input data. Rotation (and reflection) equivariance enhances robustness to variations in position and orientation, guaranteeing that the model's predictions remain consistent under rotations or reflections of the input data.

In essence, the EGNN achieves these equivariances by operating on relative positions and orientations within the graph. This means that alterations to the graph, such as translations or rotations, maintain the relative positions and orientations between nodes and edges. This preservation of relative information ensures the integrity of the data during message passing and contributes to the model's robustness and effectiveness in processing graph-structured data.

In our model, each EGNN block accepts a graph as input to update its features. Initially, we incorporate node features, including per-residue embeddings and $x, y, z$ coordinates of each residue. We experimented with optional edge features such as the distance between nodes and a binary indicator for adjacent residues connected by a peptide bond. However, these edge features did not yield improvements in prediction accuracy beyond node features and were consequently excluded from the final model.

### 1.4 Getting Started with TransFun

TransFun is an open-source tool for predicting protein functions, accessible to all individuals through GitHub. TransFun is available via the following GitHub link: https://github.com/jianlin-cheng/TransFun. The repository contains a comprehensive toolkit for working with protein structure prediction, including model training, data preprocessing, and utilities for managing and transforming datasets.

### 1.5 Installation Instructions

#### 1.5.1 Clone the Repository

First, to install TransFun on your local computer, clone the project from GitHub at your desired directory on your computer, as shown below:

```
git clone https://github.com/jianlin-cheng/
TransFun.git

cd TransFun/
```

This command clones the repository and then changes the directory to the newly created local repository folder.

#### 1.5.2 Downloading Necessary Data

TransFun uses pretrained models and additional data files that are essential for its operation. These need to be downloaded separately.

```
curl https://calla.rnet.missouri.edu/rnami-
ner/transfun/data
--output data.zip unzip data
```

These commands download the required data and model files into a zip archive and then unzip them in the current directory.

### 1.5.3 Setting Up the Conda Environment

To ensure that all dependencies required by TransFun are installed and managed properly, a Conda environment is used. This isolates the software's dependencies from other Python environments on your system.

```
conda env create -f environment.yml

conda activate transfun
```

These commands create a new Conda environment based on the specifications in the environment.yml file and then activate this environment. Make sure Conda is installed before executing these commands.

## 1.6 Using TransFun: Step-by-Step Guide

TransFun allows prediction of protein functions with either protein structures in PDB format or protein sequences in FASTA format combined with structures.

### 1.6.1 Predicting Protein Functions Using PDB Files

To predict the function of proteins based on their three-dimensional structures provided in PDB format. The software automatically extracts sequences from these PDB files.

```
python predict.py --data-path path\_to\_store
\_intermediate
\_files --ontology GO\_function\_category --
input-type
pdb --pdb-path data\/alphafold --output output
\_file
--cut-off probability\_threshold
```

### 1.6.2 Predicting Protein Functions Using FASTA and PDB Files

For cases where protein sequences are available separately from structures, TransFun can utilize both FASTA-formatted sequences and PDB files for a more comprehensive analysis.

```
python predict.py --data-path path\_to\_store
\_intermediate
\_files --ontology GO\_function\_category --
input-type fasta
--pdb-path data\/alphafold --fasta-path path
```

```
\_to\_a\_fasta
\_file--output result.txt --cut-off probabil-
ity\_threshold
```

#### 1.6.3 Command-Line Options Explained

```
--data-path: Specifies where to store interme-
diate data files.

--ontology: Indicates the GO function category
(e.g.,

cellular\_component, molecular\_function).

--input-type: Type of input (pdb or fasta).

--pdb-path: Directory path containing PDB
files.

--fasta-path: Path to a FASTA file (used only if
input-type

is fasta).

--output: Path for the output file where re-
sults will be

saved.

--cut-off: Probability threshold for predic-
tions.
```

Here, we present examples for two proteins—A0A5S9MMK5 and E9Q0S6—focusing on molecular function and cellular component, respectively. Using the commands provided below, each taking an input PDB file, we display the results. Specifically, we show the correct predictions (true positives) from TransFun along with their respective probability scores, the incorrect predictions (false positives with their corresponding scores), and the missed predictions (false negatives) in Table 2.

**Table 2**
**Predictions for protein A0A5S9MMK5 (molecular function) and protein E9Q0S6 (cellular component) are shown. True positives are indicated in blue font, false positives in red font, and false negatives in orange font**

| | | | | | |
|---|---|---|---|---|---|
| **(A) Molecular function (A0A5S9MMK5)** | | | | | |
| GO:0003690 (0.391) | GO:0000976 (0.310) | GO:1901363 (0.863) | GO:0043565 (0.672) | GO:0003677 (0.772) | GO:0003676 (0.851) |
| GO:0000981 (0.319) | GO:0001067 (0.319) | GO:0001216 (0.181) | GO:0003700 (0.580) | GO:0000987 (0.206) | GO:0005488 (0.937) |
| GO:0140110 (0.580) | GO:0001227 (0.155) | GO:0005515 (0.700) | GO:0042802 (0.136) | GO:0000978 (0.184) | GO:0097159 (0.864) |
| GO:0001228 (0.119) | GO:0001217 (0.248) | GO:0000977 (0.219) | GO:1990837 (0.361) | GO:0046872 | GO:0043169 |
| GO:0046872 | GO:0043169 | GO:0043167 | GO:0036094 | | |
| (B) Cellular component (E9Q0S6) | | | | | |
| GO:0005622 (0.901) | GO:0015629 (0.129) | GO:0043226 (0.571) | GO:0005856 (0.129) | GO:0043229 (0.519) | GO:0043232 (0.167) |
| GO:0070013 (0.127) | GO:0043231 (0.364) | GO:0030054 (0.225) | GO:0071944 (0.354) | GO:0005654 (0.127) | GO:0043233 (0.127) |
| GO:0016020 (0.316) | GO:0031974 (0.127) | GO:0120025 (0.154) | GO:0005829 (0.516) | GO:0005886 (0.316) | GO:0031981 (0.127) |
| GO:0110165 (0.992) | GO:0043227 (0.394) | GO:0043228 (0.167) | GO:0005737 (0.833) | GO:0042995 (0.167) | GO:0045202 (0.101) |
| GO:0070161 (0.121) | GO:0005634 (0.218) | GO:0031982 (0.147) | GO:0009986 | | |

```
python predict.py --data-path data --ontology
molecular
\_function --cut-off 0.01 --input-type pdb --
pdb-path
test/pdbs/ --output result.txt


python predict.py --data-path data
--ontology cellular\_component --cut-off 0.1
--input-type pdb --pdb-path test/pdbs/
--output result.txt
```

## 2 Conclusion

In this chapter, we present a book chapter detailing our research on protein function prediction (AFP) using TransFun [4]. TransFun is a structure-based method that leverages embeddings from the ESM-1B [12] pretrained protein language model and AlphaFold's predicted structures [10] to predict protein functions.

For future work, we aim to decompose and better understand the individual contributions of sequence and structural features. To the best of our knowledge, most structure-based AFP methods also incorporate sequence features. This is largely due to the widespread use of graph neural networks (GNNs) [5, 8], which typically require node-level features where sequence-derived features are prevalent. As a result, distinguishing the specific contributions of structural features has proven challenging.

To address this, we plan to utilize methods that extract features exclusively from protein structures, such as the $3Di$ alphabet [9, 17] or other structural modules, to isolate and assess the impact of structural information alone. Additionally, we aim to enhance prediction accuracy by integrating multiple modalities, including textual data and InterPro annotations, through advanced multimodal approaches and large language models.

## Acknowledgements

This work was supported in part by a grant from the US National Science Foundation (NSF grant numbers: DBI2308699, CCF2343612).

## References


1. Aleksander SA, Balhoff J, Carbon S, Cherry JM, Drabkin HJ, Ebert D et al (2023) The gene ontology knowledgebase in 2023. Genetics 224(1):iyad031
2. Ashburner M, Ball CA, Blake JA, Botstein D, Butler H, Cherry JM et al (2000) Gene ontology: tool for the unification of biology. Nat Genet 25(1):25–29
3. Boadu F, Cheng J (2024) Improving protein function prediction by learning and integrating representations of protein sequences and function labels. Bioinform. Adv. 4(1):vbae120
4. Boadu F, Cao H, Cheng J (2023) Combining protein sequences and structures with transformers and equivariant graph neural networks to predict protein function. Bioinformatics 39 (Supplement_1):i318–i325
5. Boadu F, Lee A, Cheng J (2024) Deep learning methods for protein function prediction. Proteomics. https://analyticalsciencejournals.onlinelibrary.wiley.com/doi/abs/10.1002/pmic.202300471
6. Clark WT, Radivojac P (2013) Information-theoretic evaluation of predicted ontological annotations. Bioinformatics 29(13):i53–i61
7. Gene Ontology Consortium (2004) The gene ontology (GO) database and informatics resource. Nucleic Acids Res 32(suppl_1): D258–D261
8. Gligorijević V, Renfrew PD, Kosciolek T, Leman JK, Berenberg D, Vatanen T et al (2021) Structure-based protein function prediction using graph convolutional networks. Nat Commun 12(1), 3168
9. Heinzinger M, Weissenow K, Sanchez JG, Henkel A, Mirdita M, Steinegger M, Rost B (2023) ProstT5: bilingual language model for protein sequence and structure. NAR Genom Bioinform 6(4):p.lqae150
10. Jumper J, Evans R, Pritzel A, Green T, Figurnov M, Ronneberger O et al (2021) Highly accurate protein structure prediction with AlphaFold. Nature 596(7873):583–589
11. Kulmanov M, Guzmán-Vega FJ, Duek Roggli P, Lane L, Arold ST, Hoehndorf R (2024) Protein function prediction as approximate semantic entailment. Nat Mach Intell 6(2):220–228
12. Rives A, Meier J, Sercu T, Goyal S, Lin Z, Liu J et al (2021) Biological structure and function emerge from scaling unsupervised learning to 250 million protein sequences. Proc Natl Acad Sci 118(15):e2016239118
13. Satorras VG, Hoogeboom E, Welling M (2021) E (n) equivariant graph neural networks. In: International conference on machine learning, pp 9323–9332
14. Steinegger M, Söding J (2017) MMseqs2 enables sensitive protein sequence searching for the analysis of massive data sets. Nat Biotechnol 35(11):1026–1028
15. The Gene Ontology Consortium (2023) The Gene Ontology knowledgebase in 2023. Genetics 224(1):iyad031. https://doi.org/10.1093/genetics/iyad031
16. UniProt Consortium (2019) UniProt: a worldwide hub of protein knowledge. Nucleic Acids Res 47(D1):D506–D515
17. Van Kempen M, Kim SS, Tumescheit C, Mirdita M, Lee J, Gilchrist CL et al (2024) Fast and accurate protein structure search with Foldseek. Nat Biotechnol 42(2):243–246
18. Varadi, M., Anyango, S., Deshpande, M., Nair S, Natassia C, Yordanova G et al (2022) AlphaFold protein structure database: massively expanding the structural coverage of protein-sequence space with high-accuracy models. Nucleic Acids Res 50(D1):D439–D444
19. Wang S, You R, Liu Y, Xiong Y, Zhu S (2023) NetGo 3.0: protein language model improves large-scale functional annotations. Genom. Proteomics Bioinform 21(2):349–358
20. Zhou N, Jiang Y, Bergquist TR, Lee AJ, Kacsoh BZ, Crocker AW et al (2019) The CAFA challenge reports improved protein function prediction and new functional annotations for hundreds of genes through experimental screens. Genome Biol 20(1):1–23.

# Chapter 7

# Using InterLabelGO+ for Accurate Protein Language Model-Based Function Prediction

**Chengxin Zhang, Quancheng Liu, and Lydia Freddolino**

**Abstract**

The recent availability of large language models for protein sequences has spurred the development of deep learning models for prediction of protein functions, mainly in the form of gene ontology (GO) terms. We developed InterLabelGO+, a top performing deep learning–based protein GO term prediction model in the recent fifth Critical Assessment of Function Annotation (CAFA5) challenge. InterLabelGO+ uses the ESM2 protein language model to extract sequence features, which are then used as inputs to a deep learning model that was trained under a loss function that considers the potentially complex relations between different GO terms. The deep learning–predicted GO terms are then combined with GO terms from sequence homology search to derive consensus predictions. InterLabelGO+ is available at https://seq2fun.dcmb.med.umich.edu/InterLabelGO/. In this chapter, we demonstrate procedures to perform protein GO term prediction with the InterLabelGO+ webserver and the standalone package, as well as how to retrain the model with up-to-date training data.

**Key words** Protein function prediction, Gene ontology, Protein language model, Deep learning, Sequence homology

## 1 Introduction

Proteins are the main biomacromolecules performing biological functions in living cells. As the most popular standardized vocabulary for protein functions, gene ontology (GO) terms categorize functions into three aspects: molecular function (MF), Biological Process (BP) and Cellular Component (CC). Since predicting protein functions, especially GO terms, is of great significance to the biological and biomedical community, many template-based [1–5] and machine learning–based [6–8] approaches have been developed for this purpose. With recent advances in artificial intelligence and natural language processing, almost all recently published

Authors Chengxin Zhang and Quancheng Liu have equally contributed to this chapter.

Dukka B. KC (ed.), *Large Language Models (LLMs) in Protein Bioinformatics*, Methods in Molecular Biology, vol. 2941, https://doi.org/10.1007/978-1-0716-4623-6_7,

protein GO term predictors [9–11] are at least partly based on deep learning, often making use of large language models (LLMs) such as ESM [12], ProtT5 [13], and ProteinBert [14] that are trained on self-supervised learning of protein sequences.

A typical deep learning–based protein GO prediction model [9–11] use an LLM to extract features from the query amino acid sequence. These sequence features are then input to a neural network to predict the GO term labels. The prediction of each GO term is treated as a binary classification problem, typically through the cross-entropy loss function during neural network training, having made use of a dedicated protein training set with previously known functions to use as ground truths.

We recently developed InterLabelGO+, a hybrid pipeline for accurately predicting protein GO terms by combining a deep learning model (InterLabelGO) and sequence homolog search method (AlignmentKNN) (Fig. 1) [15]. There are two main differences between InterLabelGO and most existing deep learning–based protein GO predictors. First, instead of simply using the final output of an LLM for sequence feature generation, InterLabelGO extracts sequence features from the last three layers of the ESM2 LLM to harvest richer and more informative sequence data from the embedding. Second, rather than using a loss function that considers prediction of each GO terms separately (such as the cross-entropy loss that is nearly ubiquitous in the function prediction field), InterLabelGO uses a combinatory loss function (the F1 loss

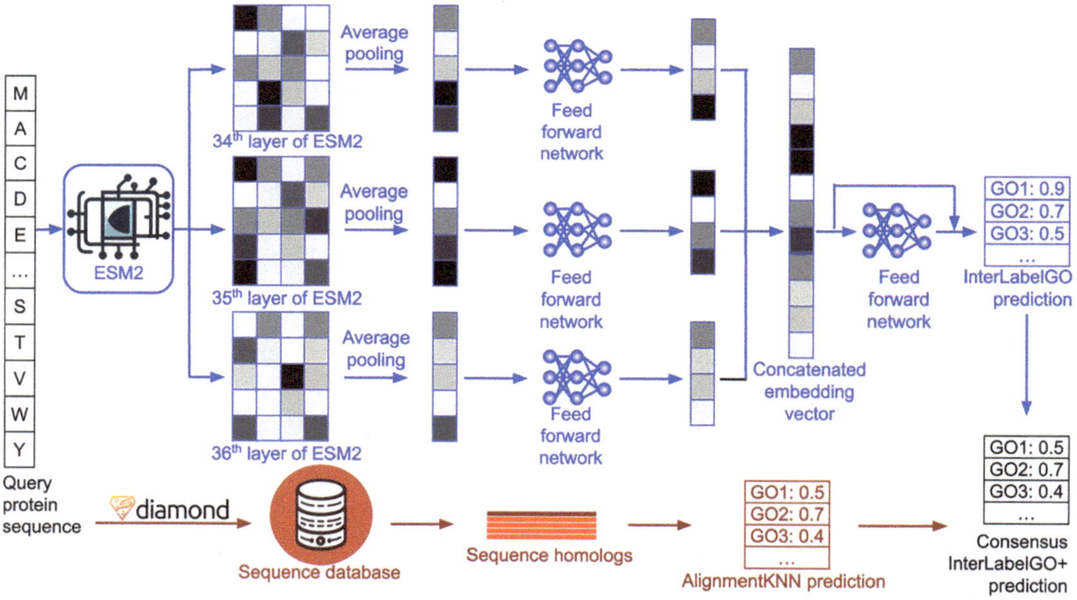


**Fig. 1** Workflow of InterLabelGO+, which combines protein GO term predictions from a deep learning–based component (InterLabelGO, blue) and a sequence homolog search-based component (AlignmentKNN, red) to obtain consensus predictions

plus the ZLPR loss [16] to consider correlation among different GO annotations as well as the information contents of each GO terms). Details of the architecture design and benchmark performance of InterLabelGO+ are reported elsewhere [15]. In this chapter, we focus on how to run the InterLabelGO+ program through the online web-server and through a standalone package.

## 2 Materials

The InterLabelGO+ web server can be run through most major web browsers, including Google Chrome, Mozilla Firefox, and Microsoft Edge. An academic email address is needed to receive notification upon job completion.

To run the standalone package of InterLabelGO+ locally, either a Linux operating system or Windows Subsystem for Linux (WSL) is needed. While predicting GO terms using the pretrained InterLabelGO model can be performed on either Central Processing Unit (CPU) or Graphics Processing Unit (GPU), Compute Unified Device Architecture (CUDA)-capable GPU is required if the user wishes to retrain the standalone InterLabelGO program. Python libraries required by InterlabelGO+ (Table 1) can be setup by conda, which is available either through Anaconda (https://www.anaconda.com/download) or Miniconda (https://docs.anaconda.com/miniconda/).

## 3 Methods

### 3.1 Using the InterLabelGO+ Webserver

#### 3.1.1 Webserver Input

The InterLabelGO+ webserver is designed for academic users who needs to perform protein function predictions for a small number of proteins. Here, we will use the Cyclin-H protein from fruit fly (UniProt accession: O76513) as an example (Fig. 2). First, the user may submit the protein of interest via https://seq2fun.dcmb.med.umich.edu/InterLabelGO/ (Fig. 2a), where the user should provide the amino acid sequence (either in FASTA format or just the sequence without the header), an academic email (Hotmail, Gmail, Yahoo, or other commercial emails are not acceptable) to track the job status, and the name of the protein. If the name of the protein is not provided, the protein name will be "protein" by default.

Upon successful job submission, a webpage link in the format of https://seq2fun.dcmb.med.umich.edu/InterLabelGO/output/IL******** is provided by the webserver (Fig. 2b). Here, "********" stands for an eight-digit number. This webpage link will be the place where the user may check the job status and view the final result. If this link does not lead to an active webpage, just check back in a few minutes. The webpage link is also sent to the user via email (Fig. 2c).

**Table 1**
**Python packages required to run the InterLabelGO+ standalone package (note that this information is also available in the `requirements.txt` file of the standalone package)**

| Package | Version |
|---|---|
| python | 3.11.5 |
| biopython | 1.82 |
| certifi | 2022.12.7 |
| charset-normalizer | 2.1.1 |
| fair-esm | 2.0.0 |
| filelock | 3.9.0 |
| fsspec | 2023.4.0 |
| idna | 3.4 |
| iterative-stratification | 0.1.7 |
| Jinja2 | 3.1.2 |
| joblib | 1.3.2 |
| MarkupSafe | 2.1.3 |
| mpmath | 1.3.0 |
| networkx | 3.0 |
| numpy | 1.24.1 |
| obonet | 1.0.0 |
| pandas | 2.1.4 |
| Pillow | 9.3.0 |
| python-dateutil | 2.8.2 |
| pytz | 2023.3.post1 |
| requests | 2.28.1 |
| scikit-learn | 1.3.2 |
| scipy | 1.11.4 |
| six | 1.16.0 |
| sympy | 1.12 |
| threadpoolctl | 3.2.0 |
| tqdm | 4.66.1 |
| triton | 2.1.0 |
| typing_extensions | 4.4.0 |
| tzdata | 2023.4 |
| urllib3 | 1.26.13 |

(continued)

**Table 1 (continued)**

| Package | Version |
|---|---|
| torch | 2.1.0+cu121 (from https://download.pytorch.org/whl/cu121) |
| torchaudio | 2.1.0+cu121 |
| torchvision | 0.16.0+cu121 |

#### 3.1.2 Webserver Output

A prediction job takes about 15 min. This wait time depends heavily on our server load and can be significantly longer if there are many prediction jobs in queue. The actual time to run the prediction is much shorter (*see* Subheading 3.2.3).

Once the prediction job is finished, the user will be notified by email of job completion (Fig. 2d). The full result will be available at the webpage link provided by Fig. 2b, d. The result webpage (Fig. 3) starts with the input sequence, followed by predicted GO terms. Predicted terms for molecular function (MF), biological process (BP), and cellular components (CC) aspects are shown separately.

For each GO aspect, we list the predicted GO terms, the InterLabelGO+ prediction scores for those terms, and the full names of the GO term in a table (Fig. 3). If both the child term and its parent term are predicted, and the parent term does not have a higher predicted scores than the child term, the parent term will not be shown in the table. Here, the InterLabelGO+ prediction score, also known as C-score in the webserver output, ranges between 0 and 1. A higher C-score indicates a higher precision, i.e., posterior probability of the predicted GO term being correct. C-scores of 0.166, 0.755, and 0.139 correspond to a precision of 0.5 for MF, BP, and CC, respectively (*see* calibration curves in Fig. 4).

Apart from the table listing predicted GO terms, we also use Graphviz [17] to draw directed acyclic graphs for the predicted terms (colored) and their parent terms (white). The predicted terms are colored in red, orange, yellow, green, cyan, blue, or purple in descending order of the prediction score.

In another table, we list the top 20 sequence homologs identified by DIAMOND in the AlignmentKNN component (Fig. 3, **right side**). Note that the sequence templates for different aspects may be different, e.g., in the case that a sequence template has GO annotations for the MF aspect but not the BP or CC aspect in the database. For each sequence template, we list its ranking, UniProt accession, alignment bit-score, sequence identity of the aligned residues, E-value, sequence length, global sequence identity normalized by the query length, and global sequence identity normalized by the template length.

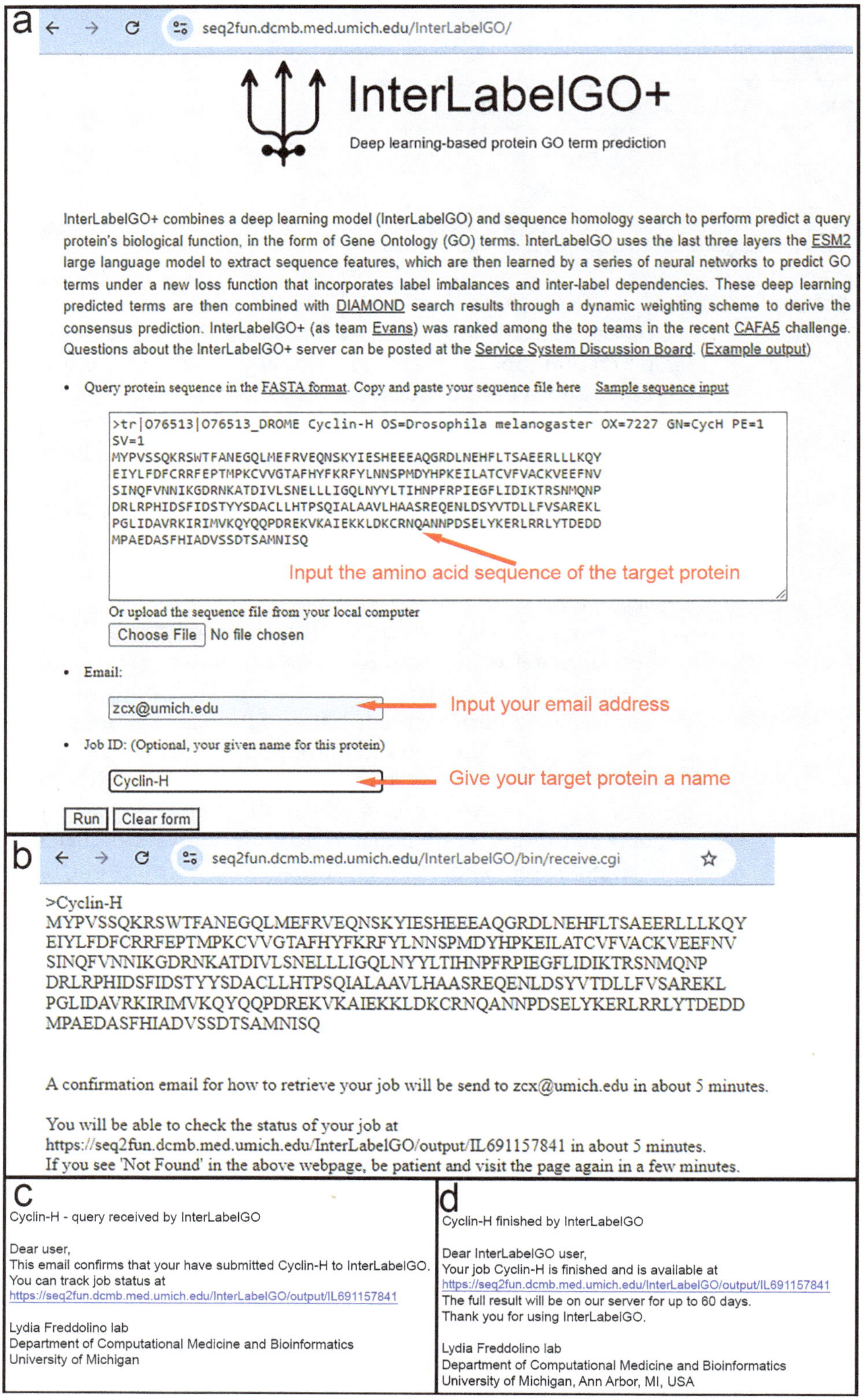


**Fig. 2** The InterLabelGO+ webserver. (**a**) Submitting a protein sequence (cyclin-H from fruit fly) to the InterLabelGO+ website. (**b**) Webpage shown after successful job submission. (**c**) Email to provide notification of successful job submission. (**d**) Email to notify the user of job completion

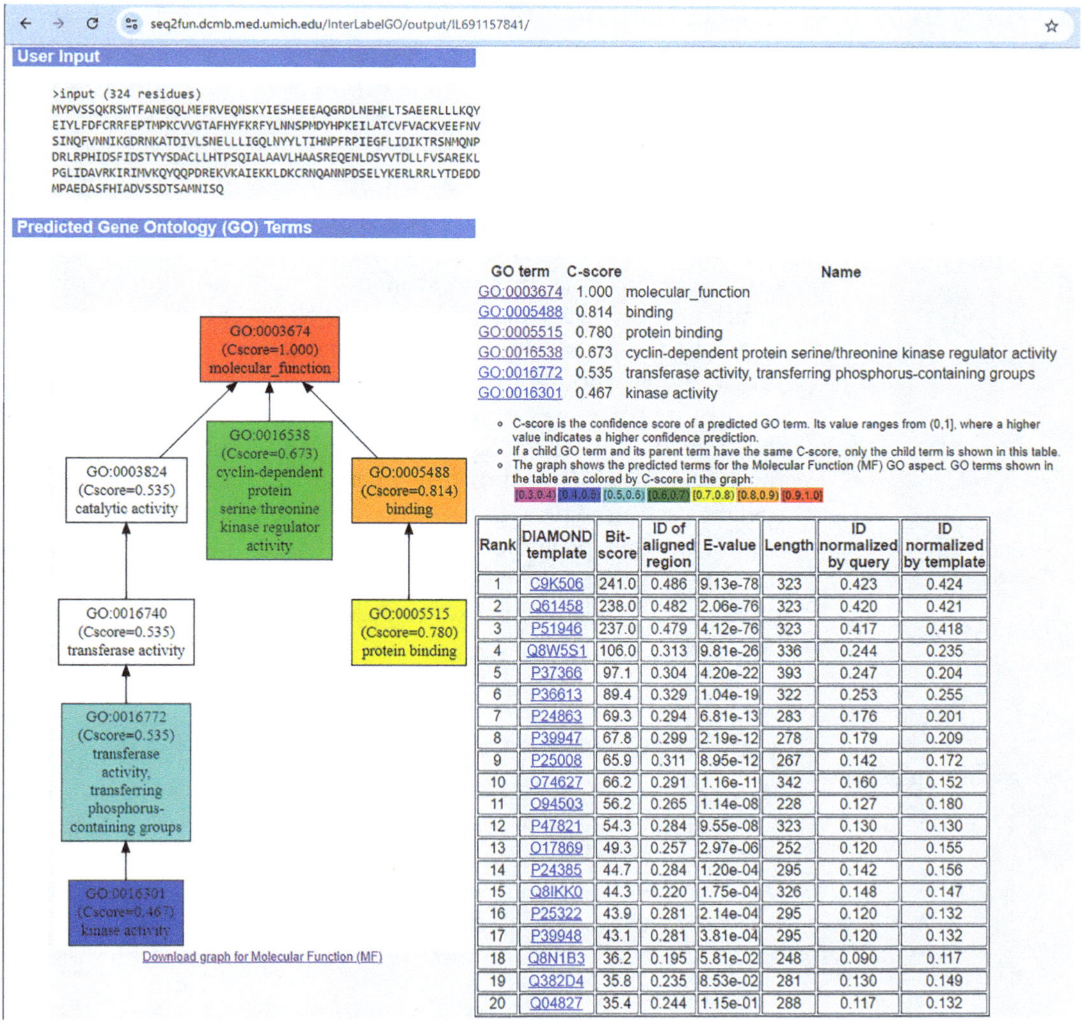


**Fig. 3** Excerpt of InterLabelGO+ webserver prediction for fruit fly cyclin-H. While the full prediction result contains function predictions for MF, BP, and CC, this figure shows predictions for only MF due to space limitations

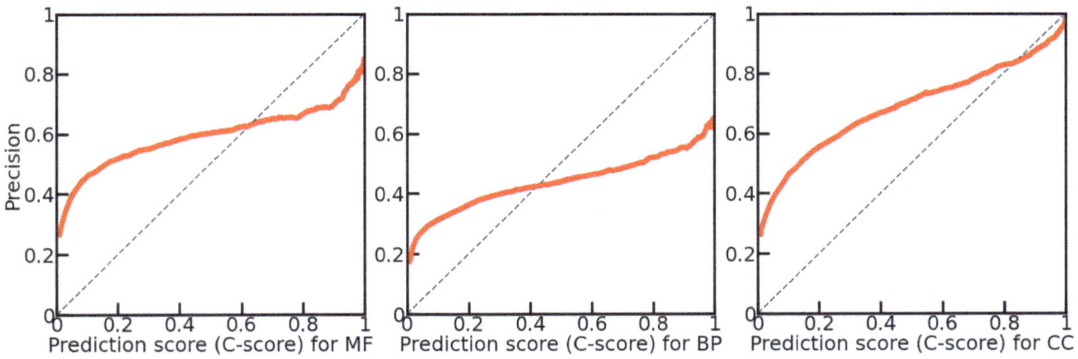


**Fig. 4** Calibration curves (red) between the InterLabelGO+ prediction score (i.e., C-score) and the precision of GO term predictions for the three aspects (MF, BP, and CC) of GO terms. These data are derived from 2041 proteins with newly annotated GO terms from December 2022 to March 2023 in the UniProt-GOA database

### 3.2 Using the InterLabelGO+ Standalone Package

Although the InterLabelGO+ webserver is user friendly, it has several limitations; thus, we also provide an InterLabelGO+ standalone package which can be applied by users whose needs exceed the usage cases supported by the webserver. The limitations of the webserver include first, the webserver requires an academic email, while the standalone package can be used by any users, including commercial ones. Second, while the webserver only process protein sequences with up to 1500 amino acids, the standalone package can work with arbitrary protein sequence lengths so long as there is sufficient physical memory. Third, the webserver only processes one sequence at a time, while the standalone package can perform bulk prediction of thousands of proteins in a single run, making the latter more useful for proteome-wide function annotation. Fourth, the standalone package allows the user to retrain the model parameters using the most recent protein function databases, or a specific version of protein function database to enable fair comparison with other programs. This retraining is not available for the user of the webserver. In the following sections, we will discuss how to run the standalone package to deploy, train, and test InterLabelGO+.

#### 3.2.1 Installing the InterLabelGO+ Standalone Package

1. Download and extract the InterLabelGO+ package:

```
$ wget https://seq2fun.dcmb.med.umich.edu/InterLabelGO/
InterLabelGO+.tar.xz
$ tar -xvf InterLabelGO+.tar.xz
$ cd InterLabelGO
```

The tar.xz compressed package is approximately 7 GB. While we also make the source code for the InterLabelGO package available at https://github.com/QuanEvans/InterLabelGO, this GitHub site does not include pretrained model parameters, which are over the file size limits of GitHub. Therefore, we recommend downloading the tar.xz compressed package instead. The files in the folder are explained by Table 2.

2. Install Python dependencies required by InterLabelGO+. Although these dependencies (Table 1) can be directly installed using `pip` (https://pypi.org/project/pip/), we strongly recommend running `pip` within a dedicated environment setup with the `conda` command:

```
$ conda create -n InterLabelGO python=3.11.5 -y
$ conda activate InterLabelGO
$ pip install -r requirements.txt
```

**Table 2**
**Files in the InterLabelGO+ standalone package**

| Filename | Description |
|---|---|
| `README.md` | Readme file for how to install and use InterLabelGO+ |
| `LICENSE` | MIT license |
| `predict.py` | Inference script to perform InterLabelGO+ prediction |
| `alignment_knn.py` | Script to perform AlignmentKNN prediction |
| `plm.py` | Script to convert FASTA sequence into ESM2 embedding |
| `train_InterLabelGO.py` | Training script |
| `prepare_data.py` | Convert data format for the training data |
| `update_data.py` | Download the latest GOA database and convert it into the required format for retraining the model |
| `settings.py` | Configuration file for paths of data files |
| `requirements.txt` | PIP script to specify python packages used by InterLabelGO+ |
| `setup_env.sh` | Script to help set up dependencies |
| `Network/` | Scripts to define InterLabelGO model architecture, loss functions, and data loading functions |
| `example/` | Folder for example files |
| `models/ZLPR_PTF1_GOF1/` | InterLabelGO pretrained model parameters |
| `Data/alignment_database/` | DIAMOND format sequence databases for AlignmentKNN |
| `Data/alignment_labels/` | GO terms for sequence databases for AlignmentKNN |
| `Data/esm_models/` | The ESM2 protein language model |
| `Data/cafa5_raw_data/` | Text format raw training set |
| `Data/network_training_data/` | InterLabelGO training dataset in numpy pickle format |

The above commands create a new conda environment called "InterLabelGO", activate the environment, and install all dependencies within that environment, making use of the explicit dependency file (`requirements.txt`) provided for InterLabelGO+. To run InterLabelGO+ in other shell sessions, make sure to run "`conda activate InterLabelGO`" beforehand.

*3.2.2 Testing the InterLabelGO+ Standalone Package*

1. To perform function inference, we will create the input FASTA format sequence file "`test/seq.fasta`", which corresponds to the Cyclin-H protein used in the previous example:

```
$ mkdir test/
$ echo ">O76513" > test/seq.fasta
$ echo
"MYPVSSQKRSWTFANEGQLMEFRVEQNSKYIESHEEEAQGRDLNEHFLTSAEERL
LLKQYEIYLFDFCRRFEPTMPKCVVGTAFHYFKRFYLNNSPMDYHPKEILATCVFV
ACKVEEFNVSINQFVNNIKGDRNKATDIVLSNELLLIGQLNYYLTIHNPFRPIE
GFLIDIKTRSNMQNPDRLRPHIDSFIDSTYYSDACLLHTPSQIALAAVLHAASRE
QENLDSYVTDLLFVSAREKLPGLIDAVRKIRIMVKQYQQPDREKVKAIEKKLDKCR
NQANNPDSELYKERLRRLYTDEDDMPAEDASFHIADVSSDTSAMNISQ"      >>
test/seq.fasta
```

While in the above example, the input file "`test/seq.fasta`" has only one sequence, the input FASTA file may contain two or more sequences, in which case predictions will be performed separately for each protein in the input FASTA sequence.

2. Run the inference script to predict the GO terms for the target sequence:

```
$ python predict.py -w test -f test/seq.fasta --use_gpu
```

The full process takes around 7 min on a modern Linux workstation without GPU and less than 1 min with GPU. In the above command, "`-w test`" specifies the folder to which output files are written. "`-f test/seq.fasta`" specifies the input FASTA file path. The flag "`—use_gpu`" requests the usage of GPU to accelerate ESM2 sequence embedding when available. The main output file in the `test/` folder is `InterLabelGO+.tsv`, whose format is explained in Fig. 5. Other output files are explained in Table 3. The procedure for running InterLabelGO+ for standard function predictions is exactly equivalent to what is shown above, just using a different query FASTA file.

If attempting to run the standalone package results in error messages complaining that a module is unavailable, such as "`ModuleNotFoundError: No module named torch`", even if the Python dependencies are correctly installed, make sure that the conda environment for InterLabelGO is activated by:

```
$ conda activate InterLabelGO
```

Please see the conda documentation (https://docs.conda.io/) for additional information.

| | A | B | C | D | E | F |
|---|---|---|---|---|---|---|
| 1 | EntryID | term | score | max_seqid | aspect | go_term_name |
| 111 | O76513 | GO:0003674 | 1 | 0.382 | MFO | molecular_function |
| 112 | O76513 | GO:0005488 | 0.814 | 0.382 | MFO | binding |
| 113 | O76513 | GO:0005515 | 0.78 | 0.382 | MFO | protein binding |
| 114 | O76513 | GO:0016538 | 0.673 | 0.382 | MFO | cyclin-dependent protein serine/threonine kinase regulator activity |
| 115 | O76513 | GO:0003824 | 0.535 | 0.382 | MFO | catalytic activity |
| 116 | O76513 | GO:0016740 | 0.535 | 0.382 | MFO | transferase activity |
| 117 | O76513 | GO:0016772 | 0.535 | 0.382 | MFO | transferase activity, transferring phosphorus-containing groups |

**Fig. 5** Excerpt from the InterLabelGO+ standalone package predictions for cyclin-H. The six columns in this Tab Separated Values (TSV) format table are the query protein name, GO term, prediction score (i.e., C-score), the average sequence identity of the top 5 templates, GO aspects (MF, BP, or CC), and description of the GO term contained on that row

**Table 3**
**Output files for testing InterLabelGO+. Files are ranked in chronological order of creation**

| Filename | Descriptionm |
|---|---|
| `InterLabelGO_pred/embed_feature/*npy` | ESM2 embedding for the query sequence |
| `InterLabelGO_pred/names.npy` | List of names of the query proteins |
| `InterLabelGO_pred/InterLabelGO.tsv` | Deep learning prediction |
| `AlignmentKNN_pred/*_blast.tsv` | DIAMOND search results |
| `AlignmentKNN_pred/seqid.tsv` | Sequence identities for top ranking templates for MF, BP and CC |
| `AlignmentKNN_pred/AlignmentKNN.tsv` | AlignmentKNN prediction |
| `InterLabelGO+.tsv` | Final GO term predictions from InterLabelGO+ |

*3.2.3 Retraining the InterLabelGO+ Standalone Package*

In this section, we detail how the InterLabelGO+ network can be retrained using the latest versions of the UniProt-GOA annotation database and Swiss-Prot protein sequence database.

1. Download the latest UniProt Gene Ontology Annotation (UniProt-GOA) database and the Swiss-Prot protein sequence database, which are approximately 20GB in total file size:

```
$ python update_data.py
```

2. Convert the format of the downloaded data files:

```
$ python prepare_data.py --make_db --ia
```

The above command will perform the following operations:

- Convert text format training set sequences and GO terms labels at "Data/cafa5_raw_data/train_*" to binary format training data at "Data/network_training_data/*".
- Create DIAMOND format sequence databases at "Data/alignment_database/" for the training data, as specified by the –make_db flag.
- Calculate the information content for each GO term for the training data and save the information contents to "Data/network_training_data/IA.txt", as specified by the –ia flag.
- Calculate the ESM2 sequence embeddings for the training sequences at "Data/embeddings/*.npy".

3. Retrain the model using the new training data:

```
$ python train_InterLabelGO.py
```

The newly trained model will overwrite the old InterLabelGO models at "models/ZLPR_PTF1_GOF1/*/model_*.pt". The full training procedure takes approximately 90 min on an NVIDIA RTX 4090 GPU.

## Acknowledgments

The authors thank Dr. Xiaoqiong Wei for insightful discussions. This work used the Advanced Cyberinfrastructure Coordination Ecosystem: Services & Support (ACCESS) program, which is supported by the National Science Foundation [2138259, 2138286, 2138307, 2137603, and 2138296]. This work was supported by the National Institute of Allergy and Infectious Diseases [AI134678 to L.F.]

### References

1. Zhang C, Freddolino PL, Zhang Y (2017) COFACTOR: improved protein function prediction by combining structure, sequence and protein-protein interaction information. Nucleic Acids Res 45(W1):W291–W299. https://doi.org/10.1093/nar/gkx366
2. Laskowski RA (2017) The ProFunc function prediction server. Methods Mol Biol 1611:75–95. https://doi.org/10.1007/978-1-4939-7015-5_7
3. Gong Q, Ning W, Tian W (2016) GoFDR: a sequence alignment based method for predicting protein functions. Methods 93:3–14. https://doi.org/10.1016/j.ymeth.2015.08.009

4. Martin DM, Berriman M, Barton GJ (2004) GOtcha: a new method for prediction of protein function assessed by the annotation of seven genomes. BMC Bioinf 5:178. https://doi.org/10.1186/1471-2105-5-178
5. Hawkins T, Chitale M, Luban S, Kihara D (2009) PFP: automated prediction of gene ontology functional annotations with confidence scores using protein sequence data. Proteins 74(3):566–582. https://doi.org/10.1002/prot.22172
6. Wang S, You R, Liu Y, Xiong Y, Zhu S (2023) NetGO 3.0: protein language model improves large-scale functional annotations. Genomics Proteomics Bioinformatics 21:349. https://doi.org/10.1016/j.gpb.2023.04.001
7. Kulmanov M, Khan MA, Hoehndorf R, Wren J (2018) DeepGO: predicting protein functions from sequence and interactions using a deep ontology-aware classifier. Bioinformatics 34(4):660–668. https://doi.org/10.1093/bioinformatics/btx624
8. Gligorijevic V, Renfrew PD, Kosciolek T, Leman JK, Berenberg D, Vatanen T, Chandler C, Taylor BC, Fisk IM, Vlamakis H, Xavier RJ, Knight R, Cho K, Bonneau R (2021) Structure-based protein function prediction using graph convolutional networks. Nat Commun 12(1):3168. https://doi.org/10.1038/s41467-021-23303-9
9. Zhu YH, Zhang C, Yu DJ, Zhang Y (2022) Integrating unsupervised language model with triplet neural networks for protein gene ontology prediction. PLoS Comput Biol 18(12): e1010793. https://doi.org/10.1371/journal.pcbi.1010793
10. Yuan Q, Xie J, Xie J, Zhao H, Yang Y (2023) Fast and accurate protein function prediction from sequence through pretrained language model and homology-based label diffusion. Brief Bioinform 24:3. https://doi.org/10.1093/bib/bbad117
11. Kulmanov M, Guzmán-Vega FJ, Roggli PD, Lane L, Arold ST, Hoehndorf R (2024) Protein function prediction as approximate semantic entailment. Nat Mach Intell 6(2): 220–228. https://doi.org/10.1038/s42256-024-00795-w
12. Lin Z, Akin H, Rao R, Hie B, Zhu Z, Lu W, Smetanin N, Verkuil R, Kabeli O, Shmueli Y, dos Santos CA, Fazel-Zarandi M, Sercu T, Candido S, Rives A (2022) Evolutionary-scale prediction of atomic level protein structure with a language model. bioRxiv:2022.2007.2020.500902. https://doi.org/10.1101/2022.07.20.500902
13. Elnaggar A, Heinzinger M, Dallago C, Rehawi G, Wang Y, Jones L, Gibbs T, Feher T, Angerer C, Steinegger M (2021) Prottrans: toward understanding the language of life through self-supervised learning. IEEE Trans Pattern Anal Mach Intell 44(10): 7112–7127. https://doi.org/10.1109/TPAMI.2021.3095381
14. Brandes N, Ofer D, Peleg Y, Rappoport N, Linial M (2022) ProteinBERT: a universal deep-learning model of protein sequence and function. Bioinformatics 38(8):2102–2110. https://doi.org/10.1093/bioinformatics/btac020
15. Liu Q, Zhang C, Freddolino L (2024) InterLabelGO+: unraveling label correlations in protein function prediction. bioRxiv:2024.2006.2026.600816. https://doi.org/10.1101/2024.06.26.600816
16. Su J, Zhu M, Murtadha A, Pan S, Wen B, Liu Y (2022) Zlpr: a novel loss for multi-label classification. arXiv preprint arXiv:220802955. https://doi.org/10.48550/arXiv.2208.02955
17. Ellson J, Gansner ER, Koutsofios E, North SC, Woodhull G (2004) Graphviz and dynagraph – static and dynamic graph drawing tools, Mathematics and visualization, pp 127-148. https://doi.org/10.1007/978-3-642-18638-7_6

# Chapter 8

# Functional Annotation of Proteomes Using Protein Language Models: A High-Throughput Implementation of the ProtTrans Model

**Ildefonso Cases, Gemma Martínez-Redondo, Rosa Fernández, and Ana M. Rojas**

## Abstract

Protein function prediction is critical for a wide range of applications in biology, spanning from functional genomics to protein design and genome evolution, among others. However, accurately predicting protein function remains a longstanding challenge in computational biology, especially for non-model organisms. Traditional methods based on sequence similarity often fail to annotate a significant proportion of proteins. The emergence of protein language models has significantly improved this process, enabling more accurate and comprehensive functional annotation. In this work, we highlight how the ProtTrans language model outperforms other tools in per-protein annotation, offering a more precise approach to predicting protein function. We also introduce functional annotation based on embedding space similarity (FANTASIA; available at https://github.com/MetazoaPhylogenomicsLab/FANTASIA), a tool developed to harness these advances for large-scale annotation of uncharacterized proteomes. We provide a detailed overview of how to use FANTASIA, interpret its outputs, and demonstrate its utility in three case studies: (a) enrichment analyses from transcriptomics data, (b) assigning novel functions to unannotated genes in model organisms, and (c) identifying genes involved in important functions in non-model organisms. These results demonstrate the potential of protein language models to advance functional annotation in diverse biological contexts.

**Key words** Protein function prediction, Protein language models, Proteome annotation, FANTASIA, Computational biology

## 1 Introduction

UniProtKB is the main repository for protein sequences, currently containing over 250 million sequences (as of the January release of UniProtKB). However, less than 0.5% of known proteins have experimental annotations [1].

Functional annotation on unknown sequences is typically predicted using sequence similarity-based methods, which rely on

Dukka B. KC (ed.), *Large Language Models (LLMs) in Protein Bioinformatics*, Methods in Molecular Biology, vol. 2941,
https://doi.org/10.1007/978-1-0716-4623-6_8,

evolutionary information from a limited set of model organisms. The rationale is that similar genes are likely to perform similar functions [2], allowing the transfer of function when sequence similarity exceeds a certain threshold. In this approach, functional information—often formalized as gene ontology (GO) terms [3]—is transferred from known to unknown sequences based on these similarity criteria. This is the standard method for investigating the coding functional landscape of genomes.

However, there are several issues with this approach. It is known that unrelated genes can perform the same function [4], while highly similar genes can have orthogonal functions [5]. Additionally, many genes do not exhibit detectable similarity. Furthermore, the dynamic structural conformation of proteins complicates the identification of evolutionary relationships, contributing to the "darkness" of certain protein functions [6]. These challenges highlight the need to decouple function prediction from evolutionary assumptions in order to annotate novel genomes more effectively.

The direct consequence of these caveats of similarity-based annotation is the limited coverage of the protein space. Even for comprehensively studied model organisms like *Caenorhabditis elegans* and *Drosophila melanogaster*, approximately 28% and 18% of their genes, respectively, remain unannotated in the reference database Uniprot [7]. The situation is even worse for non-model organisms, where roughly 50% of genes remain unannotated [8, 9]. These unannotated genes likely hold novel clues to understanding biology but remain "hidden" due to current limitations in annotation methods.

Recently, the use of large language models has emerged as a useful alternative that decouples function prediction from evolutionary information. There are many language models available that could be used to annotate proteins available here (https://huggingface.co/). Among them, the ProtTrans autoencoder language model [10], trained on 393 billion amino acids (https://github.com/agemagician/ProtTrans), is one of the most successful, as benchmarked against the CAFA3 datasets [1], and offers a promising alternative. Recent work validating the performance of ProtTrans at the level of model organisms [7] demonstrates its ability to recover functional information from transcriptomics data, making it a strong alternative to standard homology-based methods for downstream analyses.

ProtTrans is the core of the goPredSim [1] method, which is publicly available as a web server for exploring the functional annotation of individual sequences(https://embed.protein.properties/). However, it currently lacks automatic implementation for annotating whole proteomes and does not include essential steps for data preprocessing or output configuration. To address these limitations, we have developed functional annotation based on embedding space similarity (FANTASIA), an automated implementation of the goPredSim ProtTrans model with several enhancements.

The protocol we present here is twofold. First, it provides guidelines for evaluating the performance of different models in the context of function annotation at the organism-level. Second, it involves the use of a computational pipeline (FANTASIA) for large-scale annotation. The steps include (1) data collection, (2) data preprocessing, (3) computation of embeddings and distances, and (4) GO term transfer. Here, we highlight the importance of using alternative methods to homology-based function assignment.

## 2 Materials

1. This approach requires in-house computational tools, access to a high-performance computing (HPC) facility for intensive computations, ample storage capacity, and programming skills for data processing.
2. Singularity container (available at https://docs.sylabs.io/guides/3.0/user-guide/installation.html).

## 3 Methods

### *3.1 Comparison of Different ML-Based Annotation Methods: Selecting the best Suited Tool*

Performance of the different top performing machine learning methods (in CAFA datasets, *see* **Note 1**) was estimated in model organisms using different metrics (*see* **Notes 2** and **3**). The Prot-Trans (ProtT5) model was the best performer and selected to further scale-up the annotation of proteomes (*see* subheading 3.2).

### *3.2 Functional Annotation Based on Embedding Space Similarity (FANTASIA)*

FANTASIA is a large-scale implementation of the goPredSim [1] method and requires access to a workstation or a high-performance computing (HPC) facility due to the tool's high RAM requirements (*see* **Note 4**). FANTASIA (Fig. 1) (1) takes a FASTA peptide file as input (Fig. 1a), (2) preprocesses (*see* **Note 5**) the data (Fig. 1d), (3) runs GOPredSIM (with the ProtT5 model) and performs GO annotation (Fig. 1b,e and *see* **Note 6**), and (4) generates output files ready to be used as inputs for topGO enrichment methods [11] (Fig. 1c,f). If all isoforms are used as input files (Fig. 1e), then GO terms from each gene isoform will be collapsed (Fig. 1f) to produce a repertoire of GO terms per gene. Specific instructions for downloading are available here (https://github.com/MetazoaPhylogenomicsLab/FANTASIA).

#### *3.2.1 Installation from a Singularity Image, Only CPU Enabled*

The current implementation of FANTASIA as a singularity image requires basic command-line skills for execution.

1. Download the singularity image (available here https://cloud.sylabs.io/library/gemma.martinezredondo/fantasia/fantasia).

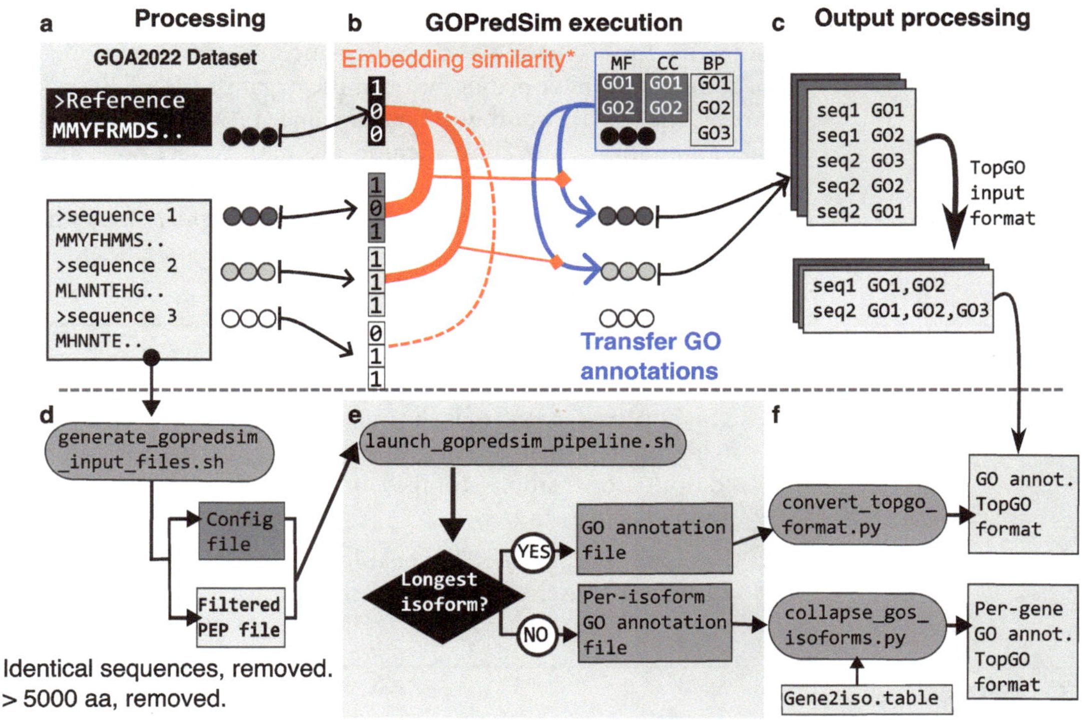


**Fig. 1** The FANTASIA pipeline. Top row: Conceptual steps. Bottom row: Implementation steps. (**a**) **Processing step:** This step takes an input file of reference sequences (from the GOA2022 dataset) with known functions (black background) and target sequences (gray backgrounds). (**b**) **GOPredSim execution:** Using the ProtTrans (T5) model, distances among embeddings from reference (black background) and target sequences are calculated. Distances are indicated as red arcs connecting the embeddings, with thicker arcs representing smaller distances (and therefore greater similarity). A dashed line indicates that the distance between the reference and sequence 3 embeddings is very large and will not be considered for the next step. A semi-supervised method then transfers GO annotations from reference sequences to the unknown ones (blue lines). For the embedding similarity(*) calculation, *see* **Note 6**. (**c**) **Output file generation:** The GO terms corresponding to each gene ontology (biological process—BP, molecular function—MF, and cellular component—CC) previously transferred to the target sequences are transformed into topGO [11] compliant format for downstream enrichment analyses. (**d**) **Pipeline execution (input file generation):** The script "generate_gopredsim_input_files.sh" first removes identical and large sequences (>5000 aa), generating the input file. Importantly, the removal of "sister species" (*see* **Note 5**) effects should be done before running this step. (**e**) **Pipeline execution (method running):** The script "launch_gopredsim_pipeline.sh" runs the GOPredSim method. This step also provides the option to use the entire repertoire of isoforms, rather than just the longest isoform, to extract potential information contributed by different isoforms. (**f**) **Pipeline execution (output production):** The output files are formatted to topGO format via "convert_topgo_format.py." If isoforms are used, an additional step collapses the annotations from all isoforms using "collapse_-go_isoforms.py," which takes information from an isoform-to-gene mapping table. The final output is a file containing all the GO terms assigned to each sequence for each gene ontology category (MF, CC, BP)

2. Once downloaded, you can execute it as follows (make sure that you have singularity installed):
   - Syntax: ./fantasia --infile protein.fasta [--outpath output_-path] [--allisoforms gene_isoform_conversion.txt] [--kee-pintermediate]. For the options explanations, *see* **Note 7**.

#### 3.2.2 Local Installation of FANTASIA, Also GPU Enabled

In some architectures where singularity cannot be used, the source code is provided for installation by users with more advanced programming skills.

1. Download the scripts and files from http://gofile.me/5wDJy/CI8zCiijR.
2. Open the "installation_guide_FANTASIA.sh" shell script (also available at https://github.com/MetazoaPhylogenomicsLab/FANTASIA) and follow the instructions.
3. Execute FANTASIA (you can check the files and options required for each script by adding -h).

## 4 Case Studies

### 4.1 Downstream GO Enrichments from RNAseq Data

Understanding the functions that are either exerted or inhibited (i.e., upregulated or downregulated) under different conditions or treatments is one of the main objectives of functional genomics. These analyses typically utilize gene expression data that undergo downstream functional enrichments of genes or pathways. For these approaches to be effective, it is necessary to have most genes functionally annotated. In the case of model organisms, most genes are annotated; however, for non-model organisms, only about 50% of their genes are annotated [8], making the discovery of novel functions or genes challenging. As these species exhibit functions that could be interesting to exploit—such as resistance to radiation and regeneration—it is important to increase the annotation coverage of these genomes. We demonstrated that ProtT5 can recover the same information as homology-based methods in differential gene expression experiments in model organisms [7] (*see* **Note 3**). In Fig. 2, we present the recovery of similar enriched terms using the reference annotations or the ProtT5-derived annotation (measured as semantic similarity) for each gene ontology for published transcriptomics experiments from *C. elegans* [12]. ProtT5-derived annotations are able to recover almost the same set of enriched terms than the reference annotation, even if we keep only the 25% top-scored annotations (Fig. 2, SS >0.8 in downregulated genes). This case illustrates how language models recover and recapitulate functional information, making them reliable choices for transcriptomic-based downstream experiments to address the functional landscapes in non-model organisms.

### 4.2 Identifying Novel Functions in Unannotated Genes from Model Organisms

Model organisms are often considered the standard for comparative genomics studies, with the assumption that most of their genes are well characterized. However, this assumption is not entirely accurate. We observed that even in extensively studied model organisms, a significant number of genes remain unannotated. For

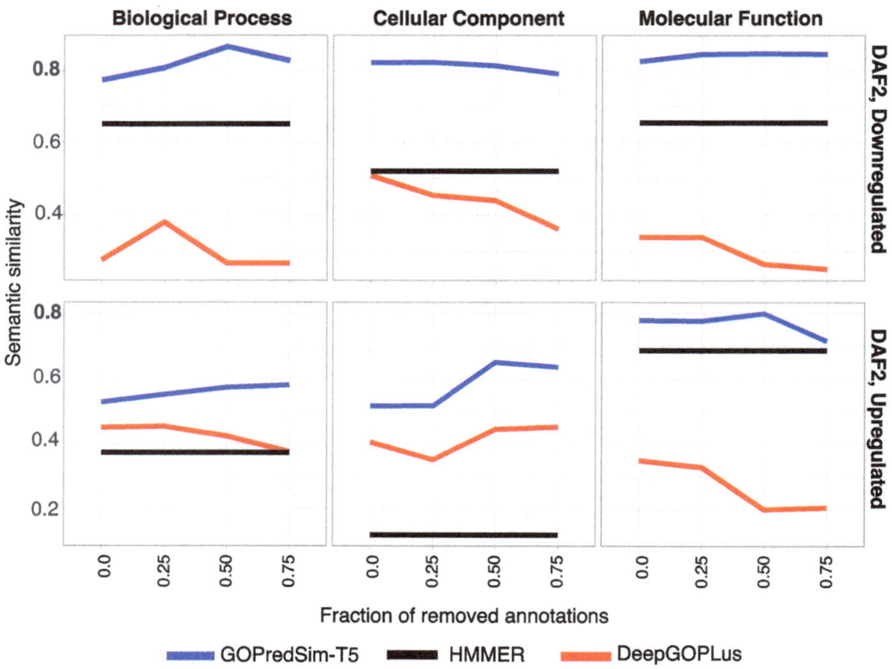


**Fig. 2** Semantic similarity from *Caenorhabditis elegans* transcriptomics data of the *Daf-2* gene in a long-lived mutant. Differentially expressed genes from a long-lived *C. elegans* mutant were retrieved from the literature [12]. To identify enriched/depleted functions, four independent topGO enrichment analyses [11] were conducted using different approaches: (i) UniProt reference annotations, (ii) ProtT5 predictions, (iii) HMMER predictions (*see* **Note 3**), and (iv) DeepGOPlus predictions. Semantic similarity (SS) [20] was then calculated for the enriched gene ontology (GO) terms, comparing each method against the reference annotations. The semantic similarity between reference/ProtT5 (blue line), HMMER/reference (black line), and DeepGOPlus/reference (red line) is depicted. The upper panel represents downregulated genes, while the lower panel represents upregulated genes. The Y-axis shows the semantic similarity values (calculated using the Wang method [25] and the Best Match Average (BMA) to handle multiple GO terms) per gene. The X-axis represents the fraction of annotations removed based on each method's score. Since the scores are not directly comparable [7], they were divided into quantiles using the R quantile function. The first quantile includes all annotations regardless of score, while the last quantile includes only the top 25% of annotations with the highest scores. Given that approximately 28% *of C. elegans* genes are unannotated, GOPredSim-ProtT5 not only annotated most proteins from the genome but also achieved the highest semantic similarity values across scores and gene ontologies in these transcriptomics experiments

example, approximately 17% of genes in *D. melanogaster* are still unannotated [7]. For instance, the CG11373 protein-coding gene lacks assigned gene ontology (GO) functions despite evidence that the protein is expressed in male germline cells, somatic cells of the testis, and adult males, as reported in the FlyBase repository (https://flybase.org/reports/FBgn0040679.htm). This lack of annotation may be attributed to the fact that the protein is predicted to be disordered, as determined by AIUPred [13]. FANTASIA, however, inferred the GO terms "meiotic cell cycle" and "signal transduction involved in cell cycle switching, mitotic to meiotic cell cycle," which are consistent with the expression data [8]. This example demonstrates how language models can retrieve valuable information to further investigate experimental findings.

### 4.3 Identifying Novel Functions in Difficult to Address Non-model Organisms

An annelid lineage underwent massive genome rearrangements during the transition from marine to non-marine ecosystems. Using transcriptomic data from several stress experiments in the earthworm *Eisenia andrei*, we were able to identify 12 stress-related genes originating during the massive genome rearrangements that may have been involved in the adaptation to life out of the sea. One of those genes could not be annotated based on homology, and was predicted to be involved in sodium–potassium cell trafficking by FANTASIA (GO:0006814, GO:0010248, GO:0071805, GO:0005391, GO:0005524, GO:0008900, GO:0016887, GO:0046872, GO:0005886, and GO:0016324) [14].

## 5 Notes

1. The Critical Assessment of protein Function Annotation algorithms (CAFA) is a competition focused on function prediction (https://biofunctionprediction.org/people/). CAFA evaluates and monitors the performance of various function prediction methods serving as the authoritative source for assessing the usability of computational tools in this context. It provides datasets of proteins, while the corresponding functions are hidden from the predictors during the competition. Participating teams then develop methods to predict these functions. The most recent analyses from CAFA3 evaluation highlighted that there is still room for improvement [15] regardless of the methods used. Remarkably, the latest competition held in 2024, hosted on Kaggle, demonstrated that all but one of the 10 top-performing teams used protein language models (https://www.kaggle.com/competitions/cafa-5-protein-function-prediction). Analyses of the competition results are still underway.
2. The objective is to evaluate performance at the model organism level using methods that have been benchmarked with the CAFA datasets. The methods selected in this benchmark [7] are deep machine–learning based (DeepGO [16] and DeepGOPlus [17] versus language models, namely ProtT5 and SeqVec [1]), while profile-based methods [18] serve as control low-coverage, high-precision methods.
3. We used the UniProtKB [19] annotations from model organisms as reference annotations. While this approach can be problematic due to the prevalence of automatic annotations, relying solely on annotations with "experimental evidence" would significantly reduce coverage. Therefore, we recommend following the CAFA guidelines regarding evidence

codes from the Gene Annotation Files (https://geneontology.org/docs/go-annotation-file-gaf-format-2.2/): EXP, IDA, IPI, IGI, IEP, TAS, or IC, which are considered "experimental" by the CAFA consortium. Performance was measured by estimating semantic similarity [20], information content [21], and precision in predicting a GO term, as developed in our previous work [7]. Additionally, we evaluated the ability to recover functional information using topGO [11] enrichment analysis on transcriptomics data to identify the most suitable tool for large-scale annotation. The code used for evaluating methods in model organisms [7], which led to the selection of the ProtT5 model, is available: https://github.com/CBBIO/func-lm and can be modified to test the performance of novel models in a transfer learning context. To ensure that GO annotations assigned to protein domains from the PFAM (Protein Families Database) correspond to the relevant regions, GO terms were extracted from http://current.geneontology.org/ontology/external2go/pfam2go.

4. RAM usage is linearly dependent on proteome size. For instance, a human-sized proteome of ~20,000 proteins will require around 50 GB of RAM.

5. The preprocessing step involves removing identical sequences using CD-HIT [22] and filtering out sequences longer than 5000 amino acids, due to a length constraint in the original ProtT5 model [10]. This process is fast, taking only a few seconds to complete. Since the pipeline was designed to annotate organisms with little presence in UniProtKB, overfitting was unlikely. However, as the databases continue to grow, the likelihood of finding sequences from these organisms, or very similar ones, will increase over time. Therefore, to minimize the risk of overfitting, it is recommended to remove sequences and embeddings from the look-up table that correspond to the target organism, if they are present. Additionally, sequences or embeddings from closely related species (e.g., mouse and human) should be excluded to minimize the "sister species" effect [7] in the predictions.

6. The annotations for transfer come from proteins with annotated GO terms in the GOA database [23] (https://geneontology.org/). The GOA2022 dataset, available in FANTASIA, contains 295,558 proteins. Embeddings (n and m) are generated from both the reference dataset and the query proteins, and the distances between them are calculated as follows:

$$d(n, m) = \sqrt{\sum_{i=1}^{1024} (n_i - m_i)^2} \quad (1)$$

Based on these distances, given a query protein, the *k*-nearest neighbor algorithm (k-NN) [24] is used to select the k closest hits from the GOA2022 reference dataset for GO term transfer (k is a tunable parameter). All annotations from the selected hits are transferred, rather than only the most frequent ones. In other words, the smaller the distance, the more similar the sequence embeddings (Fig. 1b, red arcs), making the transfer more reliable (Fig. 1b, blue line); for more details, see [1].

The computing time for GO annotation depends on the number of sequences in the proteome. Larger datasets require more time, but using GPUs with a local FANTASIA installation significantly speeds up the process. Even on CPUs, the computation takes only hours, with the longest test (~250,000 sequences) taking about 24 h. When installing the pipeline manually (outside the provided singularity container), users can choose between CPUs (default) or GPUs, depending on their system. GPU usage accelerates execution without affecting memory requirements. Memory usage for GOPredSim-ProtT5 is proportional to the number of sequences and can be estimated with the formula:

$$y = 4.4 + 0.021x \tag{2}$$

where

$y$ is the memory in gigabytes.

$x$ is the number of sequences.

7. Options for running the pipeline

   Syntax: ./fantasia --infile protein.fasta [--outpath output_-path] [--allisoforms gene_isoform_conversion.txt] [--keepintermediate].

At a given point, FANTASIA may raise a warning if your system only has CPUs or if the GPU CUDA library version is different from 11.0. You can safely ignore this message, as the singularity container is prepared to run on CPUs.

Mandatory inputs are

1. **-i/--infile** is the input protein file. For a comprehensive guide to sequence formats, refer https://www.ebi.ac.uk/jdispatcher/docs/formats/.
2. **-h/--help**: prints this Help.

The following are optional tasks:

1. -o/--outpath: (Optional) Output directory. If not provided, the input file directory will be used.

2. -a/--allisoforms: (Optional) Tab-separated conversion file specifying the correspondence between gene and isoform IDs for obtaining per-gene annotations using all isoforms.

-p/--prefix: (Optional) Prefix to add to output folders and files (e.g., the species code). If not provided, the input file name will be used.

## Acknowledgments

Ministerio de Ciencia e Innovación [MCIN/AEI/10.13039/501100011033]; PID2021-127503OB-I00 to A.M.R. and F.P.E.; GIMR, AMR, and RF acknowledge funding and support from the LifeHUB/CSIC research network (PIE-202120E047). GIMR acknowledges the support of Secretaria d'Universitats i Recerca del Departament d'Empresa i Coneixement de la Generalitat de Catalunya and ESF Investing in your future (grant 2021 FI_B 00476). R.F. acknowledges support from the following sources of funding: the Agencia Estatal de Investigación (project PID2019-108824GA-I00 funded by MCIN/AEI/10.13039/501100011033), the European Research Council (this project has received funding from the European Research Council (ERC) under the European's Union's Horizon 2020 research and innovation program (grant agreement no. 948281)), the Human Frontier Science Program (grant no. RGY0056/2022) and the Secretaria d'Universitats i Recerca del Departament d'Economia i Coneixement de la Generalitat de Catalunya (AGAUR 2021-SGR00420).

### References

1. Littmann M, Heinzinger M, Dallago C et al (2021) Embeddings from deep learning transfer GO annotations beyond homology. Sci Rep 11:1160
2. Linard B, Ebersberger I, McGlynn SE et al (2021) Ten years of collaborative progress in the quest for orthologs. Mol Biol Evol 38:3033–3045
3. Ashburner M, Ball CA, Blake JA et al (2000) Gene ontology: tool for the unification of biology. Nat Genet 25:25–29
4. Piergiorge RM, de Miranda AB, Guimarães AC et al (2017) Functional analogy in human metabolism: enzymes with different biological roles or functional redundancy? Genome Biol Evol 9:1624–1636
5. Studer RA, Robinson-Rechavi M (2009) How confident can we be that orthologs are similar, but paralogs differ? Trends Genet 25:210–216
6. Perdigão N, Heinrich J, Stolte C et al (2015) Unexpected features of the dark proteome. Proc Natl Acad Sci U S A 112:15898–15903
7. Barrios-Núñez I, Martínez-Redondo GI, Medina-Burgos P et al (2024) Decoding functional proteome information in model organisms using protein language models. NAR Genomics Bioinf 6:lqae078
8. Martínez-Redondo GI, Barrios-Núñez I, Vázquez-Valls M, et al (2024) Illuminating the functional landscape of the dark proteome across the animal tree of life through natural language processing models. https://www.biorxiv.org/content/10.1101/2024.02.28.582465v1
9. Martínez-Redondo GI, Vargas-Chávez C, Eleftheriadi K, et al (2024) MATEdb2, a collection of high-quality metazoan proteomes across the animal tree of life to speed up phylogenomic studies. https://www.biorxiv.org/content/10.1101/2024.02.21.581367v2

10. Elnaggar A, Heinzinger M, Dallago C et al (2022) ProtTrans: toward understanding the language of life through self-supervised. Learning 44:7112–7127
11. Alexa A and Rahnenfuhrer J (2023) topGO: enrichment analysis for gene ontology. https://bioconductor.org/packages/topGO/
12. Senchuk MM, Dues DJ, Schaar CE et al (2018) Activation of DAF-16/FOXO by reactive oxygen species contributes to longevity in long-lived mitochondrial mutants in *Caenorhabditis elegans*. PLoS Genet 14:e1007268
13. Erdős G, Dosztányi Z (2024) AIUPred: combining energy estimation with deep learning for the enhanced prediction of protein disorder. Nucleic Acids Res 52:W176–W181
14. Vargas-Chávez C, Benítez-Álvarez L, Martínez-Redondo GI, et al (2024) A punctuated burst of massive genomic rearrangements and the origin of non-marine annelids. https://www.biorxiv.org/content/10.1101/2024.05.16.594344v2
15. Zhou N, Jiang Y, Bergquist TR et al (2019) The CAFA challenge reports improved protein function prediction and new functional annotations for hundreds of genes through experimental screens. Genome Biol 20:244
16. Kulmanov M, Khan MA, Hoehndorf R (2018) DeepGO: predicting protein functions from sequence and interactions using a deep ontology-aware classifier. Bioinformatics 34:660–668
17. Kulmanov M, Hoehndorf R (2020) DeepGOPlus: improved protein function prediction from sequence. Bioinformatics 36:422–429
18. Johnson LS, Eddy SR, Portugaly E (2010) Hidden Markov model speed heuristic and iterative HMM search procedure. BMC Bioinformatics 11:431
19. The UniProt Consortium (2023) UniProt: the universal protein knowledgebase in 2023. Nucleic Acids Res 51:D523–D531
20. Yu G, Li F, Qin Y et al (2010) GOSemSim: an R package for measuring semantic similarity among GO terms and gene products. Bioinformatics 26:976–978
21. Fröhlich H, Speer N, Poustka A et al (2007) GOSim—an R-package for computation of information theoretic GO similarities between terms and gene products. BMC Bioinformatics 8:166
22. Li W, Godzik A (2006) Cd-hit: a fast program for clustering and comparing large sets of protein or nucleotide sequences. Bioinformatics 22:1658–1659
23. Camon E, Magrane M, Barrell D et al (2004) The gene ontology annotation (GOA) database: sharing knowledge in Uniprot with gene ontology. Nucleic Acids Res 32:D262–D266
24. Cover T, Hart P (1967) Nearest neighbor pattern classification 13:21–27
25. Wang JZ, Du Z, Payattakool R et al (2007) A new method to measure the semantic similarity of GO terms. Bioinformatics 23:1274–1281

# Chapter 9

# Advances in Language-Model-Informed Protein–Nucleic Acid Binding Site Prediction

Sumit Tarafder, Xinyu Wang, Rahmatullah Roche, and Debswapna Bhattacharya

## Abstract

Interactions between proteins and nucleic acids are essential for understanding a wide range of cellular and evolutionary processes. Recent advancements in protein language models (pLMs), trained on vast protein sequence data, have revolutionized various predictive modeling tasks, offering unprecedented scalability and generalizability. Consequently, a number of computational methods have been developed in the recent past for protein–nucleic acid binding site prediction powered by pLMs. To this end, we recently developed the EquiPNAS method that integrates pLM embeddings with E(3) equivariant deep graph neural networks for enhancing accuracy and robustness in predicting protein–DNA and protein–RNA binding sites, thereby reducing the dependency on evolutionary information. Here we present an overview of the recent protein–nucleic acid binding site prediction methods, emphasizing the recent advances in harnessing the potential of pLMs, and provide a detailed description of the EquiPNAS methodology as well as the necessary materials and procedures for the computational prediction of protein–DNA and protein–RNA binding sites.

**Key words** Protein–DNA binding site prediction, Protein–RNA binding site prediction, Language models, Graph neural networks

## 1 Introduction

Proteins interact with nucleic acids, including DNA and RNA, to perform critical cellular functions such as regulating gene expression, facilitating DNA replication, and mediating cellular signaling pathways. The precise identification of these interaction or binding sites where proteins interact with nucleic acids is fundamental to understanding these biological processes [1–5]. This understanding has significant implications not only for basic biological research but also for practical applications in drug design and therapeutic interventions. While traditional laboratory methods for identifying these binding sites are effective [6, 7], they are often time-consuming, expensive, and technically challenging, prompting

Dukka B. KC (ed.), *Large Language Models (LLMs) in Protein Bioinformatics*, Methods in Molecular Biology, vol. 2941, https://doi.org/10.1007/978-1-0716-4623-6_9,

researchers to explore computational approaches for predicting these sites based on available data. Computational methods for predicting protein–nucleic acid binding sites can be categorized into two main types: sequence-based methods [8–15] and structure-aware methods [8, 16–20]. Sequence-based methods rely solely on the amino acid sequences of proteins to make predictions about binding affinities and interaction sites. On the other hand, structure-aware methods utilize 3D structural information derived from experiments or computational predictions, resulting in more precise binding site predictions. However, these approaches frequently depend on limited experimental data from structural databases such as the Protein Data Bank (PDB) [21], which may not provide comprehensive coverage for all protein–nucleic acid interactions. The introduction of AlphaFold2 [22] and AlphaFold3 [23] significantly expands this landscape by providing highly accurate structural models for a wide range of proteins and their interactions, including with nucleic acids. Additionally, the AlphaFold Database [24] offers a comprehensive repository of these predicted structures, allowing researchers to access valuable data that was previously inaccessible. These advancements alleviate the limitations associated with reliance on PDB and enhance the predictive capabilities of protein structure modeling [25] and binding site identification.

Currently, large language models (LLMs) have emerged as transformative tools in the field of artificial intelligence [26–30], possessing the capability to comprehend and generate human language through the analysis of extensive textual datasets. Their applications extend across numerous domains, encompassing natural language processing tasks such as sentiment analysis, language translation, and content generation [31–33], thereby reshaping industries ranging from healthcare [34–39] to finance [40–44]. Notably, in the area of protein–nucleic acid binding site predictions, LLMs demonstrate significant advancements by leveraging their understanding of biological sequences and molecular interactions. The advent of protein language models (pLMs) [45–50], which are analogous to natural language processing models but specifically trained on extensive datasets of protein sequences, has opened new avenues for predicting protein characteristics and functions. While the application of pLMs in predicting protein–nucleic acid binding sites is still in its formative stages, their potential for enhancing prediction accuracy is considerable [48, 50]. Research initiatives such as EquiPPIS [51] lay critical groundwork for advancements in predicting protein–RNA binding positions.

Although numerous methods have been developed for protein–nucleic acid binding site prediction, EquiPNAS [52] is one of the first methods to leverage pLM embeddings for the effective prediction of both DNA and RNA binding protein residues, consistently surpassing state-of-the-art techniques across an array of widely utilized benchmarking datasets for protein–DNA and protein–RNA binding site prediction tasks. This innovative approach employs a specific type of graph neural network known as E(3) Equivariant Graph Neural Network (EGNN) [53], which excels at processing the intricate three-dimensional (3D) nature of protein structures. By being sensitive to positional, rotational, and orientational transformations, this model maintains robustness and accuracy, even when utilizing protein structures predicted by AlphaFold2 [22] as opposed to those resolved through experimental methods (*see* **Notes 1** and **2**).

## 2 Method

### 2.1 Overview of Existing Protein–Nucleic Acid Binding Site Prediction Methods

Protein–nucleic acid binding site prediction is a well-studied problem with numerous methods available in the literature utilizing a combination of network architectures and features as listed in Table 1. Currently, the available protein–nucleic acid binding site prediction methods can be broadly divided into sequence-based and structure-based approaches. Sequence-based methods, such as NCBRPred [9], DNAPred [10], and RNABindRPlus [12], leverage the abundance of protein sequence data to predict nucleic

**Table 1**
**A selection of state-of-the-art, deep learning–based frameworks for protein–nucleic acid binding site prediction, arranged in descending order by year of publication**

| Name | Year | Architecture | pLM_Method | Protein–DNA/RNA |
|---|---|---|---|---|
| EquiPNAS (Roche et al.) | 2024 | EGNN | ESM2 | Both |
| CLAPE (Liu et al.) | 2024 | CNN | ProtBert | Protein–DNA |
| ULDNA (Zhu et al.) | 2024 | LSTM-attention | ESM2, ProtTrans | Protein–DNA |
| EGPDI (Zheng et al.) | 2024 | EGNN+GCN-II | ESM2, ProtTrans | Protein–DNA |
| ESM-NBR (Zeng et al.) | 2023 | BiLSTM + MLP | ESM2 | Both |
| GLMsite (Song et al.) | 2023 | GVP-GNN | ProtTrans | Both |
| GraphSite (Yuan et al.) | 2022 | Graph transformer | – | Protein–DNA |
| bindEmbed21DL (Littmann et al.) | 2021 | CNN | ProtT5 | Both |
| GraphBind (Xia et al.) | 2021 | GNN | – | Both |
| NCBRPred (Zhang et al.) | 2021 | BiGRU | – | Both |
| NucleicNet (Lam et al.) | 2019 | ResNet | – | Protein–RNA |
| aaRNA (Li et al.) | 2014 | DenseNet | – | Protein–RNA |
| RNABindRPlus (Walia et al.) | 2014 | SVM | – | Protein–RNA |

acid binding sites. These methods often utilize machine learning techniques, such as hidden Markov models and bidirectional Gated Recurrent Units (BiGRUs), to capture patterns from sequence data. Although these methods are widely applicable, their reliance solely on sequence information limits their predictive accuracy, as they miss critical spatial details of protein–nucleic acid interactions. Structure-based methods, including COACH-D [16], NucBind [8], and GraphBind [19], incorporate three-dimensional structural information, which enhances prediction accuracy. By utilizing structural templates or advanced computational models like graph neural networks, structure-based methods can better capture the intricate spatial patterns essential for identifying binding sites.

In recent years, hybrid methods have emerged, combining both sequence-based and structure-based approaches to improve prediction outcomes. For instance, DNABind [54] and NABind [55] integrate machine learning with template-based methods, while NucBind [8] combines predictions from both COACH-D [16] and SVMnuc [8]. Graph-based approaches, such as GraphSite [20] and GLMSite [56], have introduced new levels of sophistication by encoding secondary structure and spatial positions of atoms. While these hybrid methods outperform some of the state-of-the-art deep learning-only based frameworks, their hybrid approaches and high dependence on the quality of the templates make them susceptible to orphan proteins with low homology depths. However, with the recent surge in the availability of various families of protein language models such as ESM and ProtTrans, an array of deep learning–based binding site prediction methods have emerged as shown in Table 1. Methods such as EquiPNAS [52], CLAPE [57], ULDNA [58], EGPDI [4], ESM-NBR [59], and bindEmbed21DL [60], utilizing these language model embeddings, have surpassed those without them in terms of prediction accuracy, reducing the overall dependency on evolutionary information for binding site prediction task [61].

### 2.2 Significance of pLMs for Protein–Nucleic Acid Binding Site Prediction

While computational methods have advanced in predicting protein and nucleic acid binding sites, significant challenges persist, particularly in effectively utilizing protein-related data. Sequence-based approaches, which depend heavily on evolutionary information, often struggle with orphan proteins and encounter difficulties with intrinsically disordered proteins (IDPs) and regions (IDRs) due to their unstable and dynamic structures [62]. Additionally, many existing techniques rely on manually curated features that demand domain-specific expertise and may miss critical biological insights [17].

In contrast, protein language models (pLMs) present notable advantages in protein–nucleic acid binding predictions. Using self-supervised learning, pLMs can capture long-range dependencies and structural information from sequences without requiring manual feature engineering [63]. These large language models, originally designed for natural language tasks like translation and question-answering, have evolved into biological language models due to the linguistic parallels between human and biological languages. Through transfer learning, these models excel at capturing the structure and function of biological molecules such as proteins, creating rich feature representation from large datasets of protein sequences, which proves to be significant for binding site prediction tasks (*see* **Notes 3** and **4**). This approach minimizes the reliance on evolutionary data and enables more accurate and generalizable embeddings [64].

However, in the context of binding site prediction, especially with nucleic acid ligands, the information derived from pLMs should account for the interactions between proteins and nucleic acids. A promising approach can be the potential incorporation of multimodal learning, where pLMs designed for DNA and RNA sequences [65, 66], or a platform like BioSeq-BLM, capable of analyzing DNA, RNA, and protein sequences using a range of biological language models [67] can be leveraged. This is especially valuable for improving prediction accuracy and efficiency when dealing with highly flexible molecules like RNA [61].

### 2.3 Overview of EquiPNAS Framework

The primary objective of the EquiPNAS method is to accurately predict nucleic acid (both DNA and RNA)-binding residues for a given input protein 3D structure as shown in Fig. 1. The prediction framework follows a structured methodology that consists of three key components. The first component is the graph representation of the protein, which involves constructing a graph in which nodes correspond to individual residues and edges denote interactions among them. This graph structure effectively captures the spatial and relational information inherent to the protein. The second component involves feature generation for both nodes (residues) and edges (interactions) within the graph. This includes the extraction of sequence-based and structure-based information, alongside the utilization of protein language model (pLM) embeddings derived from the ESM-2 model [47], thereby enriching the data set available for predictive purposes. The final component employs an E(3) equivariant graph neural network (EGNN) to process the graph representation and make binding site predictions. By leveraging the coordinate information from the input monomer in conjunction with the generated features, the EGNN performs graph node classification, estimating the probability of whether each residue acts as a binding site for the corresponding nucleic acid as annotated in Fig. 1. In the subsequent sections, each of these components will be explored in detail, illuminating their significance within the overarching EquiPNAS architecture and the enhanced predictive capabilities they provide.

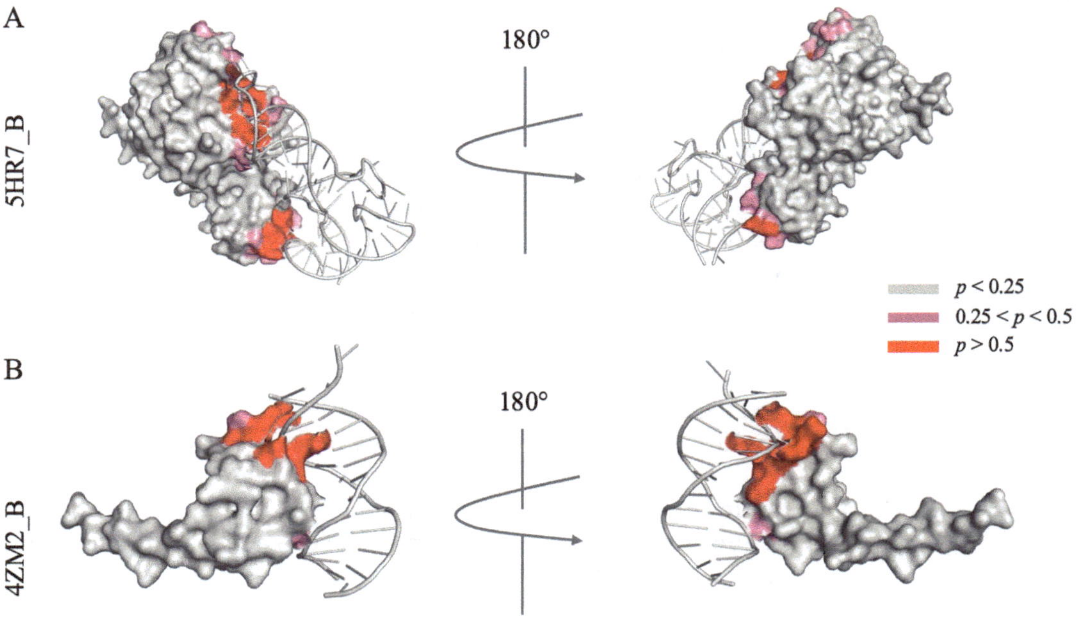


**Fig. 1** Experimental structure of (**a**) protein–RNA complex (PDB ID: 5HR7_B) and (**b**) protein–DNA complex (PDB ID: 4ZM2_B) with proteins shown in surface representation. Each residue of the protein structure is color coded according to its predicted probability score ($p$) by EquiPNAS, visualizing the correlation between EquiPNAS predictions and experimentally derived binding sites

### *2.3.1 Protein Graph Representation*

#### Protein as a Graph

In the EquiPNAS framework, proteins are represented as graphs wherein each residue is depicted as a node. An edge is established between two nodes if the Cα atoms of the corresponding residues are within a defined distance threshold, thus transforming the protein structure into an interactive network of components.

#### Interacting Residue Pairs

Residues are classified as interacting based on specific criteria. Notably, the distance between their Cα atoms must be less than or equal to 14 Å for DNA-binding interactions and 15 Å for RNA-binding interactions. Furthermore, the residues must be at least six positions apart within the protein sequence. This requirement ensures that only relevant interactions are captured, offering a focused representation of the protein's binding potential. Through this graph-based approach, EquiPNAS adeptly captures both spatial relationships and interaction dynamics, facilitating accurate predictions of P-NA binding sites.

### *2.3.2 Feature Generation*

#### Sequence-Based Features

The generation of sequence-based features is essential for the effective representation of proteins. Each amino acid is encoded using a simple binary (one-hot) encoding scheme for the 20 standard amino acids. Additionally, a Position-Specific Scoring Matrix (PSSM) is obtained through PSI-BLAST [68], with subsequent normalization of values to maintain consistent scaling. To augment the feature set, a Multiple Sequence Alignment (MSA) is generated using MMseqs2 [69] and subsequently refined by ColabFold [70] pipeline to provide distilled evolutionary information to the model. Finally, the language model embeddings have been extracted from ESM-2, a pre-trained protein language model (pLM) containing 15 billion parameters represented by a feature dimension of 5120 for each residue in the protein sequence.

#### Structure-Based Features

Structure-based features capture critical aspects of the protein's topology and spatial orientation [71]. One-hot encoding is utilized for secondary structure and solvent accessibility, indicating the exposure of each residue to the surrounding environment. Furthermore, geometric relationships—such as bond angles and the positioning of residues in three-dimensional space—are meticulously recorded. The framework also tracks the number of residues in contact and computes their spatial positions relative to the protein's center, thus refining the understanding of interaction dynamics.

#### Edge Features

For edge features, the framework calculates the ratio of sequence distance to three-dimensional (3D) distance for each pair of interacting residues. This ratio provides valuable insights into the spatial relationships between residues, significantly contributing to the overall predictive capability of the framework.

#### Coordinate Features

Finally, coordinate features leverage the 3D coordinates (x, y, z) of each residue's C$\alpha$ atom, effectively capturing the spatial arrangement within the protein structure. By integrating these diverse feature sets, EquiPNAS substantially enhances its ability to accurately predict protein–nucleic acid binding sites.

### 2.4 Architecture

The core architecture of EquiPNAS is constructed utilizing deep E (3)-equivariant graph neural networks (EGNNs), specifically tailored to predict protein–DNA and protein–RNA binding sites. These networks operate by leveraging features pertinent to the protein's structure and its spatial coordinates, such as the positions of C$\alpha$ atoms in three-dimensional space. The architecture comprises multiple layers of equivariant graph convolution layers (EGCL), which simultaneously update both the node features, and the coordinates of the protein based on the interactions (edges) among residues.

The process can be outlined as follows: (1) *Node representation*: Each node, representing a residue in the graph, encompasses features (such as the amino acid type) and coordinates (indicating its spatial position). These attributes undergo transformation across several layers of the network (2) *Feature and coordinate updates*: The layers update the node features and coordinates by utilizing edge information, including the distances between residues.

(3) *Interaction learning*: Within each layer, the model learns how residues interact based on the provided features, continuously refining its predictions to enhance accuracy.

The EGNN employs 12 layers of transformations, each characterized by 768 hidden dimensions. To mitigate the risk of overfitting, dropout regularization is applied at every layer, ensuring the model generalizes effectively beyond the training data. Ultimately, the model predicts the likelihood of each residue acting as a binding site for nucleic acids (DNA or RNA), condensing all acquired information into a singular predictive score. The training procedure of this model is conducted using ADAM optimizer [72] and cosine annealing [73] for dynamic adjustment of learning rates. The training process extends for up to 40 epochs on a high-performance NVIDIA Graphics Processing Unit (GPU), ensuring robust learning from the datasets. To assess the efficacy of the EGNN structure in enhancing prediction accuracy, baseline models devoid of equivariant updates were also trained for comparative purposes. This allows for a thorough evaluation of the model's performance enhancements attributable to the incorporation of equivariant features.

## 3 Materials

To train and evaluate a machine learning–based model for protein–DNA binding site prediction, the GraphBind [19] study's Train_573 (573 protein chains) and Test_129 (129 chains) datasets are used, along with GraphSite's [20] Test_181 (181 chains). All datasets, curated from BioLiP [74] and filtered with CD-Hit [75] for non-redundancy, span different timeframes: Train_573 (pre-2016), Test_129 (2016–2018), and Test_181 (2018–2021), with varying counts of binding and non-binding residues. For protein–RNA binding site prediction, Train_495 (495 chains) and Test_117 (117 chains) from GraphBind are utilized, also processed from BioLiP and filtered with CD-Hit. These datasets contain a higher proportion of non-binding residues compared to binding residues, necessitating the use of specialized metrics for evaluation, as detailed below.

The predictive performance of EquiPNAS is compared with an array of existing methods for protein–nucleic acid binding site prediction. The structure-aware methods such as NucBind [8], DNABind [54], GraphBind [19], and GraphSite [20] represent the state-of-the-art methodologies in the literature, with GraphBind currently being the top-performing method for protein–RNA binding site prediction. Two key metrics Receiver Operating Characteristic–Area Under the Curve (ROC-AUC) and Precision-Recall–Area Under the Curve (PR-AUC) are used for this comparison, which provides a robust, threshold-independent assessment of classification models, especially when dealing with imbalanced datasets. The ROC curve plots the true positive rate (TPR) against the false positive rate (FPR), whereas the PR curve plots precision (positive predictive value) against recall (TPR) at different thresholds. Area Under the Curve (AUC) represents the overall ability of the model to distinguish between positive and negative classes, where a ROC-AUC of 1.0 indicates perfect classification. PR-AUC, which is the area under the Precision-Recall curve, is particularly useful for evaluating models on imbalanced datasets where the positive class is rarer than the negative class (*see* **Note 5**).

## 4 Notes

1. All the structure-aware methods, such as GraphSite, GraphBind, and EquiPNAS rely on 3D structures of proteins for binding site prediction. While the use of experimental structures shows better accuracy for all the competing methods, EquiPNAS experiences a minimal performance drop when using AlphaFold2 predicted structures highlighting its robustness and generalizability with predicted input structures.
2. The accuracy of EquiPNAS predictions correlates with AlphaFold2's self-estimated accuracy, measured by pLDDT with high confidence predictions leading to better ROC-AUC and PR-AUC scores. This suggests that the self-estimated accuracy of AlphaFold2 models can reliably predict the accuracy of EquiPNAS binding site predictions, especially for highly confident AlphaFold2 structures.
3. EquiPNAS incorporates pretrained pLM embeddings from the ESM-2 model as essential sequence-based features, which play a more critical role in performance compared to evolutionary features like PSSM and MSA. Excluding pLM features results in a significant performance drop; in contrast, removing evolutionary features leads to only minor reductions in accuracy. This highlights that pLM embeddings play a major role in EquiPNAS's improved performance, whereas evolutionary features contribute only modestly.
4. The ESM-2 provides a range of pre-trained protein language models (pLMs) with sizes from 8 million to 15 billion parameters. EquiPNAS uses the largest model, esm2_t48_15B_UR50D, with 15 billion parameters by default. Interestingly, EquiPNAS with the smallest model, esm2_t6_8M_UR50D, performed the worst and the performance consistently improved as the number of parameters increased, confirming that the largest model, esm2_t48_15B_UR50D, yields the best results for both protein–DNA and protein–

RNA binding site predictions, showing the significance of large LLMs in protein bioinformatics.

5. While the ROC-AUC achieved by EquiPNAS demonstrates high accuracy, the PR-AUC results indicate poorer performance overall in terms of binding site prediction. A model with high ROC-AUC but low PR-AUC suggests that although it is effective at distinguishing between positive and negative classes, it struggles with precision, leading to a higher number of false positives, in this case, binding residue prediction. This discrepancy highlights the need for further optimization in refining the precision of EquiPNAS, especially in handling imbalanced datasets, to ensure more reliable predictions.

## Acknowledgments

This work was partially supported by the National Institute of General Medical Sciences (R35GM138146 to D.B.) and the National Science Foundation (DBI2208679 to D.B.). The authors gratefully acknowledge high-performance computing resources and technical support from the National AI Research Resource Pilot award (NAIRR240093 to D.B.).

### References

1. Ferraz RAC, Lopes ALG, Da Silva JAF, Moreira DFV, Ferreira MJN, De Almeida Coimbra SV (2021) DNA–protein interaction studies: a historical and comparative analysis. Plant Methods 17:82
2. Ofran Y, Mysore V, Rost B (2007) Prediction of DNA-binding residues from sequence. Bioinformatics 23:i347–i353
3. Yesudhas D, Batool M, Anwar M, Panneerselvam S, Choi S (2017) Proteins recognizing DNA: structural uniqueness and versatility of DNA-binding domains in stem cell transcription factors. Genes 8:192
4. Zheng M, Sun G, Li X, Fan Y (2024) EGPDI: identifying protein–DNA binding sites based on multi-view graph embedding fusion. Brief Bioinform 25:bbae330
5. Zhou J, Xu R, He Y, Lu Q, Wang H, Kong B (2016) PDNAsite: identification of DNA-binding site from protein sequence by incorporating spatial and sequence context. Sci Rep 6:27653
6. Si J, Zhao R, Wu R (2015) An overview of the prediction of protein DNA-binding sites. IJMS 16:5194–5215
7. Mittler G, Butter F, Mann M (2009) A SILAC-based DNA protein interaction screen that identifies candidate binding proteins to functional DNA elements. Genome Res 19:284–293
8. Su H, Liu M, Sun S, Peng Z, Yang J (2019) Improving the prediction of protein–nucleic acids binding residues via multiple sequence profiles and the consensus of complementary methods. Bioinformatics 35:930–936
9. Zhang J, Chen Q, Liu B (2021) NCBRPred: predicting nucleic acid binding residues in proteins based on multilabel learning. Brief Bioinform 22:bbaa397
10. Zhu Y-H, Hu J, Song X-N, Yu D-J (2019) DNAPred: accurate identification of DNA-binding sites from protein sequence by ensembled hyperplane-distance-based support vector machines. J Chem Inf Model 59:3057–3071
11. Zhang J, Ghadermarzi S, Katuwawala A, Kurgan L (2021) DNAgenie: accurate prediction of DNA-type-specific binding residues in protein sequences. Brief Bioinform 22:bbab336
12. Walia RR, Xue LC, Wilkins K, El-Manzalawy Y, Dobbs D, Honavar V (2014) RNABindRPlus: a predictor that combines machine learning

and sequence homology-based methods to improve the reliability of predicted RNA-binding residues in proteins. PLoS One 9:e97725

13. Hu J, Li Y, Zhang M, Yang X, Shen H-B, Yu D-J (2017) Predicting protein-DNA binding residues by weightedly combining sequence-based features and boosting multiple SVMs. IEEE/ACM Trans Comput Biol Bioinf 14: 1389–1398
14. Zhang J, Kurgan L. SCRIBER: accurate and partner type-specific prediction of protein-binding residues from proteins sequences
15. Yu D-J, Hu J, Yang J, Shen H-B, Tang J, Yang J-Y (2013) Designing template-free predictor for targeting protein-ligand binding sites with classifier ensemble and spatial clustering. IEEE/ACM Trans Comput Biol Bioinf 10: 994–1008
16. Wu Q, Peng Z, Zhang Y, Yang J (2018) COACH-D: improved protein–ligand binding sites prediction with refined ligand-binding poses through molecular docking. Nucleic Acids Res 46:W438–W442
17. Li S, Yamashita K, Amada KM, Standley DM (2014) Quantifying sequence and structural features of protein–RNA interactions. Nucleic Acids Res 42:10086–10098
18. Lam JH, Li Y, Zhu L et al (2019) A deep learning framework to predict binding preference of RNA constituents on protein surface. Nat Commun 10:4941
19. Xia Y, Xia C-Q, Pan X, Shen H-B (2021) GraphBind: protein structural context embedded rules learned by hierarchical graph neural networks for recognizing nucleic-acid-binding residues. Nucleic Acids Res 49:e51–e51
20. Yuan Q, Chen S, Rao J, Zheng S, Zhao H, Yang Y (2022) AlphaFold2-aware protein–DNA binding site prediction using graph transformer. Brief Bioinform 23:bbab564
21. Berman HM, Westbrook J, Feng Z, Gilliland G, Bhat TN, Weissig H, Shindyalov IN, Bourne PE (2000) The Protein Data Bank. Nucleic Acids Res 28(1):235–242
22. Jumper J, Evans R, Pritzel A et al (2021) Highly accurate protein structure prediction with AlphaFold. Nature 596:583–589
23. Abramson J, Adler J, Dunger J et al (2024) Accurate structure prediction of biomolecular interactions with AlphaFold 3. Nature 630: 493–500
24. Varadi M, Anyango S, Deshpande M et al (2022) AlphaFold protein structure database: massively expanding the structural coverage of protein-sequence space with high-accuracy models. Nucleic Acids Res 50:D439–D444
25. Moussad B, Roche R, Bhattacharya D (2023) The transformative power of transformers in protein structure prediction. Proc Natl Acad Sci USA 120:e2303499120
26. Minaee S, Mikolov T, Nikzad N, Chenaghlu M, Socher R, Amatriain X, Gao J (2024) Large language models: a survey
27. Zhao WX, Zhou K, Li J, et al (2023) A survey of large language models
28. Zhou C, Li Q, Li C, et al (2023) A comprehensive survey on Pretrained foundation models: a history from BERT to ChatGPT
29. Liu S, Guo B, Fang C, Wang Z, Luo S, Zhou Z, Yu Z (2024) Enabling resource-efficient AIoT system with cross-level optimization: a survey. IEEE Commun Surv Tutor 26:389–427
30. Dong Q, Li L, Dai D, et al (2024) A survey on in-context learning
31. Chowdhery A, Narang S, Devlin J, et al (2022) PaLM: scaling language modeling with pathways
32. Touvron H, Lavril T, Izacard G, et al (2023) LLaMA: open and efficient foundation language models
33. OpenAI, Achiam J, Adler S, et al (2024) GPT-4 Technical Report
34. Singhal K, Azizi S, Tu T, et al (2022) Large language models encode clinical knowledge
35. Esteva A, Chou K, Yeung S, Naik N, Madani A, Mottaghi A, Liu Y, Topol E, Dean J, Socher R (2021) Deep learning-enabled medical computer vision. npj Digit Med 4:5
36. Tomašev N, Harris N, Baur S et al (2021) Use of deep learning to develop continuous-risk models for adverse event prediction from electronic health records. Nat Protoc 16:2765–2787
37. Yim J, Chopra R, Spitz T et al (2020) Predicting conversion to wet age-related macular degeneration using deep learning. Nat Med 26:892–899
38. Aydın Ö, Karaarslan E (2022) OpenAI ChatGPT generated literature review: digital twin in healthcare. SSRN J. https://doi.org/10.2139/ssrn.4308687
39. Yang G, Liu X, Shi J, Wang Z, Wang G (2024) TCM-GPT: efficient pre-training of large language models for domain adaptation in traditional Chinese medicine. Comput Methods Prog Biomed Update 6:100158
40. Shah RS, Chawla K, Eidnani D, Shah A, Du W, Chava S, Raman N, Smiley C, Chen J, Yang D (2022) WHEN FLUE MEETS FLANG: benchmarks and large pre-trained language model for financial domain

41. Shah A, Chava S (2023) Zero is not hero yet: benchmarking zero-shot performance of LLMs for financial tasks
42. Chen ZZ, Ma J, Zhang X, Hao N, Yan A, Nourbakhsh A, Yang X, McAuley J, Petzold L, Wang WY (2024) A survey on large language models for critical societal domains: finance, healthcare, and law
43. Yang H, Liu X-Y, Wang CD (2023) FinGPT: Open-Source Financial Large Language Models
44. Kalyan KS (2024) A survey of GPT-3 family large language models including ChatGPT and GPT-4. Nat Lang Process J 6:100048
45. Elnaggar A, Heinzinger M, Dallago C, et al (2020) ProtTrans: towards cracking the language of life's code through self-supervised learning. https://doi.org/10.1101/2020.07.12.199554
46. Brandes N, Ofer D, Peleg Y, Rappoport N, Linial M (2022) ProteinBERT: a universal deep-learning model of protein sequence and function. Bioinformatics 38:2102–2110
47. Lin Z, Akin H, Rao R et al (2023) Evolutionary-scale prediction of atomic-level protein structure with a language model. Science 379:1123–1130
48. Chowdhury R, Bouatta N, Biswas S et al (2022) Single-sequence protein structure prediction using a language model and deep learning. Nat Biotechnol 40:1617–1623
49. Rives A, Meier J, Sercu T et al (2021) Biological structure and function emerge from scaling unsupervised learning to 250 million protein sequences. Proc Natl Acad Sci USA 118:e2016239118
50. Ferruz N, Schmidt S, Höcker B (2022) ProtGPT2 is a deep unsupervised language model for protein design. Nat Commun 13:4348
51. Roche R, Moussad B, Shuvo MH, Bhattacharya D (2023) E(3) equivariant graph neural networks for robust and accurate protein-protein interaction site prediction. PLoS Comput Biol 19:e1011435
52. Roche R, Moussad B, Shuvo MH, Tarafder S, Bhattacharya D (2024) EquiPNAS: improved protein–nucleic acid binding site prediction using protein-language-model-informed equivariant deep graph neural networks. Nucleic Acids Res 52:e27–e27
53. Satorras, V. G., Hoogeboom, E., & Welling, M. (2021, July). E (n) equivariant graph neural networks. In International conference on machine learning (pp. 9323-9332). PMLR.
54. Liu R, Hu J (2013) DNABind: a hybrid algorithm for structure-based prediction of DNA-binding residues by combining machine learning-and template-based approaches. Proteins 81:1885–1899
55. Jiang Z, Shen Y-Y, Liu R (2023) Structure-based prediction of nucleic acid binding residues by merging deep learning- and template-based approaches. PLoS Comput Biol 19:e1011428
56. Song Y, Yuan Q, Zhao H, Yang Y (2023) Accurately identifying nucleic-acid-binding sites through geometric graph learning on language model predicted structures. Brief Bioinform 24:bbad360
57. Liu Y, Tian B (2024) Protein–DNA binding sites prediction based on pre-trained protein language model and contrastive learning. Brief Bioinform 25:bbad488
58. Zhu Y-H, Liu Z, Liu Y, Ji Z, Yu D-J (2024) ULDNA: integrating unsupervised multi-source language models with LSTM-attention network for high-accuracy protein–DNA binding site prediction. Brief Bioinform 25:bbae040
59. Zeng W, Lv D, Liu X, Chen G, Liu W, Peng S (2023) ESM-NBR: fast and accurate nucleic acid-binding residue prediction via protein language model feature representation and multi-task learning. In: 2023 IEEE international conference on bioinformatics and biomedicine (BIBM). pp 76–81
60. Littmann M, Heinzinger M, Dallago C, Weissenow K, Rost B (2021) Protein embeddings and deep learning predict binding residues for various ligand classes. Sci Rep 11:23916
61. Wang B, Li W (2024) Advances in the application of protein language modeling for nucleic acid protein binding site prediction. Genes 15:1090
62. Järvelin AI, Noerenberg M, Davis I, Castello A (2016) The new (dis)order in RNA regulation. Cell Commun Signal 14:9
63. Baek M, McHugh R, Anishchenko I, Baker D, DiMaio F (2022) Accurate prediction of nucleic acid and protein-nucleic acid complexes using RoseTTAFoldNA 2022.09.09.507333
64. Shen Y, Chen Z, Mamalakis M, He L, Xia H, Li T, Su Y, He J, Wang YG (2024) A fine-tuning dataset and benchmark for large language models for protein understanding. https://doi.org/10.48550/arXiv.2406.05540
65. Ji Y, Zhou Z, Liu H, Davuluri RV (2021) DNABERT: pre-trained bidirectional encoder representations from transformers model for DNA-language in genome. Bioinformatics 37:2112–2120

66. Zhang Y, Lang M, Jiang J et al (2023) Multiple sequence alignment-based RNA language model and its application to structural inference. Nucleic Acids Res gkad1031:e103
67. Li H-L, Pang Y-H, Liu B (2021) BioSeq-BLM: a platform for analyzing DNA, RNA and protein sequences based on biological language models. Nucleic Acids Res 49:e129
68. Altschul S (1997) Gapped BLAST and PSI-BLAST: a new generation of protein database search programs. Nucleic Acids Res 25: 3389–3402
69. Steinegger M, Söding J (2017) MMseqs2 enables sensitive protein sequence searching for the analysis of massive data sets. Nat Biotechnol 35:1026–1028
70. Mirdita M, Schütze K, Moriwaki Y, Heo L, Ovchinnikov S, Steinegger M (2022) ColabFold: making protein folding accessible to all. Nat Methods 19:679–682
71. Jing B, Eismann S, Suriana P, Townshend RJL, Dror R (2020) Learning from protein structure with geometric vector perceptrons. In: International conference on learning representations
72. Kingma DP, Ba J (2017) Adam: a method for stochastic optimization. https://doi.org/10.48550/arXiv.1412.6980
73. Loshchilov I, Hutter F (2017) SGDR: stochastic gradient descent with warm restarts. https://doi.org/10.48550/arXiv.1608.03983
74. Yang J, Roy A, Zhang Y (2013) BioLiP: a semi-manually curated database for biologically relevant ligand–protein interactions. Nucleic Acids Res 41:D1096–D1103
75. Huang Y, Niu B, Gao Y, Fu L, Li W (2010) CD-HIT suite: a web server for clustering and comparing biological sequences. Bioinformatics 26:680–682

# Chapter 10

# Practical Applications of Language Models in Protein Sorting Prediction: SignalP 6.0, DeepLoc 2.1, and DeepLocPro 1.0

**Henrik Nielsen**


## Abstract

Protein sorting prediction, or protein subcellular location prediction, is an important task in bioinformatics. It is also a task where protein language models have proven their value in practice. SignalP 6.0 (for prediction of signal peptides), DeepLoc 2.1 (for prediction of subcellular location and membrane association in eukaryotes), and DeepLocPro 1.0 (for prediction of subcellular location in prokaryotes) are all based on large pretrained protein language models. This has led to substantial improvements over previous methods, especially for rare classes where available labeled data are limited.


**Key words** Protein sorting, Protein subcellular locations, Protein language models, Signal peptides

## 1 Introduction

Cells have a number of distinct compartments, each with their characteristic set of proteins. Eukaryotic cells have several membrane-bound organelles such as the nucleus, the endoplasmic reticulum (ER), and the mitochondria [1]. Prokaryotes generally do not have organelles (although there are exceptions, such as the thylakoids in cyanobacteria), but still, proteins can belong to the cytoplasm, the plasma membrane, the cell wall, or the extracellular environment. For diderm (i.e., most Gram-negative) bacteria, there are two more possible locations, namely periplasm and outer membrane. So how does the cell know where to put which proteins? The Nobel Prize in physiology or medicine was in 1999 awarded to Günter Blobel "for the discovery that proteins have intrinsic signals that govern their transport and localization in the cell" [2].

But how do *we* know which proteins belong to which compartments? Since the experimental determination of subcellular locations of proteins is slow and costly compared to DNA sequencing, there is a great interest in *protein sorting prediction*, i.e., the

Dukka B. KC (ed.), *Large Language Models (LLMs) in Protein Bioinformatics*, Methods in Molecular Biology, vol. 2941,
https://doi.org/10.1007/978-1-0716-4623-6_10,

inference of subcellular locations based on sequences. This has been a target in bioinformatics since the beginning of the field, see Ref. [3] for a historical review. Many machine learning techniques have been employed in this task, and pretrained protein language models are among them. The advantage of this technique is that the pretrained model has "seen" many examples of proteins and their intrinsic sorting signals and therefore is able to generalize from a relatively small set of experimentally labeled examples in the downstream task.

In 2019, one of the early published applications of protein language models used subcellular location prediction in eukaryotic proteins (ten categories) as one of its two downstream tasks [4]. This was done with the ELMo architecture [5] based on long short-term memory recurrent neural networks. The more computationally efficient transformer architecture [6] was used by the same group 2 years later on the same task [7], comparing several different pretrained protein language models and reporting the best performance for the 3B-parameters ProtT5 [8].

### 1.1 Signal Peptides and SignalP

A signal peptide (SP) is the N-terminal part of a protein that is targeted to the secretory pathway in both prokaryotes and eukaryotes (*see*, however, **Note 1**). In eukaryotes, a protein with an SP will be targeted to the endoplasmic reticulum (ER) membrane and be cotranslationally translocated across the membrane. In prokaryotes, translocation takes place across the cytoplasmic membrane (inner membrane in diderm bacteria), and the process can happen during or after translation.

SPs are generally described as having three regions: an N-terminal n-region of variable length characterized by positive charge, a central h-region of typically 7–15 hydrophobic residues, and a C-terminal c-region of typically 3–7 polar residues. Positions −1 and −3 relative to the cleavage site are occupied by small uncharged residues, most often Alanine. SPs of Gram-positive bacteria tend to be longer than those of Gram-negative bacteria, which in turn tend to be longer than eukaryotic SPs [9, 10].

SignalP was in 1995 the first machine learning-based SP prediction method to be made available as a web server [11]. It has gone through six major versions—see Ref. [12] for a historical account—and has gained a very wide usage. The papers about SignalP have been cited more than 30,000 times in total according to Google Scholar. While SignalP 5.0 from 2019 [13] was the first version to be based on deep learning, version 6.0 from 2022 [14] is the first version to be based on a pretrained protein language model, ProtBERT [8].

Conventionally, we think of prokaryotes as being simpler than eukaryotes, but regarding SPs, the opposite is the case. Prokaryotes (both bacteria and archaea) have two types of translocon (Sec and Tat) and three types of signal peptidase (I, II, and III). This gives rise to five types of SPs (not six, since signal peptidase III substrates are only transported through the Sec translocon).

Prokaryotic lipoproteins have SPs that are cleaved by signal peptidase II [15]. They do not have a standard c-region but instead a motif termed a *lipobox* and a cysteine right after the cleavage site as the first residue of the mature protein. During cleavage, a lipid group is covalently attached to this cysteine, anchoring the protein to the membrane.

Those SPs that direct their proteins toward the Tat translocon have a special motif at the border between the n- and h-regions containing two consecutive arginines [16]. This is how the Tat translocon got its name—it is short for **t**win **a**rginine **t**ranslocation. It differs from the Sec translocon by being able to transport proteins that are already folded into their native 3D structure. Tat SPs differ from Sec SPs not only by the presence of the motif, they also tend to be longer and less hydrophobic [17].

Finally, signal peptidase III cleaves some very short SPs that do not contain an h-region; instead there is a hydrophobic region *after* the cleavage site [18]. Proteins cleaved by signal peptidase III typically belong to type IV pili and are always transported by the Sec translocon.

Previous versions of SignalP were only able to predict Sec/SPI SPs (the only type found at the eukaryotic ER membrane), and in order to predict SPII and Tat substrates, users had to consult other predictors such as LipoP [19] and TatP [20]. SignalP 5.0 changed this by including prediction of Sec/SPII and Tat/SPI SPs, but it could not predict Tat/SPII or Sec/SPIII SPs. Experimentally known examples of these two are extremely rare in UniProt [21] or PROSITE [22]—as a matter of fact, we failed to find even one example of Tat/SPII by our usual data retrieval protocol. When making the dataset for SignalP 6.0, we discovered that the PROSITE pattern for the Tat motif (PS51318) had a postprocessing rule that prevented simultaneous matching to the lipoprotein motif (PS51257), which we regarded as an error, since there is evidence for the existence of such proteins [23, 24]. By combining the two motifs and relabeling some proteins, we built a set of 36 Tat/SPII sequences.

This is an extremely small set for training one of the classes in a deep learning method, but we found that with a pretrained language model, it worked. We also tried to retrain the SignalP 5.0 architecture on the SignalP 6.0 dataset, and while it had comparable performances on the large classes of SPs, it failed to predict the rare classes Tat/SPII (33 examples after homology partitioning) and Sec/SPIII (70 examples after homology partitioning), see Fig. 2a of the SignalP 6.0 paper [14]. The ability to predict rare classes was the main advantage of the language model technology in SignalP 6.0, but there were two other substantial improvements.

The first additional improvement was better performance for sequences with low similarity to the training set. The partitioning of the SignalP 6.0 data to avoid leakage between training, validation, and test sets was done with the algorithm that later became known as GraphPart [25]. In contrast to the homology reduction algorithms of Hobohm et al. [26] that removes sequences until no pair with a similarity above the threshold remains in the data set, GraphPart strives to keep as many data points as possible while ensuring that no pair of sequences *between* partitions have similarity above the threshold. In many cases (including SignalP 6.0), this cannot be done without removing some data points. We used the removed sequences to estimate the how the performance of the finished SignalP 6.0 depended on the degree of similarity to the training set, and found that while the retrained SignalP 5.0 performed on par with SignalP 6.0 at high similarity values, it was substantially worse at similarity values below 60% (see Fig. 2c of the SignalP 6.0 paper), showing that the pretrained language model improves generalization to distantly related sequences.

The other additional improvement was independence of the organism group label: Ever since SignalP 1.0, the user has been asked to specify which group of organisms the sequences came from; eukaryotes, Gram-positive bacteria, Gram-negative bacteria, or (from version 5.0) archaea. In SignalP 1–4, this was implemented by keeping the data sets for each group separate, while SignalP 5.0 was trained on the combined data set with an extra input informing the network about the organism group. In SignalP 6.0, we also initially had such an extra input, but when we made an experiment where the group labels were randomly shuffled, we found that SignalP 6.0 performance values were unaffected, while performance of the retrained SignalP 5.0 suffered (*see* Supplementary Fig. 4 of the SignalP 6.0 paper). Apparently, the pretrained language model is able to infer the peculiarities of each organism group directly from the sequence. This has two advantages: SignalP 6.0 can now be applied to sequences of unknown origin, and it can be applied to groups of bacteria that are not easily classified into the Gram-positive/Gram-negative divide (such as genus *Deinococcus* that stain Gram-positive even though they are diderm, class Mollicutes that stain Gram-negative even though they are monoderm, and class Negativicutes which are diderm and Gram-negative even though they belong to phylum Bacillota, f.k.a. Firmicutes).

One new feature in SignalP 6.0 that is not attributable to the language model is the region assignment. SignalP 6.0 assigns n-, h-, and c-regions in predicted SPs by training the conditional random field (CRF) layer to maximize the differences in amino acid composition between the regions (the architecture of the CRF layer is shown in Supplementary Fig. 8 of the SignalP 6.0 paper). Region predictions can be useful for investigating properties of groups of SPs. As an example, we used SignalP 6.0 on a library of synthetic

SPs that were found to be either functional or non-functional in *Bacillus subtilis* [27]. In the original work, the authors did not find any discriminating properties between the two groups using traditional sequence analysis, but region predictions show a significant difference in n-region net charge and hydrophobicity between the groups (*see* Supplementary Fig. 5 of the SignalP 6.0 paper), revealing possible factors that contribute to in vivo functionality.

### 1.2 Eukaryotic Subcellular Locations and DeepLoc

The presence or absence of a SP is not the whole story about the localization of a protein. It is important to stress that the presence of a SP does *not* necessarily mean that the protein is secreted to the extracellular environment, it only means that it enters the secretory pathway. In all kinds of organisms, the protein could have one or more transmembrane helices downstream of the SP and therefore be retained in the membrane [28]. In eukaryotes, the protein could also be retained in one of the compartments that belong to the secretory pathway: the ER, the Golgi apparatus, or the lysosome/vacuole [29]; or it could be anchored to the outer face of the cytoplasmic membrane by a glycosylphosphatidylinositol (GPI) group [30].

A computational method that can predict the location of a protein based on its amino acid sequence is therefore highly desirable. Many such methods have been published, starting with PSORT in 1992 [31]. Some of the methods use homology information, i.e., they search a database of annotated proteins using BLAST or another sequence similarity search software and use the annotations of the found hits to infer the location of the query sequence. This inference can be direct (using the location of the best hit as the predicted location) or indirect (using some annotation of the homologues found as input to a machine learning method). The annotations used in indirect homology inference can be, e.g., keywords or functional descriptions [32], titles and/or abstracts of literature references [33], or Gene Ontology terms [34, 35].

Homology-based methods, or hybrid methods containing homology-based components, often present the best measured performances, but the performance depends critically on the source of the query protein. Organisms that have been subject to intense research will naturally tend to have more high-quality annotations, so proteins from those and their close relatives will find more close homologues with richer annotations from which to make predictions, while predictions for less well-studied organisms will suffer from lack of annotations of close homologues. This is typically not taken into consideration when reporting the predictive performances of such methods. Another drawback is that such methods will not be able to distinguish between very closely related proteins that differ in the presence or absence of a sorting signal, and they will not be able to predict the effects of small mutations that destroy or create a sorting signal.

DeepLoc 1.0, launched in 2017 [36], was a ten-category subcellular location predictor based on deep learning (convolutional and recurrent neural networks), but not language models. The input sequences were encoded either by the BLOSUM62 substitution matrix or by profiles. Profile encoding, where each position in an input sequence is represented by a vector of the frequencies of the 20 amino acids at that position in a multiple alignment, has been used extensively in protein sequence-based prediction methods for many years, beginning with secondary structure prediction in 1993 [37]. It should be stressed that the use of profiles does not make a method homology-based, since the annotations of the matching sequences are not used. We found that the BLOSUM62 encoding gave an accuracy four percentage points lower than profile encoding, but still better than the benchmark methods (which included homology-based methods). Profile encoding is, however, computationally heavy; in DeepLoc 1.0 we found that it was taking around ten times as long as the rest of the prediction, even though we used precomputed Pfam [38] profiles to speed up the calculation. Consequently, the web server of DeepLoc 1.0 allowed only 50 sequences per submission for profile encoding, where 500 sequences were allowed for BLOSUM62.

DeepLoc 2.0 from 2022 [39] was based on transformer language models (ESM1b [40] and ProtT5 [8]), which improved the performance, especially for the small categories where prediction is particularly hard (such as ER, lysosome/vacuole, Golgi, and peroxisome). Additionally, the language model encoding made profiles obsolete, allowing for a faster prediction than the profile mode. Two added features in DeepLoc 2.0 were multiple locations and sorting signal prediction.

Several proteins function in more than one subcellular location. When training DeepLoc 1.0, we experimented with adding an 11th category for "Cytoplasm+Nucleus" (the by far most abundant category of multi-localized proteins) but it turned out to be nearly impossible to predict, so in the end we discarded all multi-localized proteins from the data set. To remedy this situation, DeepLoc 2.0 is trained as a multi-label method. The average number of labels per protein in the DeepLoc 2.0 training data is 1.27, implying that a substantial minority of proteins are muti-localized.

The other added feature was multi-label prediction of sorting signals. We included nine sorting signals: SPs; transmembrane helices; transit peptides for mitochondria, chloroplasts, and chloroplast thylakoids; nuclear localization and export signals; peroxisomal targeting signals; and GPI anchor propeptides. It should be stressed that there is not necessarily a 1:1 correspondence between localizations and sorting signals. For example, some proteins are imported into mitochondria without a transit peptide [41], and for peroxisomal proteins we only predict the C-terminal PTS1 signal, not the N-terminal PTS2 [42].

For DeepLoc 2.0, we did not do nested cross-validation (as for SignalP 6.0) but just plain five-fold cross-validation on the UniProt-derived data set. In addition, we had an external test set consisting of sequences from the Human Protein Atlas (HPA) [43]. Note that the HPA test set had fewer classes than the UniProt set it did not include extracellular (due to the experimental procedure), plastid (because it is a human dataset), or peroxisome and lysosome/vacuole (because there were too few examples with good confidence and low similarity to the UniProt dataset).

Most of the locations predicted by DeepLoc should actually be regarded as two locations, because proteins can be inside an organelle or associated with its membrane. For example, the ER category comprises both ER lumen and ER membrane proteins, which are rather different with respect to the sorting process and the involved sorting (retention) signals. In DeepLoc 1.0, this was addressed by adding a binary prediction of whether each query protein was membrane-bound or soluble. In DeepLoc 2.0, this prediction was missing.

This was the reason for updating DeepLoc to version 2.1 in 2024 [44]. Not only did we want to reinstate the "membrane or soluble" prediction, we also wanted to make it more fine-grained by distinguishing between transmembrane proteins, peripheral membrane proteins, and lipid-anchored membrane proteins. Since there were a substantial number of proteins in the data set belonging to two of these categories (e.g., soluble and peripheral), we decided to make this classification multilabel as well. The membrane type prediction was made with the same architecture as DeepLoc 2.0, using ESM1b and ProtT5. At the time, ESM2 had been published [45], so we also tried two versions of ESM2, one with 650 M parameters (like ESM1b) and the other with 15B parameters. The improvements over ESM1b and ProtT5 were at best marginal for ESM2-650 M (and ESM2-15B would have been too computationally heavy to deploy in the web server), so we decided to stick with ESM1b and ProtT5. In this way, only a single per-position representation for a query protein needs to be computed for the subcellular location and membrane protein type prediction tasks.

The data for DeepLoc 2.1 were derived from UniProt. We split the data set into five partitions using GraphPart [25] and set aside one partition as a fixed test set, using four-fold cross-validation on the rest.

### 1.3 Prokaryotic Subcellular Locations and DeepLocPro

Also in prokaryotes, the presence of a SP does *not* necessarily mean that the protein is secreted. In diderm bacteria, the protein could be retained in the periplasm, or be inserted into the outer membrane as a β-barrel transmembrane protein [46]. In cyanobacteria, there is also the possibility that the protein could be imported into the thylakoids [47]. In Gram-positive bacteria and archaea, the protein could be attached to the cell wall [48]. In addition, there is the aforementioned possibility of transmembrane helices downstream in the sequence. It was reported in 1994 that SPs were rarely found in bacterial cytoplasmic membrane proteins [49], but it is uncertain whether that result holds today. On the other hand, the absence of a SP does not necessarily mean that the protein is cytoplasmic. There are several pathways for SP-independent secretion, especially in Gram-negative bacteria where types I, III, IV, and VI secretion happen without a SP [50].

For these reasons, multi-category protein sorting prediction is also of great interest in prokaryotes (see Ref. [51] for an overview). The very first published version of PSORT in 1991 was intended for Gram-negative bacteria [52]. Later, the development of PSORT forked, with WoLF PSORT from 2007 for eukaryotic sequences [53] and PSORTb from 2003 for prokaryotic sequences [54]. PSORTb version 3.0 from 2010 [55] is a homology-based method which is still regarded as state-of-the-art. Although some newer methods have been published, implementations are typically not functional or no longer available to the public [56–59]. From the PSORTb group came in 2020 PSORTm [60], a version of PSORTb 3.0 that is optimized for metagenomic data. However, it does not represent a retraining of the underlying method, so it is not expected to perform better on full-length proteins. Recently, two language-model-based methods have been described in the literature [61, 62], but they have not been made available to the scientific community.

Accordingly, we perceived a need for a new non-homology-based method using pretrained protein language models. That method is DeepLocPro 1.0 [63]. Unlike DeepLoc 2.1, DeepLoc-Pro 1.0 is a single-label multiclass method. We considered making a multi-label method but found that the amount of multi-localized proteins is much lower than in eukaryotes (only 3% of the labeled data set had more than one label), so in the end, we decided to ignore this minority of proteins.

Since the number of membranes is so low in prokaryotes (one in archaea and monoderm bacteria, two in diderm bacteria, and three in cyanobacteria), there is not the same need for a separate membrane association prediction as in eukaryotes. Still, a user might want to know whether a prokaryotic membrane protein is integral, peripheral, or lipid anchored, but that is a feature to be added in future versions of DeepLocPro. The same is true for thylakoid membrane and lumen proteins, which are not included in this version.

Benchmarking DeepLocPro against PSORTb 3.0 was complicated, because PSORTb 3.0 includes a BLAST-based homology module with direct transfer of annotation. Since the annotation database used in PSORTb 3.0 is derived from UniProt exactly like our training and test sets, this method could reach an artificially high performance value simply by retrieving the annotations of exactly the same sequences that we used for testing. The solution was to empty the BLAST database used in our installation of PSORTb 3.0 in order to disable the homology module, see details in the supplementary materials of the DeepLocPro paper.

## 2 Materials

### 2.1 Amino Acid Sequences

SignalP, DeepLoc, and DeepLocPro require amino acid sequences in FASTA format. They will not produce useful predictions from nucleotide sequences (*see* **Note 2**). For DeepLoc and DeepLocPro, sequences should be complete, i.e., not fragments, since many sorting signals are located in the N- or C-terminal regions. SignalP only considers the first 70 residues of each input sequence, so it is important that this part of the sequence is complete and has the correct start codon (*see* **Note 3**).

### 2.2 Web Servers

The web servers are found here:

- https://services.healthtech.dtu.dk/services/SignalP-6.0/
- https://services.healthtech.dtu.dk/services/DeepLoc-2.1/
- https://services.healthtech.dtu.dk/services/DeepLocPro-1.0/

In addition, SignalP 6.0 and DeepLocPro 1.0 have mirror sites at the company BioLib:

- https://dtu.biolib.com/SignalP-6
- https://ku.biolib.com/DeepLocPro

which can be used if the server load is high at the dtu.dk site. DeepLoc 2.1 does not have a mirror site. The web servers grant free access to all users, including commercial users.

### 2.3 Downloadable Packages

If you want to predict on large datasets, we recommend downloading the software packages for command line use on your own hardware. The packages are free for academic institutions, while there is a license fee for commercial users. For SignalP 6.0 and DeepLoc 2.1, academic users should click "Downloads" at the web server pages, choose the latest version, and fill out the form to accept the license. For DeepLocPro, academic users should go to https://github.com/Jaimomar99/deeplocpro. Commercial users are asked to contact health-software@dtu.dk.

In the case of SignalP 6.0, there is a choice between "Fast" and "Slow" versions, see "model mode" in Subheading 3.1.1. The "Slow" version takes six times as long as to run, and the download is also six times as large as the "Fast" version. If you want the opportunity to choose between the two versions via a command line option, you need to download both packages.

Note that the programs, being based on large pretrained language models, run much faster on a system with a CUDA-compatible GPU. For SignalP 6.0, there is a script included in the package to convert the model files into GPU versions. For DeepLoc 2.1 and DeepLocPro 1.0, you can make the predictions run on GPU with the "-d cuda" command line option.

## 3 Methods

The interface to the three web servers is quite similar—in the "Submission" tab, you can paste your sequence(s) in the window or upload a file with the sequences, see an example in Fig. 1. Then, there are a few options and a "Submit" button. After you click the "Submit" button, you will be informed about whether your job is queued or actively being processed. On that page, there is an option to enter your email address to be informed when the job finishes. In the following, the options and the output will be explained in detail.

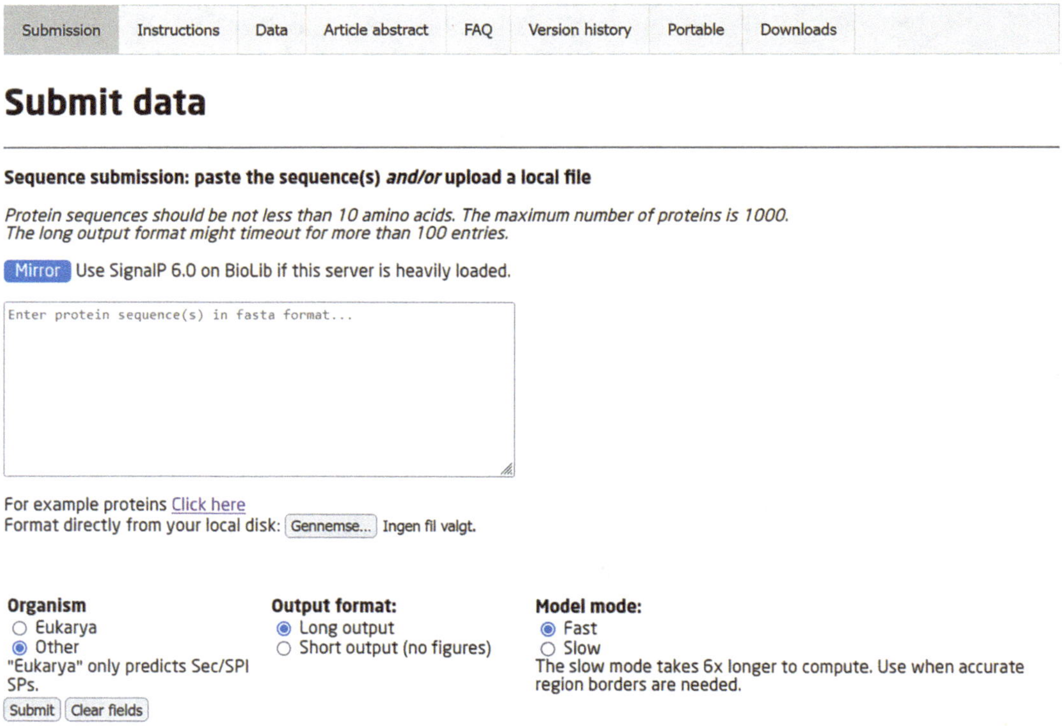


**Fig. 1** The submission tab of the SignalP 6.0 server page

### 3.1 Using SignalP 6.0

On the web server, paste or upload your sequences and click "Submit" (*see* Fig. 1). On the command line, run "signalp6 -ff *input.fasta* -od *output-directory.*"

#### 3.1.1 Options

**Organism** As described in Subheading 1.1, SignalP 6.0 is independent of the organism group label, in the sense that we did not see any deterioration in performance when we randomized the group labels. When SignalP 6.0 still offers the user a choice between "*Eukarya*" (eukaryotes) and "*Other,*" it is solely to suppress false predictions of SPs that are not of the Sec/SPI type in eukaryotes (*see*, however, **Note 4**). If the origin of the sequences is unknown, choose "*Other.*" In the command line interface, the group is indicated by "-org eukarya" or "-org other."

**Output Format** The "*Long*" output format supplies, for each sequence, a plot showing the probabilities of each region for each position in the sequence (see below). If you submit a large number of sequences, use the "*Short*" format without figures instead. In the command line interface, the output format is given by the "-fmt" option which can take the values "txt," "png," "eps," "all," or "none." The last two values correspond to "*Long*" and "*Short*" format, while the three first are in-between possibilities (making all textual output, or additionally making one of the plot types).

**Model Mode** This can be "*Fast*" and "*Slow.*" SignalP 6.0 is trained with three-fold nested cross-validation, i.e., the dataset has been divided into three partitions (without any sequence being >30% identical to any sequence in the other two partitions), and each partition has been used as a test set while the other two partitions have been used as training and validation in both combinations. In total, that gives six trained networks. Since we used fine-tuning, the ProtBERT models are not identical between these six. The "*Slow*" mode runs all six models sequentially and averages the output probabilities. The "*Fast*" mode is based on so-called model distillation, where one network is trained to emulate the output of the "*Slow*" mode (*see* Supplementary Note 1 in the SignalP 6.0 publication). Thus, the "*Fast*" mode takes only one sixth of the time taken by the "*Slow*" mode. The discrimination performance between non-SPs and the five SP types is very close to the original (*see* Supplementary Table 12 in the SignalP 6.0 publication), but the cleavage site performance will suffer. If precise cleavage sites and region borders are important, use the "*Slow*" mode.

#### 3.1.2 Output

When using the "*Long*" output format, there will be a table and a plot for each sequence in the input. The table shows the per-sequence probabilities for the five SP types and "Other" (non-SP). For predicted SPs, the cleavage site is also indicated. The plot (see Figs. 2, 3, 4, 5, and 6) shows the region probabilities

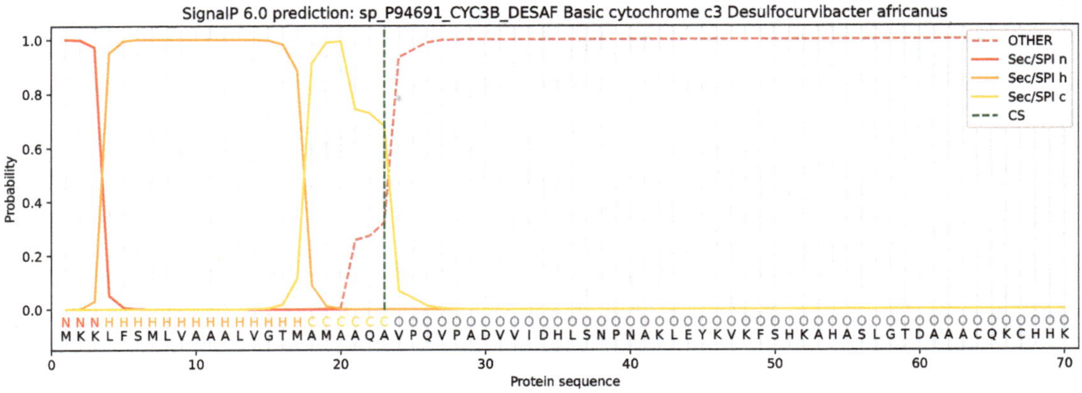


**Fig. 2** An example of a plot from SignalP 6.0, showing a predicted Sec/SPI signal peptide with n-, h-, and c-regions

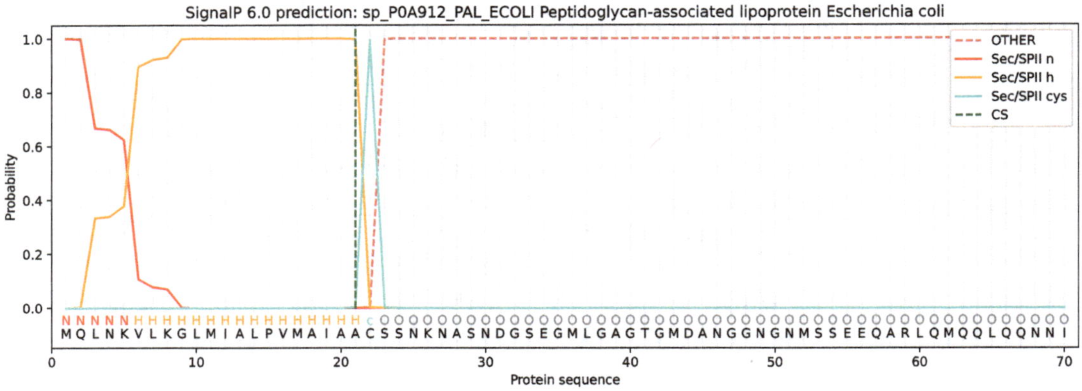


**Fig. 3** An example of a plot from SignalP 6.0, showing a predicted Sec/SPII signal peptide with n- and h-regions and a lipid-modified cysteine

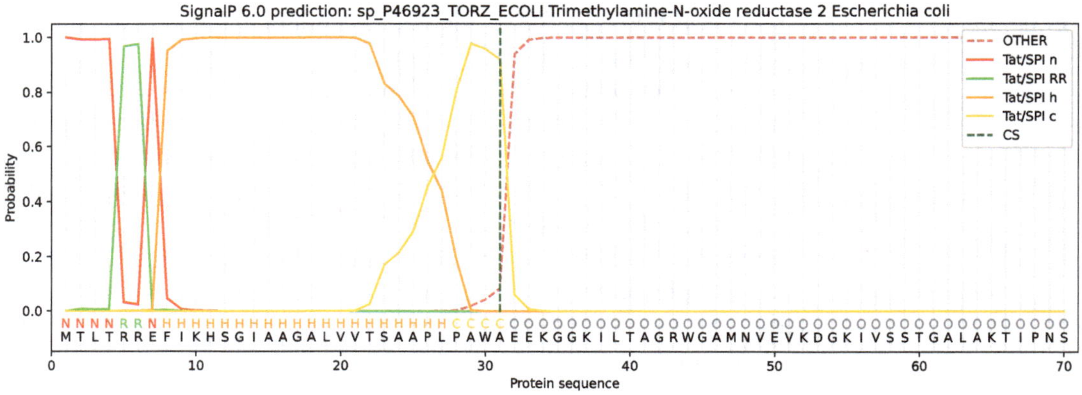


**Fig. 4** An example of a plot from SignalP 6.0, showing a predicted Tat/SPI signal peptide with n-, h-, and c-regions and the twin-arginine motif ("RR")

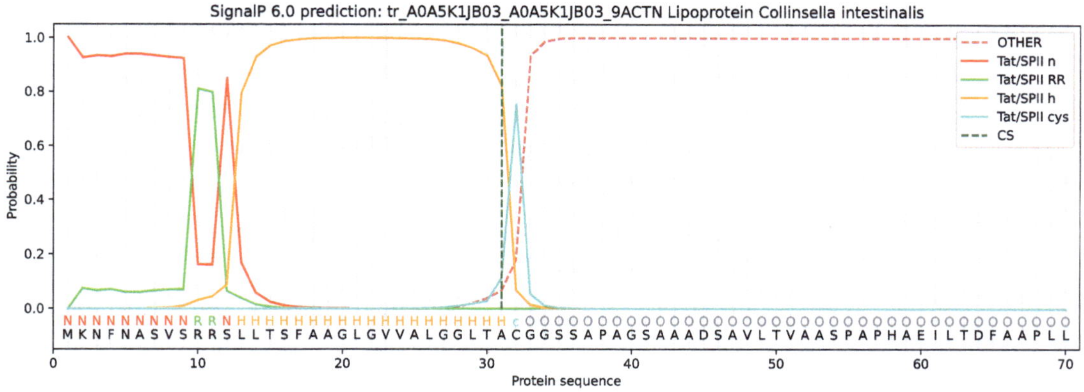


**Fig. 5** An example of a plot from SignalP 6.0, showing a predicted Tat/SPII (Tat-lipo) signal peptide, having both a twin-arginine motif ("RR") and a lipid-modified Cys

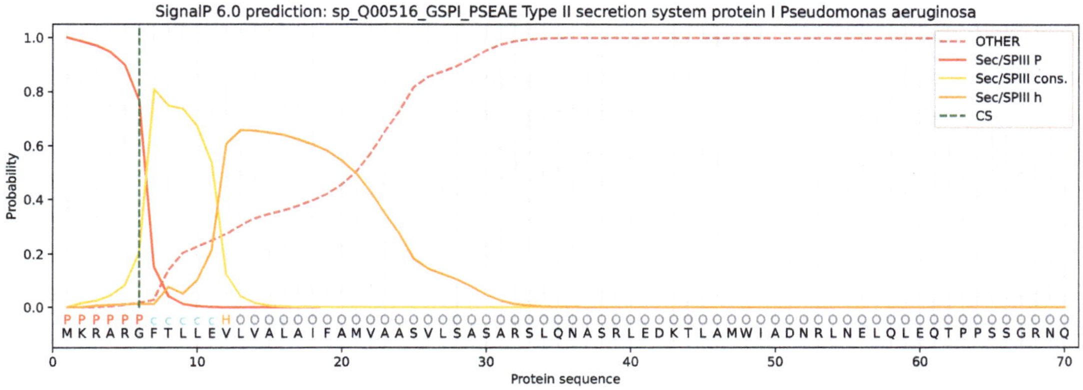


**Fig. 6** An example of a plot from SignalP 6.0, showing a predicted Sec/SPIII signal peptide with consensus region and hydrophobic region *after* the cleavage site

for each position in the sequence (truncated to 70 residues). In the bottom of the plot, the sequence is indicated, and above the sequence is the assignment of the regions. For SPI-cleaved SPs (Fig. 2), there are n-, h-, and c-regions assigned (*see* **Note 5**). The cleavage site is per definition the last position of the c-region and is marked by a vertical dashed line. Note that the cleavage site does not have its own state in the CRF layer, so the cleavage site probability is defined as the marginal probability of "c-region" in the cleavage position (*see* **Note 6**).

SPII-cleaved lipoprotein SPs have no c-region, but their n- and h-regions are indicated (Fig. 3). In addition, the lipid-modified cysteine is marked by a peak right after the cleavage site (*see* **Note 7**). Tat/SPI SPs have an annotated twin-arginine motif (Fig. 4) within their n-region. Note the long h-region, which is characteristic of Tat SPs. An example of the rare Tat/SPII type is seen in Fig. 5, showing both the twin-arginine motif and the lipid-modified cysteine. Finally, the short Sec/SPIII prepilin SPs are not divided into regions, but after their cleavage site, there is a consensus region and a hydrophobic region (see Fig. 6).

The full output from SignalP 6, attained via the "*Long*" output format or the "-fmt all" option, comprises five files pertaining to the whole prediction plus three files per sequence. The five files available per prediction are:

- JSON Summary: used for defining the output that the user sees on the webpage.
- Prediction summary: a tab-delimited file with one line per sequence, containing the probabilities for the five SP types and "Other" (non-SP). For predicted SPs, the cleavage site is also indicated.
- Processed entries fasta: a FASTA sequence file containing the sequences of those proteins that had predicted SPs, with the SP removed. This can be useful for certain types of downstream analysis.
- Processed entries gff3: a GFF3 (generic feature format version 3 [64]) file showing the length of each predicted SP.
- Region predictions gff3: a GFF3 file showing the start and end positions of each region in each predicted SP.

And the three files available for each sequence are:

- PNG: the plot shown on the webpage in Portable Network Graphics format.
- EPS: the plot in Encapsulated PostScript, a text-based vector graphics format.
- Tabular: the data for the plot in a tab-delimited format with one line per position. It contains the probabilities for each region of each SP type.

Additionally, the web server gives an option for downloading all of these files as a ZIP archive.

### 3.2 Using DeepLoc 2.1

On the web server, paste or upload your sequences and click "Submit." On the command line, run "deeploc2 -f *input.fasta* -o *output-directory*."

#### 3.2.1 Options

**Model** This can be "*High-quality (Slow)*" or "*High-throughput (Fast)*." These options select the ProtT5 (3B parameters) or ESM1b (650M parameters) pretrained models, respectively. The "*Slow*" model takes approximately two to four times as long as the "*Fast*" model (*see* Supplementary Table 1 of the DeepLoc 2.0 and 2.1 papers [39, 44]). In the command line interface, the model is selected by "-m Accurate" or "-m Fast."

**Output Format** The "*Long*" output format supplies, for each sequence, a logo-like attention plot (see below). If you submit a large number of sequences, use the "*Short*" format without figures instead. In the command line interface, plots are requested by the "-p" option.

### 3.2.2 Output

If the "Long" output is chosen, there will, for each sequence, be a header, two tables, and a figure. The header shows predicted localizations, membrane association, and sorting signals. Since DeepLoc 2.1 is a multi-label method, there can be more than one label predicted in each of these categories. "Localizations" and "membrane association" will always have at least one label, while "signals" can be empty.

Figure 7 shows the header and the tables for the Thy-1 antigen from rat, which is correctly predicted to localize to the cell membrane as a lipid-anchored protein. However, the prediction is not perfect, since there are no predicted signals, while there should have been a SP and a GPI anchor. The chosen labels in the "Localization" and "Membrane association" categories are marked by a green background. The intensity of the green color reflects the difference between the probability score and its associated threshold. The higher the probability difference, the darker the green color. The threshold values can be displayed by clicking the button labeled "Probability thresholds" at the top of the output page.

If none of the labels in "Localization" or "Membrane association" scores above its threshold, the label that is closest to its threshold is picked as the prediction. In that case, there will be no green background, and the confidence in the prediction is low.

Below the tables, there is a logo-like attention plot showing, using the letters of the input sequence, which parts of the sequence were most important for the localization prediction (*see* Fig. 8). The output probabilities are computed as weighted sums of scores pertaining to every position in the sequence, and these weights are

sp_P01830_THY1_RAT
**Predicted localizations:** Cell membrane
**Predicted membrane association:** Lipid anchor
**Predicted signals:**

| Localization | Cytoplasm | Nucleus | Extracellular | Cell membrane | Mitochondrion | Plastid | Endoplasmic reticulum | Lysosome/ Vacuole | Golgi apparatus | Peroxisome |
|---|---|---|---|---|---|---|---|---|---|---|
| Probability | 0.1207 | 0.0555 | 0.4058 | 0.8029 | 0.1195 | 0.0031 | 0.3417 | 0.1773 | 0.2141 | 0.0803 |

| Membrane association | Peripheral | Transmembrane | Lipid anchor | Soluble |
|---|---|---|---|---|
| Probability | 0.1480 | 0.4440 | 0.9490 | 0.3160 |

**Fig. 7** A header and two tables showing DeepLoc 2.1 prediction results for the UniProt sequence P01830/THY1_RAT

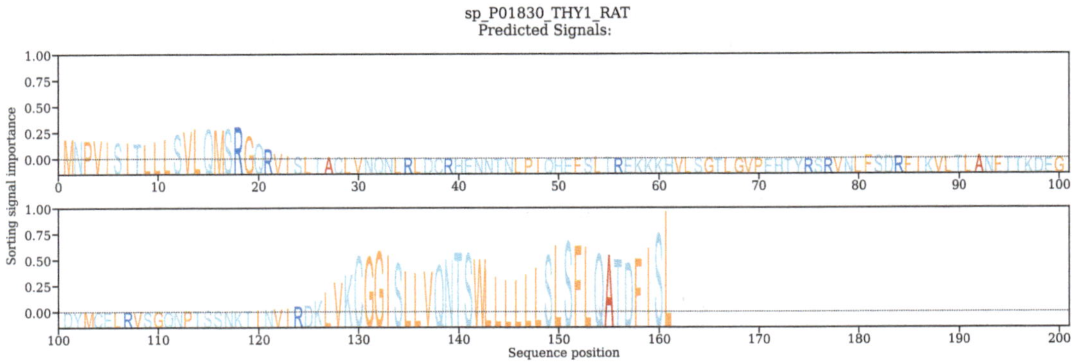


**Fig. 8** Example of the logo-like attention plot from DeepLoc 2.1, showing results for the UniProt sequence P01830/THY1_RAT

called alpha values or attention scores. In the model, the alpha values per definition sum to 1 over the entire sequence, so they are rather small numbers. In the plot, they are rescaled so that the highest scoring position in the sequence is set to 1, making it easier to compare attention plots from sequences of varying lengths.

In the attention plot for Thy-1 from rat (Fig. 8), it is evident that two regions are important; from the N-terminus (1) to approximately 20, and from approximately 125 to the C-terminus (161). These correspond well to the experimentally known SP (1–19) and propeptide (131–161), with the ω-site of the lipid anchor being the cysteine in position 130. So, even though no signals were predicted, it is clear that DeepLoc 2.1 has in some sense captured these two signals.

Even if there are predicted signals, DeepLoc 2.1 will only predict their presence, not their exact positions, but the attention plot can help point out where the signals are. Note, however, that there is a general tendency to have higher attention in the N- and C-termini, so a high attention value in the first or last handful of residues is not necessarily a sorting signal.

For cleaved signals, there are alternative specialized predictors which will predict the exact positions of their cleavage sites: SignalP 6.0 for SPs, TargetP 2.0 for N-terminal mitochondrial, plastid and thylakoid transit peptides [65], and NetGPI 1.1 for C-terminal GPI anchor propeptides [66]. (SignalP 6.0 and NetGPI 1.1 predict the SP and GPI anchor of rat Thy-1 correctly.) For transmembrane helices, there are two language model-based methods available, DeepTMHMM [67] and TMbed [68]. Regarding DeepLoc 2.1's prediction of transmembrane helices, *see* **Note 8**.

The remaining three signals—nuclear localization signals (NLS), nuclear export signals (NES), and C-terminal peroxisomal targeting signals (PTS1)—are not cleaved, and not structurally defined like transmembrane helices, so it is hard to define their limits in the sequences. We used NLSdb [69] for NLS and NES definitions, and "motif" features of UniProt for defining PTS1 sequences. In a quantitative comparison of the correlation between attention values and annotated sorting signals (Table 4 of the DeepLoc 2.0 paper), the PTS1 signal was found as efficiently as SPs, while NLS and NES were much harder to find.

Note that the three predictions of DeepLoc 2.1 (localization, membrane association, and sorting signal) are independently trained, even though they are made by the same language models. Since they do not depend on each other, they can sometimes disagree. This can be the case if one of the two soluble-only localizations ("cytoplasm" or "extracellular") is predicted alongside a membrane association other than "soluble," or if the "cell membrane" localization is predicted together with "soluble." In such cases, the results should be taken with caution. The sorting signal prediction can also disagree with the other two predictions. For example, we have seen cases of "peroxisomal targeting signal" together with "cytoplasm." Here, it is advisable to check the performances of the two categories in the DeepLoc 2.0 paper; "peroxisomal targeting signal" has a markedly higher performance (MCC 0.90) than the localization "peroxisome" (MCC 0.56), so in this case, it is probably best to believe the sorting signal prediction and regard the protein as peroxisomal (in addition to cytoplasmic).

### 3.3 Using DeepLocPro 1.0

On the web server, paste or upload your sequences and click "Submit." On the command line, run "deeplocpro -f *input.fasta* -o *output-directory*."

#### 3.3.1 Options

**Organism Group** The options here are "*Any*" (the default), "*Archaea*," "*Gram negative*," and "*Gram positive*" (*see* **Note 9**). The model is in itself agnostic to the organism group; the effect of this option is simply to suppress the biologically implausible predictions of "outer membrane" and "periplasmic" if the data are from archaea and Gram-positive bacteria. Such predictions will be remapped to "extracellular." In the command line interface, the group is indicated by the "-g" option with the arguments "any," "archaea," "positive," or "negative."

**Output Format** The "*Long*" output format supplies, for each sequence, a logo-like attention plot (see below). If you submit a large number of sequences, use the "*Short*" format without figures instead. In the command line interface, plots are requested by the "-p" option.

**sp_P0ADC1_LPTE_ECOLI**
**Predicted localization:** Outer Membrane

| Localization | Cell wall & surface | Extracellular | Cytoplasmic | Cytoplasmic Membrane | Outer Membrane | Periplasmic |
|---|---|---|---|---|---|---|
| Probability | 0 | 0 | 0.0001 | 0.0213 | 0.9783 | 0.0002 |

**Fig. 9** DeepLocPro 1.0 prediction results for the UniProt sequence P0ADC1/LPTE_ECOLI

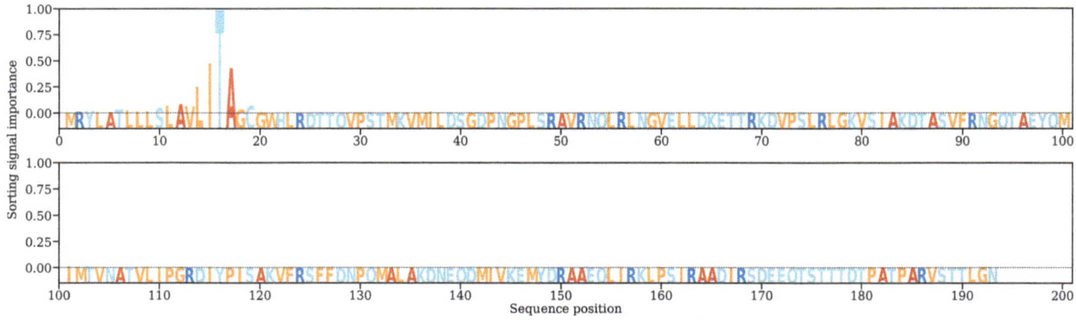


**Fig. 10** Example of the logo-like attention plot from DeepLocPro 1.0, showing results for the UniProt sequence P0ADC1/LPTE_ECOLI

*3.3.2 Output*

The DeepLocPro 1.0 output is very similar to the DeepLoc 2.1 output, but it is in several respects simpler. There are no sorting signal or membrane type predictions, Additionally, there is no table of prediction thresholds, since DeepLocPro 1.0, being a single-label method, simply takes the highest scoring category to be the prediction.

There is a table and an attention plot for each sequence. In the example shown, the LPS-assembly lipoprotein LptE from *Escherichia coli* is correctly predicted to be localized to the outer membrane. In the table (Fig. 9), the background color is green with an intensity proportional to the probability of each location. If there are two locations with a visibly green background, it is not a prediction of multiple locations, but a low-confidence prediction where DeepLocPro 1.0 is in doubt about the correct location.

In the attention plot (Fig. 10), it seems that DeepLocPro 1.0 puts emphasis on the lipobox of the signal peptide (which is of the Sec/SPII type and covers positions 1–18).

# 4 Notes

1. Sometimes, especially in introductory textbooks [1], the term "signal peptide" is used in a broader sense, meaning any sorting signal embedded in the amino acid sequence of a protein. However, the definition of SP used in this chapter corresponds

to the usage in most of the scientific literature [9, 10], as well as in UniProt [70] and the Sequence feature ontology [71].

2. The servers have no detection feature for nucleotide sequences; they will treat such sequences as proteins exclusively consisting of Ala, Cys, Gly, and Thr, leading to nonsensical output without a warning.
3. Since SignalP predicts an N-terminal signal, it is dependent on a correct start codon assignment. A start codon assigned too far downstream will cut part of the true amino acid sequence, while a start codon assigned too far upstream will add arbitrary sequence to the N-terminus, in both cases making it difficult to recognize a possible SP.
4. When we say that the only SP type in eukaryotes is Sec/SPI, that statement pertains to the translocation system at the ER membrane. In mitochondria and (especially) plastids, Tat SPs can occur [16].
5. Occasionally, it can happen that some positions within a long SP are predicted as "Other." In other words, the SP does not follow the grammar defined by the CRF layer in SignalP 6.0. In those cases, the region assignments in the GFF3 file will not make sense, since the regions will be annotated as overlapping. If this happens, we recommend taking the predicted SP with caution.
6. The "cleavage site probability" is *not* the probability that the SP is cleaved, but a measure of confidence in the predicted placement of the cleavage site. A low probability means that cleavage might occur in another position.
7. Occasionally, there are predicted Sec/SPII or Tat/SPII SPs that do not have a cysteine immediately after the cleavage site. These predictions are false; a prokaryotic lipoprotein without an N-terminal cysteine has never been observed in nature. For a simpler computational model, it would be easy to avoid these predictions that do not correspond to any example in the dataset, but for a language model, there is apparently a tendency to extrapolate beyond the training data, sometimes in unwanted directions (analogous to the "hallucinations" observed in ChatGPT and related models). Such sequences are most probably actually Sec/SPI or Tat/SPI SPs.
8. In multispanning transmembrane proteins, it was only the *first* (most N-terminal) transmembrane segment that was labeled as a sorting signal for DeepLoc 2.0 training, since this has been shown to be important for protein sorting in the secretory pathway. Therefore, DeepLoc 2.1 should not be used as a transmembrane topology predictor.

9. "Gram negative" and "Gram positive" should in this context be interpreted as diderm and monoderm, respectively. Class Mollicutes should thus be regarded as "Gram positive" even though they stain Gram-negative.

## Acknowledgments

Thanks go to all coauthors of the SignalP 6.0, DeepLoc 2.1, and DeepLocPro 1.0 publications: Felix Teufel, José Juan Almagro Armenteros, Alexander Rosenberg Johansen, Magnús Halldór Gíslason, Silas Irby Pihl, Konstantinos D. Tsirigos, Ole Winther, Søren Brunak, Gunnar von Heijne, Marius Thrane Ødum, Vineet Thumuluri, and Jaime Moreno. In addition, I want to thank Peter Wad Sackett, who is responsible for keeping the services.healthtech.dtu.dk website up and running.

### References

1. Hillis DM, Heller HC, Hacker SD et al (2020) Life: the science of biology, 12th edn. Sinauer Associates/ Macmillan, Sunderland
2. The Nobel Prize in Physiology or Medicine (1999). https://www.nobelprize.org/prizes/medicine/1999/summary/. Accessed 19 Oct 2024
3. Nielsen H, Tsirigos KD, Brunak S, von Heijne G (2019) A brief history of protein sorting prediction. Protein J 38:200–216. https://doi.org/10.1007/s10930-019-09838-3
4. Heinzinger M, Elnaggar A, Wang Y et al (2019) Modeling aspects of the language of life through transfer-learning protein sequences. BMC Bioinformatics 20:723. https://doi.org/10.1186/s12859-019-3220-8
5. Peters ME, Neumann M, Iyyer M et al (2018) Deep contextualized word representations. arXiv 1802.05365 [cs.CL]. https://doi.org/10.48550/arXiv.1802.05365
6. Vaswani A, Shazeer N, Parmar N et al (2017) Attention is all you need. In: Guyon I, von Luxburg U, Bengio S et al (eds) Advances in neural information processing systems. Curran Associates, Inc
7. Stärk H, Dallago C, Heinzinger M, Rost B (2021) Light attention predicts protein location from the language of life. Bioinforma Adv 1:vbab035. https://doi.org/10.1093/bioadv/vbab035
8. Elnaggar A, Heinzinger M, Dallago C et al (2022) ProtTrans: toward understanding the language of life through self-supervised learning. IEEE Trans Pattern Anal Mach Intell 44:7112–7127. https://doi.org/10.1109/TPAMI.2021.3095381
9. von Heijne G (1990) The signal peptide. J Membr Biol 115:195–201. https://doi.org/10.1007/BF01868635
10. Owji H, Nezafat N, Negahdaripour M et al (2018) A comprehensive review of signal peptides: structure, roles, and applications. Eur J Cell Biol 97:422–441. https://doi.org/10.1016/j.ejcb.2018.06.003
11. Nielsen H, Brunak S, Engelbrecht J, von Heijne G (1997) Identification of prokaryotic and eukaryotic signal peptides and prediction of their cleavage sites. Protein Eng Des Sel 10:1–6. https://doi.org/10.1093/protein/10.1.1
12. Nielsen H, Teufel F, Brunak S, von Heijne G (2024) SignalP: the evolution of a web server. In: Lisacek F (ed) Protein bioinformatics. Springer, New York, pp 331–367. https://doi.org/10.1007/978-1-0716-4007-4_17
13. Almagro Armenteros JJ, Tsirigos KD, Sønderby CK et al (2019) SignalP 5.0 improves signal peptide predictions using deep neural networks. Nat Biotechnol 37:420–423. https://doi.org/10.1038/s41587-019-0036-z
14. Teufel F, Almagro Armenteros JJ, Johansen AR et al (2022) SignalP 6.0 predicts all five types of signal peptides using protein language models. Nat Biotechnol 40:1023–1025. https://doi.org/10.1038/s41587-021-01156-3

15. von Heijne G (1989) The structure of signal peptides from bacterial lipoproteins. Protein Eng Des Sel 2:531–534. https://doi.org/10.1093/protein/2.7.531
16. Berks BC (2015) The twin-arginine protein translocation pathway. Annu Rev Biochem 84: 843–864. https://doi.org/10.1146/annurev-biochem-060614-034251
17. Cristóbal S, de Gier J-W, Nielsen H, von Heijne G (1999) Competition between sec- and TAT-dependent protein translocation in Escherichia coli. EMBO J 18:2982–2990. https://doi.org/10.1093/emboj/18.11.2982
18. Henche A-L, van Wolferen M, Ghosh A, Albers S-V (2014) Dissection of key determinants of cleavage activity in signal peptidase III (SPaseIII) PibD. Extremophiles 18:905–913. https://doi.org/10.1007/s00792-014-0675-4
19. Juncker AS, Willenbrock H, von Heijne G et al (2003) Prediction of lipoprotein signal peptides in gram-negative bacteria. Protein Sci 12:1652–1662. https://doi.org/10.1110/ps.0303703
20. Bendtsen JD, Nielsen H, Widdick D et al (2005) Prediction of twin-arginine signal peptides. BMC Bioinformatics 6:167. https://doi.org/10.1186/1471-2105-6-167
21. The UniProt Consortium (2023) UniProt: the universal protein knowledgebase in 2023. Nucleic Acids Res 51:D523–D531. https://doi.org/10.1093/nar/gkac1052
22. Sigrist CJA, de Castro E, Cerutti L et al (2013) New and continuing developments at PROSITE. Nucleic Acids Res 41:D344–D347. https://doi.org/10.1093/nar/gks1067
23. Hutchings MI, Palmer T, Harrington DJ, Sutcliffe IC (2009) Lipoprotein biogenesis in gram-positive bacteria: knowing when to hold 'em, knowing when to fold 'em. Trends Microbiol 17:13–21. https://doi.org/10.1016/j.tim.2008.10.001
24. Storf S, Pfeiffer F, Dilks K et al (2010) Mutational and bioinformatic analysis of haloarchaeal lipobox-containing proteins. Archaea 2010:410975. https://doi.org/10.1155/2010/410975
25. Teufel F, Gíslason MH, Almagro Armenteros JJ et al (2023) GraphPart: homology partitioning for biological sequence analysis. NAR Genomics Bioinforma 5:lqad088. https://doi.org/10.1093/nargab/lqad088
26. Hobohm U, Scharf M, Schneider R, Sander C (1992) Selection of representative protein data sets. Protein Sci 1:409–417. https://doi.org/10.1002/pro.5560010313
27. Wu Z, Yang KK, Liszka MJ et al (2020) Signal peptides generated by attention-based neural networks. ACS Synth Biol 9:2154–2161. https://doi.org/10.1021/acssynbio.0c00219
28. von Heijne G (1988) Transcending the impenetrable: how proteins come to terms with membranes. Biochim Biophys Acta 947:307–333. https://doi.org/10.1016/0304-4157(88)90013-5
29. Harter C, Wieland F (1996) The secretory pathway: mechanisms of protein sorting and transport. Biochim Biophys Acta BBA – Rev Biomembr 1286:75–93. https://doi.org/10.1016/0304-4157(96)00003-2
30. Ferguson MAJ, Williams AF (1988) Cell-surface anchoring of proteins via glycosyl-phosphatidylinositol structures. Annu Rev Biochem 57:285–320. https://doi.org/10.1146/annurev.bi.57.070188.001441
31. Nakai K, Kanehisa M (1992) A knowledge base for predicting protein localization sites in eukaryotic cells. Genomics 14:897–911. https://doi.org/10.1016/S0888-7543(05)80111-9
32. Lu Z, Szafron D, Greiner R et al (2004) Predicting subcellular localization of proteins using machine-learned classifiers. Bioinformatics 20:547–556. https://doi.org/10.1093/bioinformatics/btg447
33. Briesemeister S, Blum T, Brady S et al (2009) SherLoc2: a high-accuracy hybrid method for predicting subcellular localization of proteins. J Proteome Res 8:5363–5366. https://doi.org/10.1021/pr900665y
34. Briesemeister S, Rahnenführer J, Kohlbacher O (2010) Going from where to why – interpretable prediction of protein subcellular localization. Bioinformatics 26:1232–1238. https://doi.org/10.1093/bioinformatics/btq115
35. Wan S, Mak M-W, Kung S-Y (2017) FUEL-mLoc: feature-unified prediction and explanation of multi-localization of cellular proteins in multiple organisms. Bioinformatics 33:749–750. https://doi.org/10.1093/bioinformatics/btw717
36. Almagro Armenteros JJ, Sønderby CK, Nielsen H, Winther O (2017) DeepLoc: prediction of protein subcellular localization using deep learning. Bioinformatics 33:3387–3395. https://doi.org/10.1093/bioinformatics/btx431
37. Rost B, Sander C (1993) Improved prediction of protein secondary structure by use of sequence profiles and neural networks. Proc Natl Acad Sci 90:7558–7562. https://doi.org/10.1073/pnas.90.16.7558

38. Mistry J, Chuguransky S, Williams L et al (2021) Pfam: the protein families database in 2021. Nucleic Acids Res 49:D412–D419. https://doi.org/10.1093/nar/gkaa913
39. Thumuluri V, Almagro Armenteros JJ, Johansen AR et al (2022) DeepLoc 2.0: multi-label subcellular localization prediction using protein language models. Nucleic Acids Res 50: W228–W234. https://doi.org/10.1093/nar/gkac278
40. Rives A, Meier J, Sercu T et al (2021) Biological structure and function emerge from scaling unsupervised learning to 250 million protein sequences. Proc Natl Acad Sci 118: e2016239118. https://doi.org/10.1073/pnas.2016239118
41. Schneider K, Zimmer D, Nielsen H et al (2021) iMLP, a predictor for internal matrix targeting-like sequences in mitochondrial proteins. Biol Chem 402:937–943. https://doi.org/10.1515/hsz-2021-0185
42. Montilla-Martinez M, Beck S, Klümper J et al (2015) Distinct pores for peroxisomal import of PTS1 and PTS2 proteins. Cell Rep 13: 2126–2134. https://doi.org/10.1016/j.celrep.2015.11.016
43. Thul PJ, Åkesson L, Wiking M et al (2017) A subcellular map of the human proteome. Science 356:eaal3321. https://doi.org/10.1126/science.aal3321
44. Ødum MT, Teufel F, Thumuluri V et al (2024) DeepLoc 2.1: multi-label membrane protein type prediction using protein language models. Nucleic Acids Res 52:W215–W220. https://doi.org/10.1093/nar/gkae237
45. Lin Z, Akin H, Rao R et al (2023) Evolutionary-scale prediction of atomic-level protein structure with a language model. Science 379:1123–1130. https://doi.org/10.1126/science.ade2574
46. Duong F, Eichler J, Price A et al (1997) Biogenesis of the gram-negative bacterial envelope. Cell 91:567–573. https://doi.org/10.1016/S0092-8674(00)80444-4
47. Liberton M, Austin II JR, Berg RH, Pakrasi HB (2011) Insights into the complex 3-D architecture of thylakoid membranes in the unicellular cyanobacterium Cyanothece sp. ATCC 51142. Plant Signal Behav 6:566. https://doi.org/10.4161/psb.6.4.14946
48. Mazmanian SK, Liu G, Ton-That H, Schneewind O (1999) Staphylococcus aureus sortase, an enzyme that anchors surface proteins to the cell wall. Science 285:760–763. https://doi.org/10.1126/science.285.5428.760
49. Broome-Smith JK, Gnaneshan S, Hunt LA et al (1994) Cleavable signal peptides are rarely found in bacterial cytoplasmic membrane proteins. Mol Membr Biol 11:3–8. https://doi.org/10.3109/09687689409161023
50. Desvaux M, Hébraud M, Talon R, Henderson IR (2009) Secretion and subcellular localizations of bacterial proteins: a semantic awareness issue. Trends Microbiol 17:139–145. https://doi.org/10.1016/j.tim.2009.01.004
51. Nielsen H (2024) Protein sorting prediction. In: Journet L, Cascales E (eds) Bacterial secretion systems: methods and protocols. Springer, New York, pp 27–63. https://doi.org/10.1007/978-1-0716-3445-5_2
52. Nakai K, Kanehisa M (1991) Expert system for predicting protein localization sites in gram-negative bacteria. Proteins Struct Funct Bioinforma 11:95–110. https://doi.org/10.1002/prot.340110203
53. Horton P, Park K-J, Obayashi T et al (2007) WoLF PSORT: protein localization predictor. Nucleic Acids Res 35:W585–W587. https://doi.org/10.1093/nar/gkm259
54. Gardy JL, Spencer C, Wang K et al (2003) PSORT-B: improving protein subcellular localization prediction for gram-negative bacteria. Nucleic Acids Res 31:3613–3617. https://doi.org/10.1093/nar/gkg602
55. Yu NY, Wagner JR, Laird MR et al (2010) PSORTb 3.0: improved protein subcellular localization prediction with refined localization subcategories and predictive capabilities for all prokaryotes. Bioinformatics 26:1608–1615. https://doi.org/10.1093/bioinformatics/btq249
56. Goldberg T, Hecht M, Hamp T et al (2014) LocTree3 prediction of localization. Nucleic Acids Res 42:W350–W355. https://doi.org/10.1093/nar/gku396
57. Wang X, Zhang J, Li G-Z (2015) Multi-location gram-positive and gram-negative bacterial protein subcellular localization using gene ontology and multi-label classifier ensemble. BMC Bioinformatics 16:S1. https://doi.org/10.1186/1471-2105-16-S12-S1
58. Wan S, Mak M-W, Kung S-Y (2017) Gram-LocEN: interpretable prediction of subcellular multi-localization of gram-positive and gram-negative bacterial proteins. Chemom Intell Lab Syst 162:1–9. https://doi.org/10.1016/j.chemolab.2016.12.014
59. Grasso S, van Rij T, van Dijl JM (2020) GP4: an integrated gram-positive protein prediction pipeline for subcellular localization mimicking bacterial sorting. Brief Bioinform 22:bbaa302. https://doi.org/10.1093/bib/bbaa302
60. Peabody MA, Lau WYV, Hoad GR et al (2020) PSORTm: a bacterial and archaeal protein

subcellular localization prediction tool for metagenomics data. Bioinformatics 36:3043–3048. https://doi.org/10.1093/bioinformatics/btaa136
61. Singh G, Tyagi R, Singh A et al (2022) Protein language model for prediction of subcellular localization of protein sequences from Gram-negative bacteria (ProtLM.SCL). bioRxiv 2022.12.16.520742. https://doi.org/10.1101/2022.12.16.520742
62. Arora I, Kummer A, Zhou H et al (2024) Mtx-COBRA: subcellular localization prediction for bacterial proteins. Comput Biol Med 171:108114. https://doi.org/10.1016/j.compbiomed.2024.108114
63. Moreno J, Nielsen H, Winther O, Teufel F (2024) Predicting the subcellular location of prokaryotic proteins with DeepLocPro. Bioinformatics 40:btae677. https://doi.org/10.1093/bioinformatics/btae677
64. Stein L (2024) Specifications/gff3.md at master. The-Sequence-Ontology/Specifications. In: GitHub. https://github.com/The-Sequence-Ontology/Specifications/blob/master/gff3.md. Accessed 27 Oct 2024
65. Almagro Armenteros JJ, Salvatore M, Emanuelsson O et al (2019) Detecting sequence signals in targeting peptides using deep learning. Life Sci Alliance 2:e201900429. https://doi.org/10.26508/lsa.201900429
66. Gíslason MH, Nielsen H, Almagro Armenteros JJ, Johansen AR (2021) Prediction of GPI-anchored proteins with pointer neural networks. Curr Res Biotechnol 3:6–13. https://doi.org/10.1016/j.crbiot.2021.01.001
67. Hallgren J, Tsirigos KD, Pedersen MD et al (2022) DeepTMHMM predicts alpha and beta transmembrane proteins using deep neural networks. bioRxiv. https://doi.org/10.1101/2022.04.08.487609
68. Bernhofer M, Rost B (2022) TMbed: transmembrane proteins predicted through language model embeddings. BMC Bioinformatics 23:326. https://doi.org/10.1186/s12859-022-04873-x
69. Bernhofer M, Goldberg T, Wolf S et al (2018) NLSdb – major update for database of nuclear localization signals and nuclear export signals. Nucleic Acids Res 46:D503–D508. https://doi.org/10.1093/nar/gkx1021
70. UniProt Help: Signal Peptide (2024). http://www.uniprot.org/help/signal. Accessed 1 Nov 2024
71. SO_0000418 (2024). http://purl.obolibrary.org/obo/SO_0000418. Accessed 1 Nov 2024

# Chapter 11

# CNN-Meth: A Tool to Accurately Predict Lysine Methylation Sites Using Evolutionary Information-Based Protein Modeling

**Austin Spadaro, Alok Sharma, and Iman Dehzangi**

## Abstract

Lysine methylation is a crucial posttranslational modification influencing both histone and nonhistone protein functions. Disruptions in lysine methyltransferase activity are linked to numerous diseases, including various cancers and developmental disorders. Accurate identification of lysine methylation sites is essential for early diagnosis and therapeutic development. Here, we present CNN-Meth, a newly developed Web-based utility that employs a convolutional neural network (CNN) to predict lysine methylation sites. CNN-Meth leverages evolutionary, structural, and physicochemical data alongside binary encoding for its training process. Evolutionary and structural features used to build CNN-Meth are extracted using protein modeling, which works similarly to using Protein Language Models (PLM). Unlike traditional approaches that rely on manually extracted features, CNN-Meth uses CNNs for automated feature extraction, ensuring minimal information loss. This novel methodology enhances prediction accuracy, achieving 96.0% Accuracy, 85.1% Sensitivity, 96.4% Specificity, and a Matthew's Correlation Coefficient (MCC) of 0.65. This demonstrates the possible effectiveness of using PLM to predict Methylation sites as a future direction. The CNN-Meth tool and its source code are readily accessible at https://github.com/MLBC-lab/CNN-Meth, providing a robust resource for researchers and clinicians.

**Key words** Posttranslational modification, Methylation, Automated feature extraction, Convolutional neural network, Evolutionary features, Structural features, Physicochemical features

## 1 Introduction

Posttranslational modifications (PTMs) are enzymatic changes in protein after translation. These modifications can alter protein structure, impacting their roles within various biological processes [1, 2]. PTMs are implicated in a wide range of biological functions, including protein folding, signal transduction, and the regulation of gene expression. Conversely, aberrant PTMs are associated with various diseases, such as cancer, neurodegenerative disorders, and cardiovascular diseases, underscoring the importance of accurately identifying PTM sites [3, 4].

Dukka B. KC (ed.), *Large Language Models (LLMs) in Protein Bioinformatics*, Methods in Molecular Biology, vol. 2941, https://doi.org/10.1007/978-1-0716-4623-6_11,

With its reversible nature, lysine methylation emerges as a promising candidate from a drug design perspective [5]. The dynamic regulation of lysine methylation by lysine methyltransferases (KMTs) and demethylases (KDMs) underscores its potential impact on drug development. Identifying lysine methylation sites is essential for understanding disease mechanisms and developing targeted therapies.

Traditionally, lysine methylation sites have been identified using experimental techniques such as mass spectrometry [6]. While these methods provide high accuracy, they are labor-intensive, time-consuming, and costly, making them impractical for large-scale studies. Consequently, there is a growing need for computational methods that can predict lysine methylation sites efficiently and accurately.

Over the past few years, several machine learning-based tools have been developed to predict lysine methylation sites. Notable among them are PLMLA, iLM-2 L, MethK, mLysPTMpred, predML-Site, and ilRMethyl [7–12]. These tools typically rely on handcrafted features derived from protein sequences, such as amino acid composition, secondary structure, and evolutionary information. For instance, PLMLA, developed by Shi et al., leverages a combination of group weight, amino acid composition, and secondary structure features extracted from the tool PSIPRED 2.0 to predict lysine methylation. It employs a Support Vector Machine (SVM) classifier to identify methylation sites based on these features. PLMLA demonstrated moderate success in methylation site prediction, but like many early tools, it relies on predefined features, limiting the complexity of the patterns it can capture from protein sequences.

Similarly, iLM-2 L, proposed by Ju et al., also utilizes an SVM classifier but introduces a more sophisticated feature representation by incorporating k-gap amino acid pairs and pseudo-amino acid composition into its feature set. This model employs a two-level classification approach, allowing it to predict whether a lysine residue is methylated and infer the degree of methylation. By incorporating sequence-based features that capture more nuanced relationships within the protein sequence, iLM-2 L improved upon earlier methods, demonstrating better performance in predicting methylation degrees. However, like PLMLA, iLM-2 L is constrained by its reliance on handcrafted features, which may not fully encapsulate the complexities of lysine methylation, potentially leading to information loss during the feature extraction process.

These approaches highlight a common trend in early machine learning-based PTM prediction tools—while they achieved promising results, they often relied on predefined, manually extracted features, which may overlook subtle but important aspects of the underlying biological processes. For example, MethK, a tool designed by Lee et al., predicts lysine methylation for both histone and nonhistone proteins. It combines sequence-based features and evolutionary data derived from PSSM matrices, further demonstrating the importance of incorporating multiple layers of biological information in PTM prediction. Despite these innovations, limitations persist in these tools due to their manual feature extraction processes, which can inadvertently miss relevant data patterns. The need for more comprehensive and automated methods became apparent as methylation site prediction models sought higher accuracy and deeper insights into the biological mechanisms driving these modifications.

CNN-Meth is a novel tool designed to predict lysine methylation sites using a convolutional neural network (CNN) [13]. This tool uses deep learning to automatically extract features from protein sequences, incorporating evolutionary, structural, and physicochemical data. Using CNNs, CNN-Meth avoids the limitations associated with handcrafted feature extraction, ensuring minimal information loss and capturing complex patterns in the data. Our approach combines multiple types of information, including Position-Specific Scoring Matrices (PSSM), secondary structure predictions, physicochemical properties, and binary encoding of amino acids, to create a comprehensive feature set.

Our evolutionary and structural protein modeling is similar to using large language models (LLMs), which have emerged as promising and powerful paradigms for sequence modeling. In this approach, a very large transformer model is trained on a large natural language dataset. The transformer model comprises neural networks, including an encoder and a decoder, both of which utilize self-attention mechanisms. The resulting transformer is able to encode a given sequence using all the information that it learned during the training process. LLMs are also being used to encode proteins through Protein Language Models (PLMs). In this case, a large transformer is trained using large unlabeled protein datasets such as UniProt. PLMs have been successfully used to predict different PTMs, such as Succinylation [14, 15], Sumoylation [16], Glycation [17], Phosphoglycerylation [18], and many more [19–21].

Although we do not directly use PLMs, we use the PSSM as the output of PSI-BLAST for protein encoding. PSI-BLAST conducts a sequence alignment on a very large protein dataset and encodes the level of similarity as the PSSM. In other words, PSSM is also an encoding of a given protein sequence with respect to a large protein dataset. Using PSSM in this study can be considered as a proof of concept in using PLMs for methylation site prediction. CNN-Meth has been trained and validated on a large benchmark dataset, achieving superior performance compared to existing methods. The tool demonstrates high accuracy, sensitivity, and specificity, making it a robust resource for researchers and clinicians.

The promising performance of CNN-Meth also demonstrates the effectiveness of modeling using large protein datasets. Therefore, this study can be considered as a proof of concept that shows the possible promising performance of PLMs to accurately predict methylation sites as the next step. CNN-Meth is publicly available as a Web-based utility (at: https://mlbclab.camden.rutgers.edu/research/), providing easy access for the scientific community. This paper presents the details of CNN-Meth, including the datasets, feature extraction methods, model architecture, implementation, and evaluation results.

## 2 Materials and Methods

This section details the methods and data used to design and implement CNN-Meth.

### 2.1 Dataset

For this analysis, we used data extracted from the Protein Lysine Modification Database (PLMD), which has since been transitioned to CPLM 4.0 [22–24]. This data is publicly available at https://cplm.biocuckoo.cn/. This dataset contains various types of lysine PTMs that have been experimentally verified and is the most extensive publicly available collection of confirmed lysine methylation sites. The lysine methylation dataset consists of 6322 methylation sites and 133,293 non-methylation sites. To reduce bias within our model, CD-HIT was used to remove sequences with greater than 40% sequential similarity, resulting in a dataset with 4913 methylation sites and 108,958 non-methylation sites [25].

Given the significant imbalance between positive and negative samples, a balancing strategy was employed to improve model performance [26]. The training data was balanced to a ratio of 3:1 by down-sampling the negative samples using K-nearest neighbor (KNN) [27]. This down-sampling strategy maximizes the distance between selected training data points, preserving as much nonredundant information as possible. As a result, the final balanced dataset contains 4913 positive and 14,739 negative samples. For the reader's convenience, the number of samples in our dataset at the beginning, after removing those with over 40% sequential similarities, and after the down-sampling is presented in Table 1.

**Table 1**
**Positive and negative samples**

| | All | CD-HIT | Balanced | Test |
|---|---|---|---|---|
| Positive | 6322 | 4913 | 4913 | 983 |
| Negative | 133,293 | 108,958 | 14,739 | 21,617 |

### 2.2 Feature Representation

The feature representation for CNN-Meth was extracted from sequences of 15 upstream and 15 downstream amino acids from the target lysine [26]. This window size was chosen based on its demonstrated effectiveness in previous studies. A mirroring technique was used to ensure a consistent window size for sequences where the lysine is near the termini depicted in Fig. 1 [28].

**Evolutionary Features** Evolutionary information was extracted using Position-Specific Scoring Matrices (PSSM), calculated with PSI-BLAST [29]. PSI-BLAST aligned a given sequence against a very large protein dataset (in this case, nonredundant protein dataset (NR) containing over 80 million proteins). This modeling is similar to using PLMs for protein modeling. PSSMs provide a matrix of substitution probabilities, capturing evolutionary conservation at each position in the sequence. Instead of using handcrafted features from PSSMs, CNN-Meth inputs the entire PSSM matrix directly into the convolutional neural network, allowing for automated feature extraction.

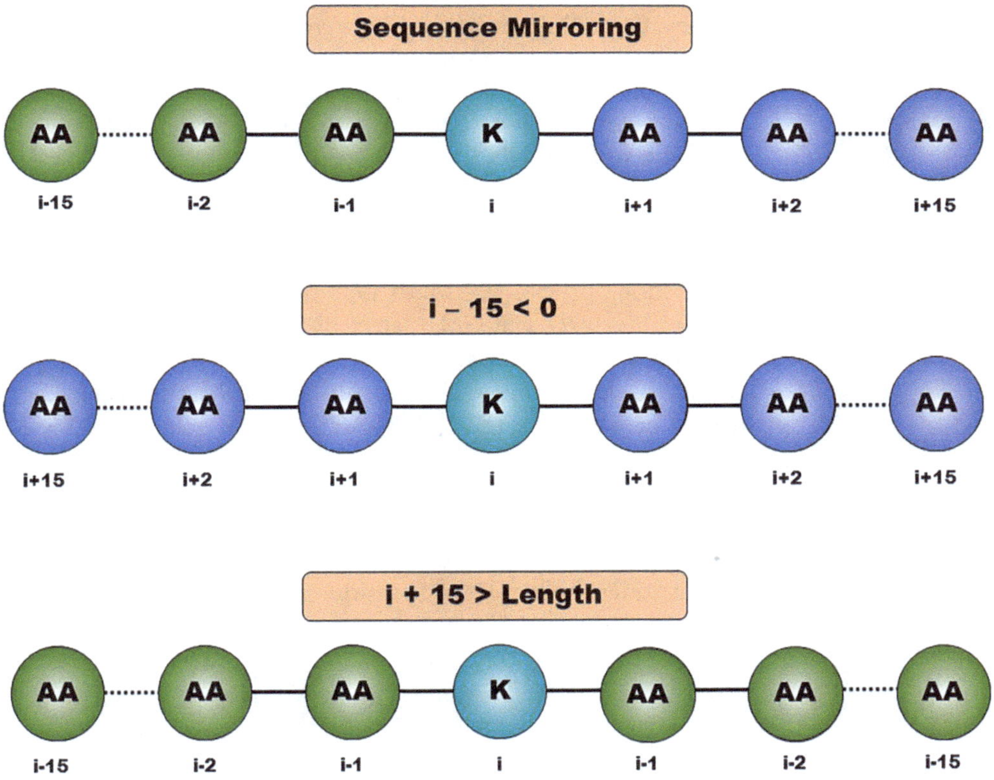


**Fig. 1** Polypeptide mirroring

**Structural Properties** Structural features were calculated using SPIDER2 and include accessible surface area (ASA), torsion angles ($\theta$, $\tau$, $\psi$, $\varphi$), and secondary structure states (helix, sheet, coil) [30–36]. These properties provide critical information about the protein's three-dimensional structure and local folding patterns. The input to SPIDER2 is the PSSM of a given protein. Therefore, it also uses PSSM modeling for feature encoding. Therefore, it can also be considered a modeling similar to PLMs. It is important to note that despite similarities, we are fully aware of the differences between PSSM and PLM protein modeling. Therefore, using PSSM here can solely be considered as a proof of concept for using PLMs in the future and not showing their absolute effectiveness in methylation site prediction.

**Physicochemical Properties** Each amino acid's physicochemical attributes were sourced from the AAIndex database, with 55 selected properties shown to be effective in previous studies [37–39]. This results in a 31x55 feature matrix for each sample, which captures a comprehensive set of physicochemical characteristics.

**Binary Encoding** Amino acids were also encoded as 20-element binary vectors, where each element represents the presence of a specific amino acid. This encoding creates a 31x20 feature matrix for each sequence window, facilitating the input of sequence information into the CNN.

### 2.3 Web Server

CNN-Meth is accessible as an online tool and is publicly available at https://mlbclab.camden.rutgers.edu/research/ (*see* **Note 1**). A screenshot of the CNN-Meth Web utility is shown in Fig. 2. The Web utility takes a PSSM matrix file as an input (*see* **Note 2**). A PSSM file represents a protein sequence's evolutionary profile, capturing each amino acid's substitution probabilities at every position. Users can use the PSI-BLAST tool to generate a PSSM file from a FASTA file, which aligns the protein sequence against a database to compute the PSSM [29] (*see* **Note 3**). The required PSSM file format is shown in Fig. 3.

The uploaded file should contain the entire protein sequence that includes the target lysine residue (*see* **Note 4**). The file name of the uploaded PSSM file must be the row number of the target lysine followed by ".pssm" (e.g., 124.pssm) (*see* **Note 5**). It is important to ensure that a valid PSSM file is uploaded using this naming convention, as an invalid file will result in an incorrect prediction from the Web utility. Once a valid PSSM file is provided, CNN-Meth will automatically handle feature extraction, including generating the corresponding SPD3 file, based on the provided PSSM file (*see* **Notes 6–8**). This streamlined process ensures that users can easily obtain accurate predictions for lysine methylation sites.

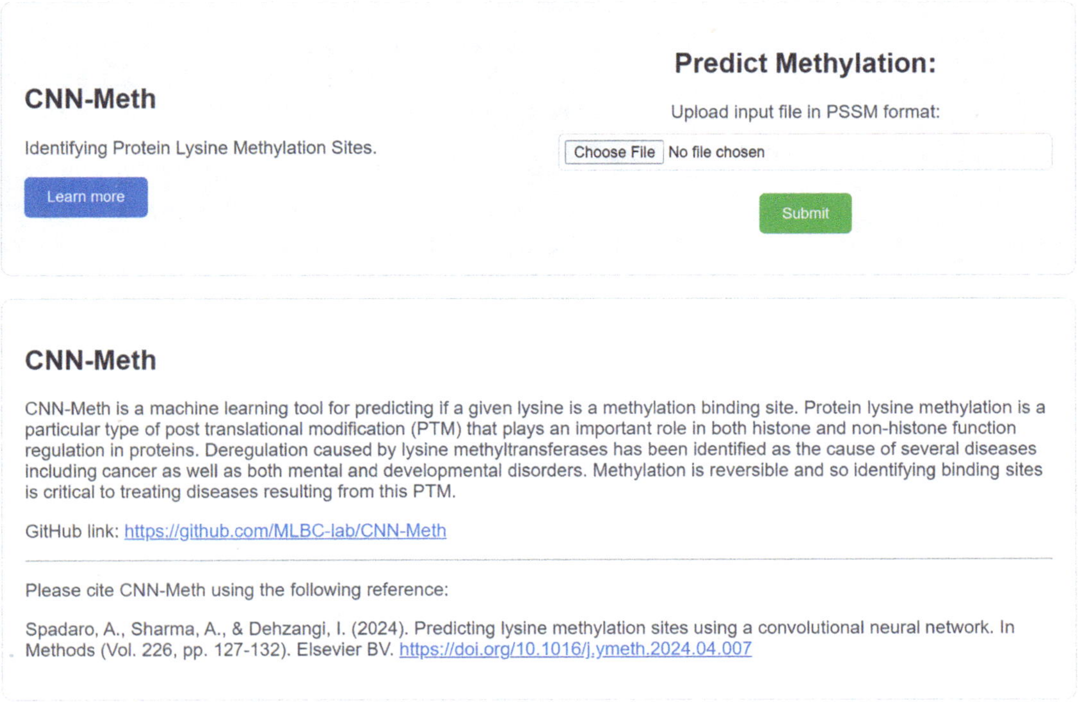


**Fig. 2** CNN-Meth Web utility

```
        A  R  N  D  C  Q  E  G  H  I  L  K  M  F  P  S  T  W  Y  V    A   R   N   D   C   Q   E   G   H   I   L   K   M   F   P   S   T   W   Y   V
 1 M   -1 -2 -2 -3 -2  0 -2 -3 -2  1  2 -2  6  0 -3 -2 -1 -2 -1  1    0   0   0   0   0   0   0   0   0   0   0   0 100   0   0   0   0   0   0   0  0.35 0.00
 2 T    0 -1  0 -1 -1 -1 -1 -2 -2 -1 -1 -1 -1 -2 -1  2  5 -3 -2  0    0   0   0   0   0   0   0   0   0   0   0   0   0   0   0   0 100   0   0   0  0.47 0.00
 3 E   -1  0  0  2 -4  2  6 -2  0 -4 -3  1 -2 -4 -1  0 -1 -3 -2 -3    0   0   0   0   0   0 100   0   0   0   0   0   0   0   0   0   0   0   0   0  0.74 0.00
 4 F   -2 -3 -3 -4 -3 -4 -4 -3 -1  0  0 -3  0  7 -4 -3 -2  1  3 -1    0   0   0   0   0   0   0   0   0   0   0   0   0 100   0   0   0   0   0   0  0.74 0.00
 5 R   -1  5  0 -1 -4  4  1 -2  0 -3 -2  2 -1 -3 -2 -1 -1 -3 -2 -3    0  65   0   0   0  35   0   0   0   0   0   0   0   0   0   0   0   0   0   0  0.59 0.00
 6 Y   -2 -2 -2 -3 -3 -2 -2 -3  2 -2 -1 -2 -1  3 -3 -2 -2  2  7 -1    0   0   0   0   0   0   0   0   0   0   0   0   0   0   0   0   0   0 100   0  0.87 0.00
 7 A    4 -2 -2 -2  0 -1 -1  0 -2 -1 -2 -1 -1 -2 -1  1  0 -3 -2  0  100   0   0   0   0   0   0   0   0   0   0   0   0   0   0   0   0   0   0   0  0.40 0.00
 8 S    1 -1  1  0 -1  0  0  0 -1 -3 -3  0 -2 -3 -1  4  2 -3 -2 -2    0   0   0   0   0   0   0   0   0   0   0   0   0   0   0 100   0   0   0   0  0.46 0.00
 9 V    0 -3 -3 -4 -1 -2 -3 -4 -4  3  1 -3  1 -1 -3 -2  0 -3 -1  4    0   0   0   0   0   0   0   0   0   0   0   0   0   0   0   0   0   0   0 100  0.35 0.00
10 S    1 -1  1  0 -1  0  0  0 -1 -3 -3  0 -2 -3 -1  4  2 -3 -2 -2    0   0   0   0   0   0   0   0   0   0   0   0   0   0   0 100   0   0   0   0  0.46 0.00
11 L   -2 -2 -4 -4 -1 -2 -3 -4 -3  2  4 -3  2  0 -3 -3 -1 -2 -1  1    0   0   0   0   0   0   0   0   0   0 100   0   0   0   0   0   0   0   0   0  0.46 0.00
12 I   -1 -3 -4 -4 -1 -3 -4 -4 -4  5  2 -3  1  0 -3 -3 -1 -3 -2  3    0   0   0   0   0   0   0   0   0 100   0   0   0   0   0   0   0   0   0   0  0.39 0.00
13 T    0 -1  0 -1 -1 -1 -1 -2 -2 -1 -1 -1 -1 -2 -1  2  5 -3 -2  0    0   0   0   0   0   0   0   0   0   0   0   0   0   0   0   0 100   0   0   0  0.47 0.00
14 N   -2 -1  5  4 -3  0  1 -1  0 -4 -4  0 -3 -4 -2  0  0 -4 -3 -3    0   0  65  35   0   0   0   0   0   0   0   0   0   0   0   0   0   0   0   0  0.65 0.00
15 S    1 -1  1  0 -1  0  0  0 -1 -3 -3  0 -2 -3 -1  4  2 -3 -2 -2    0   0   0   0   0   0   0   0   0   0   0   0   0   0   0 100   0   0   0   0  0.46 0.00
16 S    0  3  0 -1 -2  0  0 -1 -1 -3 -3  1 -2 -3 -1  3  1 -3 -2 -2    0  35   0   0   0   0   0   0   0   0   0   0   0   0   0  65   0   0   0   0  0.37 0.00
17 K   -1  2  0 -1 -3  1  1 -2 -1 -3 -3  5 -2 -3 -1  0 -1 -3 -2 -3    0   0   0   0   0   0   0   0   0   0   0 100   0   0   0   0   0   0   0   0  0.52 0.00
18 D   -2 -2  1  6 -4  0  2 -1 -1 -4 -4 -1 -3 -4 -2  0 -1 -5 -3 -4    0   0   0 100   0   0   0   0   0   0   0   0   0   0   0   0   0   0   0   0  0.93 0.00
19 A    3 -2 -2 -2 -1 -1 -1 -1 -2  1 -1 -1  0 -2 -1  0  0 -3 -2  2   65   0   0   0   0   0   0   0   0   0   0   0   0   0   0   0   0   0   0  35  0.24 0.00
20 I   -1 -3 -4 -4 -1 -3 -4 -4 -4  5  2 -3  1  0 -3 -3 -1 -3 -2  3    0   0   0   0   0   0   0   0   0 100   0   0   0   0   0   0   0   0   0   0  0.39 0.00
```

**Fig. 3** A sample PSSM file format

### 2.4 *Algorithm*

To build CNN-Meth, we utilized a CNN because it can automatically extract relevant features from raw data, eliminating the need for manual feature engineering. CNNs are well-suited for tasks where spatial hierarchies in data are essential, such as image recognition or sequence-based [18, 40, 41]. The key advantage of CNNs lies in their convolutional layers, which apply filters to input data, enabling the model to capture intricate patterns and relationships between neighboring features [42, 43]. By progressively learning abstract representations, CNNs can automatically identify relevant biological signals, such as evolutionary, structural, and physicochemical properties, without the limitations associated with predefined feature sets [44–46]. This makes them ideal for the lysine methylation prediction problem, where multiple types of complex data interact, and manual feature extraction can be prone to information loss.

The specific architecture of CNN-Meth consists of two convolutional layers, each with 64 filters and a [3] window size, followed by max-pooling layers with a [2] window size. These convolutional and pooling layers extract and condense the relevant features from the input data. Afterward, the model passes through five fully connected layers containing 50 neurons, enabling representation learning. To prevent overfitting, two dropout layers with a dropout rate of 0.2 are included [47–49]. Finally, the output layer consists of a single neuron with a sigmoid activation function, which performs binary classification by predicting whether a lysine site may become methylated.

The model was trained using the Adam optimization algorithm and binary cross-entropy loss function. The evaluation involved five-fold and ten-fold cross-validation and an independent test set to ensure robust performance evaluation. The results of cross-validation are shown in Table 2. Table 3 presents a comparison of CNN-Meth being evaluated on the independent test against methods proposed in previous studies.

**Table 2**
**The results achieved for fivefold and tenfold cross-validation**

| Cross validation | Accuracy | Specificity | Sensitivity | MCC |
|---|---|---|---|---|
| 5-fold | 88.4% | 92.6% | 78.7% | 0.69 |
| 10-fold | 90.1% | 93.0% | 81.7% | 0.74 |

**Table 3**
**Comparing CNN-Meth with previous studies found in the literature on the independent test set**

| Model | Accuracy | Specificity | Sensitivity | MCC |
|---|---|---|---|---|
| PLMLA | 82.6% | 83.5% | 83.0% | 0.66 |
| iLM-2 L | 85.3% | 91.9% | 76.5% | 0.69 |
| MethK (histone) | 81.4% | 80.3% | 85.6% | 0.56 |
| MethK (nonhistone) | 87.2% | 88.7% | 69.1% | 0.44 |
| mLysPTMpred | 83.7% | – | – | – |
| predML-Site | 84.2% | – | – | – |
| **CNN-Meth** | **96.0%** | **96.4%** | **85.1%** | **0.65** |

## 3 Notes

1. CNN-Meth is accessible as an online tool and is publicly available at https://mlbclab.camden.rutgers.edu/research/,
2. The Web utility takes a PSSM matrix file as an input. An invalid PSSM will result in an inaccurate prediction.
3. Using PSSM for protein modeling can be considered similar to protein modeling using PLMs.
4. Promising CNN-Meth prediction performance demonstrates the possible effectiveness of PLMs in enhancing methylation site prediction as a future direction.
5. Users can use the publicly available tool PSI-BLAST to generate a PSSM file from a FASTA file.
6. The entire PSSM file is required to generate secondary structure features. The provided PSSM file must be generated against the entire protein sequence that contains the target lysine.
7. The target lysine in question must be specified by the naming convention of the passed in PSSM file. The file name of the uploaded file must be the row number of the target lysine followed by ".pssm". For example, if the target lysine is the 123rd amino acid in the protein sequence, the file name must be 123.pssm.
8. Only the PSSM file is required by CNN-Meth. Other features required will automatically be generated by referencing the provided PSSM file.

### References

1. Bhat KP, Ümit Kaniskan H, Jin J, Gozani O (2021) Epigenetics and beyond: targeting writers of protein lysine methylation to treat disease. Nat Rev Drug Discov 20(4):265–286
2. Luo M (2018) Chemical and biochemical perspectives of protein lysine methylation. Chem Rev 118(14):6656–6705
3. Wang Y-C, Peterson SE, Loring JF (2014) Protein post-translational modifications and regulation of pluripotency in human stem cells. Cell Res 24(2):143–160
4. Bai B, Vanderwall D, Li Y, Wang X, Poudel S, Wang H, Dey KK, Chen P-C, Yang K, Peng J (2021) Proteomic landscape of Alzheimer's disease: novel insights into pathogenesis and biomarker discovery. Mol Neurodegener 16(1):55
5. Ramazi S, Zahiri J (2021) Post-translational modifications in proteins: resources, tools and prediction methods. Database 2021:baab012
6. Bremang M, Cuomo A, Agresta AM, Stugiewicz M, Spadotto V, Bonaldi T (2013) Mass spectrometry-based identification and characterisation of lysine and arginine methylation in the human proteome. Mol BioSyst 9(9):2231–2247
7. Shi S-P, Qiu J-D, Sun X-Y, Suo S-B, Huang S-Y, Liang R-P (2012) PLMLA: prediction of lysine methylation and lysine acetylation by combining multiple features. Mol BioSyst 8(5):1520–1527
8. Ju Z, Cao J-Z, Gu H (2015) iLM-2L: a two-level predictor for identifying protein lysine methylation sites and their methylation degrees by incorporating K-gap amino acid pairs into Chou's general PseAAC. J Theor Biol 385:50–57
9. Lee T-Y, Chang C-W, Lu C-T, Cheng T-H, Chang T-H (2014) Identification and characterization of lysine-methylated sites on histones

and non-histone proteins. Comput Biol Chem 50:11–18

10. Hasan MAM, Ahmad S (2018) Mlysptmpred: multiple lysine ptm site prediction using combination of svm with resolving data imbalance issue. Nat Sci 10(9):370–384
11. Ahmed S, Rahman A, Hasan MAM, Rahman J, Islam MKB, Ahmad S (2021) predML-site: predicting multiple lysine PTM sites with optimal feature representation and data imbalance minimization. IEEE/ACM Trans Comput Biol Bioinform 19(6):3624–3634
12. Ali SD, Tayara H, Chong KT (2022) Interpretable machine learning identification of arginine methylation sites. Comput Biol Med 147: 105767
13. Spadaro A, Sharma A, Dehzangi I (2024) Predicting lysine methylation sites using a convolutional neural network. Methods 226:127–132
14. Pokharel S, Pratyush P, Heinzinger M, Newman RH, Kc DB (2022) Improving protein succinylation sites prediction using embeddings from protein language model. Sci Rep 12(1):16933
15. Pokharel, S., P. Pratyush, M. Heinzinger, R. Newman, K. Dukka, *LMSuccSite:* improving protein succinylation sites prediction using embeddings from protein language model. 2022
16. Palacios AV, Acharya P, Peidl AS, Beck MR, Blanco E, Mishra A, Bawa-Khalfe T, Pakhrin SC (2024) SumoPred-PLM: human SUMOylation and SUMO2/3 sites prediction using pre-trained protein language model. NAR Genomics Bioinf 6(1):lqae011
17. Liu Y, Liu Y, Wang G-A, Cheng Y, Bi S, Zhu X (2022) BERT-Kgly: a bidirectional encoder representations from transformers (BERT)-based model for predicting lysine glycation site for Homo sapiens. Front Bioinf 2:834153
18. Chandra A, Sharma A, Dehzangi I, Tsunoda T, Sattar A (2023) Predicting Phosphoglycerylation with transformer features and deep learning. In: 2023 IEEE Asia-Pacific conference on computer science and data engineering (CSDE). IEEE
19. Soylu NN, Sefer E (2024) DeepPTM: protein post-translational modification prediction from protein sequences by combining deep protein language model with vision transformers. Curr Bioinforma 19(9):810–824
20. Ahmed, F., A. Sharma, S. Shatabda, and I. Dehzangi, DeepPhoPred: accurate deep learning model to predict microbial phosphorylation. Proteins: structure, function, and bioinformatics, 2024
21. Meng L, Chen X, Cheng K, Chen N, Zheng Z, Wang F, Sun H, Wong K-C (2024) TransPTM: a transformer-based model for non-histone acetylation site prediction. Brief Bioinform 25(3):bbae219
22. Xu H, Zhou J, Lin S, Deng W, Zhang Y, Xue Y (2017) PLMD: an updated data resource of protein lysine modifications. J Genet Genomics 44(5):243–250
23. Liu Z, Cao J, Gao X, Zhou Y, Wen L, Yang X, Yao X, Ren J, Xue Y (2011) CPLA 1.0: an integrated database of protein lysine acetylation. Nucleic Acids Res 39(suppl_1):D1029–D1034
24. Zhang W, Tan X, Lin S, Gou Y, Han C, Zhang C, Ning W, Wang C, Xue Y (2022) CPLM 4.0: an updated database with rich annotations for protein lysine modifications. Nucleic Acids Res 50(D1):D451–D459
25. Huang Y, Niu B, Gao Y, Fu L, Li W (2010) CD-HIT suite: a web server for clustering and comparing biological sequences. Bioinformatics 26(5):680–682
26. Dipta SR, Taherzadeh G, Ahmad MW, Arafat ME, Shatabda S, Dehzangi A (2020) SEMal: accurate protein malonylation site predictor using structural and evolutionary information. Comput Biol Med 125:104022
27. Beckmann M, Ebecken NF, Pires de Lima BS (2015) A KNN undersampling approach for data balancing. J Intell Learn Syst Appl 7(04): 104–116
28. Ahmad MW, Arafat ME, Taherzadeh G, Sharma A, Dipta SR, Dehzangi A, Shatabda S (2020) Mal-light: enhancing lysine malonylation sites prediction problem using evolutionary-based features. IEEE Access 8: 77888–77902
29. Bhagwat M, Aravind L (2008) Psi-blast tutorial, Comparative genomics, pp 177–186
30. Heffernan R, Paliwal K, Lyons J, Dehzangi A, Sharma A, Wang J, Sattar A, Yang Y, Zhou Y (2015) Improving prediction of secondary structure, local backbone angles and solvent accessible surface area of proteins by iterative deep learning. Sci Rep 5(1):11476
31. Yang Y, Heffernan R, Paliwal K, Lyons J, Dehzangi A, Sharma A, Wang J, Sattar A, Zhou Y (2017) SPIDER2: a package to predict secondary structure, accessible surface area, and main-chain torsional angles by deep neural networks, Prediction of protein secondary structure, pp 55–63
32. Kang HS, Kurochkina NA, Lee B (1993) Estimation and use of protein backbone angle probabilities. J Mol Biol 229(2):448–460

33. Wood MJ, Hirst JD (2005) Protein secondary structure prediction with dihedral angles. Proteins: Structure, Function, and Bioinformatics 59(3):476–481
34. Dor O, Zhou Y (2007) Real-SPINE: an integrated system of neural networks for real-value prediction of protein structural properties. Proteins: Structure, Function, and Bioinformatics 68(1):76–81
35. Xue B, Dor O, Faraggi E, Zhou Y (2008) Real-value prediction of backbone torsion angles. Proteins: Structure, Function, and Bioinformatics 72(1):427–433
36. Rost B (2001) Protein secondary structure prediction continues to rise. J Struct Biol 134(2–3):204–218
37. Kawashima S, Kanehisa M (2000) AAindex: amino acid index database. Nucleic Acids Res 28(1):374–374
38. Kawashima S, Ogata H, Kanehisa M (1999) AAindex: amino acid index database. Nucleic Acids Res 27(1):368–369
39. Tomii K, Kanehisa M (1996) Analysis of amino acid indices and mutation matrices for sequence comparison and structure prediction of proteins. Protein Eng Des Sel 9(1):27–36
40. Bilal A, Jourabloo A, Ye M, Liu X, Ren L (2017) Do convolutional neural networks learn class hierarchy? IEEE Trans Vis Comput Graph 24(1):152–162
41. Anwar SM, Majid M, Qayyum A, Awais M, Alnowami M, Khan MK (2018) Medical image analysis using convolutional neural networks: a review. J Med Syst 42:1–13
42. Li Z, Liu F, Yang W, Peng S, Zhou J (2021) A survey of convolutional neural networks: analysis, applications, and prospects. IEEE Trans Neural Networks Learn Syst 33(12): 6999–7019
43. O'shea K, Nash R (2015) An introduction to convolutional neural networks. arXiv preprint arXiv:1511.08458
44. Min S, Lee B, Yoon S (2017) Deep learning in bioinformatics. Brief Bioinform 18(5): 851–869
45. Chandra A, Sharma A, Dehzangi I, Tsunoda T, Sattar A (2023) PepCNN deep learning tool for predicting peptide binding residues in proteins using sequence, structural, and language model features. Sci Rep 13(1):20882
46. Manavi F, Sharma A, Sharma R, Tsunoda T, Shatabda S, Dehzangi I (2023) CNN-Pred: prediction of single-stranded and double-stranded DNA-binding protein using convolutional neural networks. Gene 853:147045
47. Li H, Li J, Guan X, Liang B, Lai Y, Luo X (2019) Research on overfitting of deep learning. In: 2019 15th international conference on computational intelligence and security (CIS). IEEE
48. Rice L, Wong E, Kolter Z (2020) Overfitting in adversarially robust deep learning. In: International conference on machine learning. PMLR
49. Salman S, Liu X (2019) Overfitting mechanism and avoidance in deep neural networks. arXiv preprint arXiv:1901.06566

# Chapter 12

# Predicting the Pathogenicity of Human Protein Variants: Not Only a Matter of Residue Labeling

**Matteo Manfredi, Gabriele Vazzana, Giulia Babbi, Elisa Bertolini, Castrense Savojardo, Pier Luigi Martelli, and Rita Casadio**

## Abstract

The pathogenicity of human variants is an important annotation feature that may help in understanding, at a molecular level, the propensity for a human being to develop a certain disease or pathology. Recently, protein sequence embedding associated with machine and/or deep learning has been proven useful in improving results in this area. Different aspects of pathogenic variants can help in understanding the molecular mechanisms of the disease at a molecular level. These include solvent accessibility in the folded gene, the effect on the protein stability, and eventually the perturbation on interaction networks important for biological processes. Here, we describe how, once a variant is predicted "pathogenic", other important structural and functional properties can be derived computationally at the same website (https://bioinformaticsweeties.biocomp.unibo.it/), including the protein structure, if not available. All the properties can help to understand variant effects within the complex context of the cell environment.

**Key words** Single residue variations, Pathogenicity, Machine learning, Deep learning, Protein embedding, Suite of tools, Solvent accessibility, Protein stability, Structure prediction, Human proteome

## 1 Introduction

The introduction of protein Language Models (pLMs), somehow inspired by natural Language Models [1, 2], has been adopted in the development of several tools as a unique and powerful representation of protein sequences along with machine learning and more recently deep learning [3] methods. In the '90, it was noticed that inference of protein structural properties, such as secondary structure from sequence, improved if the protein was represented by a sequence profile, computed after sequence alignment with similar proteins in the family, sharing structural templates [4–6]. Nowadays, the introduction of embeddings to represent a protein sequence stands upon the same notion, with the difference that the background knowledge may include all the known protein

Dukka B. KC (ed.), *Large Language Models (LLMs) in Protein Bioinformatics*, Methods in Molecular Biology, vol. 2941, https://doi.org/10.1007/978-1-0716-4623-6_12,

sequences that have been derived from next-generation sequencing experiments (several billion for the last ESMFold language) [7–11]. Each residue in the protein sequence can carry along the information derived from the background database and this has proven useful to improve the performance of different predictors, including the computation of protein functional and structural features and of protein–protein interactions along with the development of deep learning schemes for training the algorithms on given selected properties [9, 12–16]. More recently, embedding procedures have been adopted by ESMFold for structure prediction from protein sequences, and for distantly remote homology search [10, 17, 18]. In the following, we will guide the end user to solve a specific problem where, after detecting the presence of pathogenic variants in the protein sequence, he/she may follow a given workflow to endow its protein with other relevant physico-chemical features. We will use as a test case a gene, AGPS (https://www.uniprot.org/uniprotkb/O00116/entry), a fairly well-characterized human enzyme, whose variants are associated with a form of rhizomelic chondrodysplasia punctata (Rhizomelic chondrodysplasia punctata 3 (RCDP3)). Our analysis with state-of-the-art predictors characterizes the physico-chemical properties of the pathogenic variants, and frames the protein within Reactome Pathways (https://curator.reactome.org/), from where all the possible protein–ligand and protein–protein interactions can be derived. This can help to verify putative hypotheses on the molecular mechanisms of the genotype–phenotype relationships.

## 2 Predictors Relevant to Pathogenicity Aspects at Bioinformatic Sweeties

Bioinformatic Sweeties (Fig. 1a) is a suite of in-house developed tools based on machine and/or deep learning and derived databases for characterizing proteins and their variants. Here, the end user can start an analysis without any installation, login, or code skills [19]. Among the tools and databases, we will focus on (1) E-SNPs&GO for predicting the pathogenicity of protein variants [20]; (2) E-pRSA [16], for predicting residue Relative Solvent Accessibility, including also (3) ISPRED-SEQ [14] for predicting protein–protein interactions; (4) INPS for predicting protein stability upon variation starting from protein sequence [21]; (5) Alpha&ESMhFolds, a database for the comparison of structural models predicted with ESMFold and AlphaFold2 for 42,942 human reference proteins, which allows to retrieve the precomputed structure of the protein and map the protein variants for further studies [22]; and (6) DAR, a database mapping disease-associated enzyme into Reactome pathways to derive possible relevant interactions among biological processes [23] (Fig. 1b). E-SNPs&GO, E-pRSA, ISPRED-SEQ, and ESMFold adopt

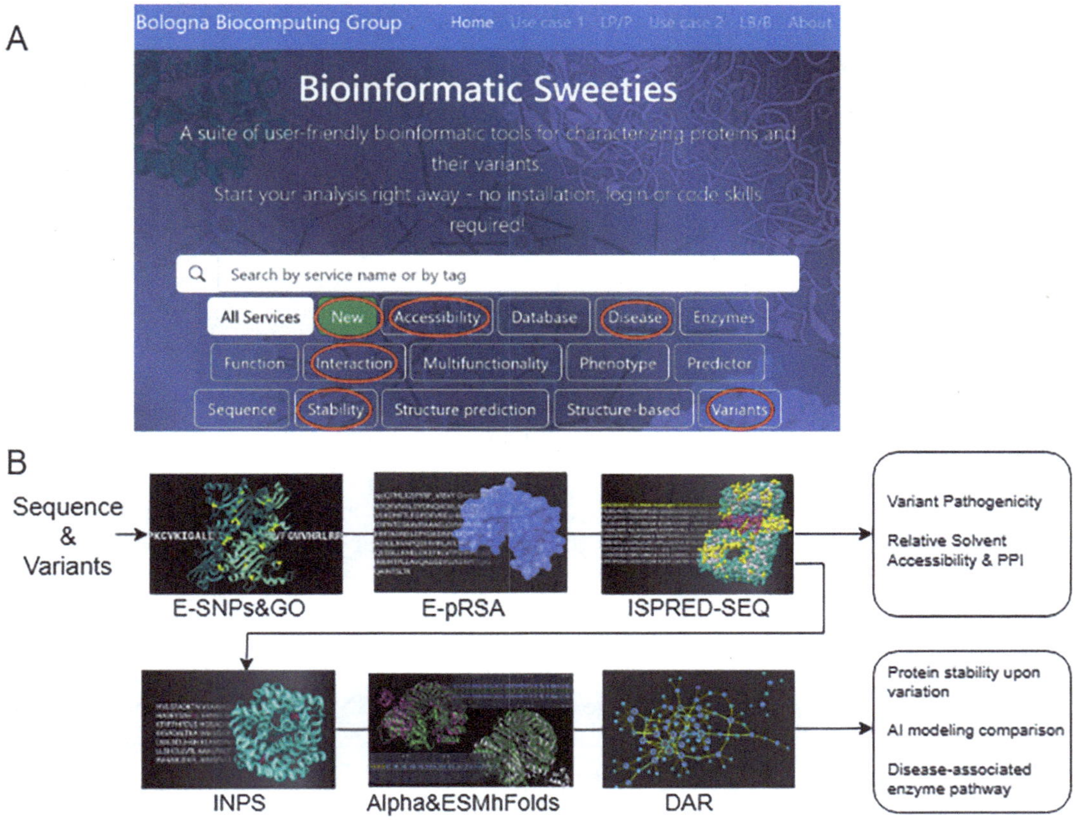


**Fig. 1** Bioinformatic Sweeties. The suite hosts the tools and databases developed by the Biocomputing Group for different purposes (Fig. 1a). Red ovals encircle tags to press in order to have direct access to the tools and databases adopted to solve the specific problem at hand (Fig. 1a)

embedding procedures for protein sequence representations. All the details relevant to understanding tools and database developments can be easily retrieved directly on the suite, along with the reference papers, and help pages for their usage (https://bioinformaticsweeties.biocomp.unibo.it/).

## 3 Methods

Here, as an example, we focus on the AGPS gene, whose product Alkyldihydroxyacetonephosphate synthase peroxisomal, is an enzyme which is known to catalyze "the exchange of the acyl chain in acyl-dihydroxyacetonephosphate (acyl-DHAP) for a long chain fatty alcohol, yielding the first ether-linked intermediate, i.e. alkyl-dihydroxyacetonephosphate (alkyl-DHAP), in the pathway of ether lipid biosynthesis" (EC:2.5.1.26, short name Alkyl-DHAP synthase, see for details and supporting literature the protein Swiss-Prot/UniProt file O00116, https://www.uniprot.org/

uniprotkb/O00116/entry). According to OMIM, the gene is located in chromosome 2 (2q31.2) and its pathogenic mutations, both homozygous and compound heterozygous, promote an alkyl-DHAP synthase deficiency in the protein, and are associated with a form of rhizomelic chondrodysplasia punctata (Rhizomelic chondrodysplasia punctata 3 (RCDP3)), a peroxisomal disorder characterized by disproportionately short stature primarily affecting the proximal parts of the extremities (for details and references see the OMIM site at https://www.omim.org/entry/600121)). Protein pathogenic variations are derived from the UniProt file, filtering out those of uncertain annotation. Similarly, benign variations were retrieved for the sake of comparing predictions. The genetics, biochemistry, and physiology of the protein have been extensively characterized. Still, the protein structure is lacking. A reason to select this protein is that while applying our predictors to understand the pathogenicity of the protein at a molecular level, we will be able to score the efficiency of our computations.

### 3.1 Predicting the Pathogenicity of SRVs

Following our flowchart (Fig. 1), we will start by asking whether all the pathogenic variations known to be associated with the pathology are correctly predicted. In Bioinformatic Sweeties [19] we select under the tag "variant" E-SNPs&GO to predict whether the variants are or are not pathogenic. To this aim, we also add some neutral natural variations not disease-related. All the variants are downloaded from the O00116 UniProt entry, where the variant origin is also detailed. E-SNPs&GO [20] is a fast and accurate method that, given an input protein sequence and a single residue variation or more, can predict whether the variation/s is/are related to diseases or not. Protein representation relies on an encoding scheme based on protein language models (ProtT5 [9], ESM-1v [7]), and Anc2Vec [24] for Gene Ontology (GO) terms. The predictor implements a Support Vector Machine [25] for classifying the variant as pathogenic. A reliability index is associated with each prediction, scoring the prediction confidence [20]. In Fig. 2, the results of the predictions match what we expected, supporting the confidence of the predictor. One should consider however that when experimental validation is lacking, only the reliability score can rate the quality of the prediction, which could still be wrong.

### 3.2 Predicting the Solvent Exposure of Residues

An interesting feature of the variations can be their relative solvent exposure [26]. This may help in understanding whether the mutated residues are or are not exposed to the polar solvent and eventually are or are not part of possible interaction sites. In Bioinformatic Sweeties [19], we can now select the tag "Accessibility" and adopt E-pRSA [16], a predictor based on embedding and deep learning to predict whether a residue is or is not at the protein surface. The same website implements ISPRED-SEQ [14], based

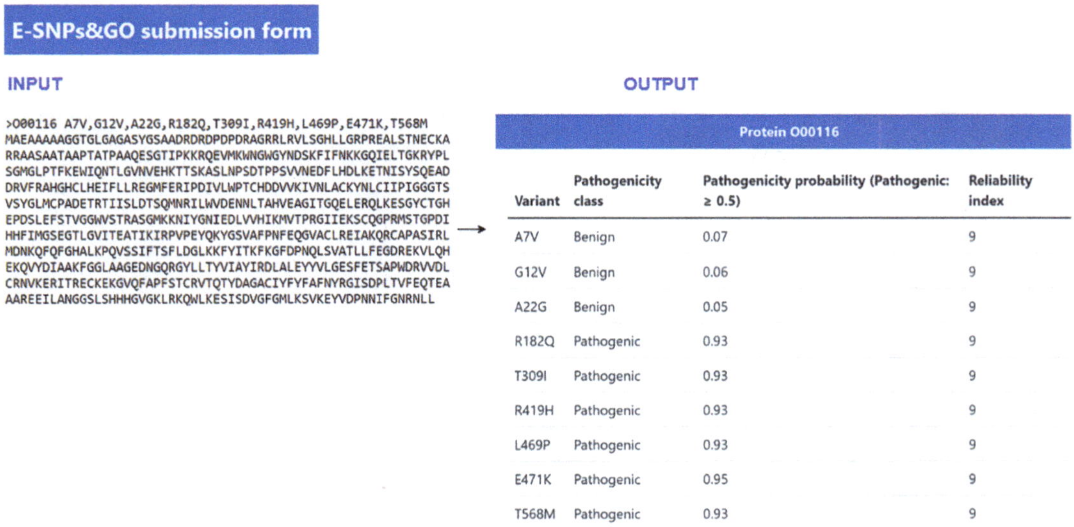


| Variant | Pathogenicity class | Pathogenicity probability (Pathogenic: ≥ 0.5) | Reliability index |
|---|---|---|---|
| A7V | Benign | 0.07 | 9 |
| G12V | Benign | 0.06 | 9 |
| A22G | Benign | 0.05 | 9 |
| R182Q | Pathogenic | 0.93 | 9 |
| T309I | Pathogenic | 0.93 | 9 |
| R419H | Pathogenic | 0.93 | 9 |
| L469P | Pathogenic | 0.93 | 9 |
| E471K | Pathogenic | 0.95 | 9 |
| T568M | Pathogenic | 0.93 | 9 |

**Fig. 2** E-SNPs&GO: Prediction of protein variant pathogenicity. This is a predictor based on the embeddings of protein sequences and Gene Ontology terms (GO, https:/www.geneontology.org/) to compute with a machine learning method (support vector machine) the pathogenicity or not of protein variants [20]. The website allows the simultaneous prediction of different variants, and alongside the predicted pathogenicity or not, it gives the associated confidence (measured as the Reliability index of the prediction). Details on the method and its implementation can be found at the corresponding website, https:/bioinformaticsweeties.biocomp.unibo.it/pmanager/resource/E-SNPs&GO

again on protein embedding and deep learning to predict the likelihood that a residue is or is not an interaction site. Figure 3 shows the sequence feature view of the prediction. According to the predictor, 58% of the protein residues (the protein sequence comprises 658 residues) are exposed with 77 interaction sites, confirming the high number of interacting proteins (https://www.uniprot.org/uniprotkb/O00116/entry#interaction). The region including R419H is highlighted indicating that the residue is buried and it is not a protein–protein interaction site. None of the pathogenic variations are in exposed regions and in interaction sites (Table 1).

### 3.3 Predicting Effects on the Protein Stability

A related property to a pathogenic variation in a protein region is whether the substitution may or not affect the protein stability. In Bioinformatic Sweeties we have available INPS [21], which is a classical predictor based on machine learning and sequence profile as input to the Support Vector Machine (SVM). It will be soon complemented with another predictor, not yet released, based on embedding techniques and deep learning [28]. INPS computes the change in free energy (DDG) associated with variants in the protein (Fig. 4). According to INPS, only three out of the six pathogenic variants affect the protein stability.

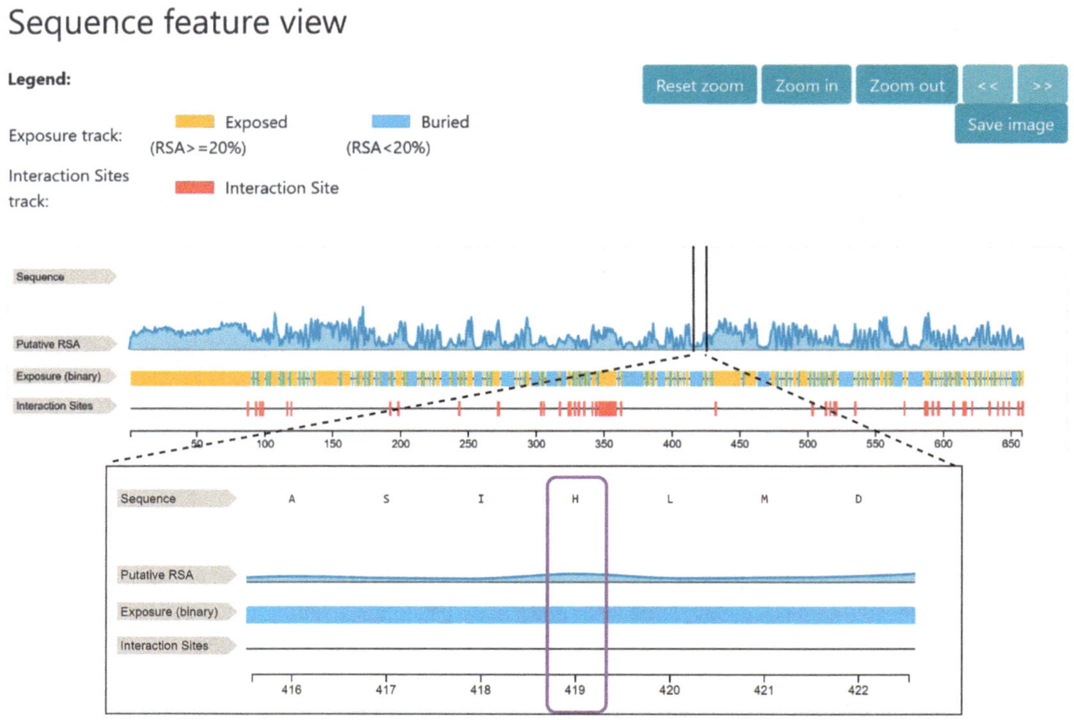


**Fig. 3** Output of E-pRSA and ISPRED-SEQ. Embedding schemes of protein sequences in E-pRSA [16] improve the prediction of relative solvent accessibility. Furthermore, the same procedure improves the prediction of interaction sites [14]. Both predictors are running on the same website. The output allows the association of exposed residues with interaction sites. Orange and blue colors identify exposed and buried regions along the protein sequence. Red color marks interaction sites. In the figure, the region including the pathogenic variant R419H is highlighted indicating that the residue is buried, and it is not a protein–protein interaction site. For details see: https:/bioinformaticsweeties.biocomp.unibo.it/pmanager/resource/E-pRSA

### 3.4 Comparing Results of Different Tools

In Table 1, we sum up the results of our prediction. Considering the six known pathogenic variations associated with RCDP3, we found that they are correctly predicted as pathogenic, they are buried, and not involved in interaction sites. We previously defined the threshold value between neutral, destabilizing, and stabilizing variations after comparing it with other state-of-the-art predictors for the same task [27].

For all variants, the associated disease is Rhizomelic chondrodysplasia punctata 3 (RCDP3). Specifically, R182Q and E471K are associated with a reduction of the protein level and R419H to a loss of enzyme activity (see the UniProt file O00116, https://www.uniprot.org/uniprotkb/O00116/entry).

### 3.5 Mapping SRVs on Predicted 3D Structures

We used our database of models, Alpha&ESMhFolds [22], (in Bioinformatic Sweeties under the tag "New") to derive the pairwise models of our protein. Both methods agree on the putative structure (TM-score is 0.88 [29]). With the model/s which can be

**Table 1**
**Predicting pathogenic natural variants in ADAS_HUMAN (O00116), associated with the disease Rhizomelic chondrodysplasia punctata type 3 (RCDP3) and their properties**

| Variant | E-SNPs&GO[a] | E-pRSA[b] | ISPRED-SEQ[c] | INPS[d] |
|---|---|---|---|---|
| A7V | Benign (0.07) | E (0.36) | N | 0.84 |
| G12D | Benign (0.09) | E (0.4) | N | 0.14 |
| A22G | Benign (0.05) | E (0.44) | N | −0.39 |
| R182Q | Pathogenic (0.93) | B (0.08) | N | **−1.32** |
| T309I[e] | Pathogenic (0.93) | B (0.08) | N | −0.98 |
| R419H[e] | Pathogenic (0.93) | B (0.14) | N | **−1.05** |
| L469P | Pathogenic (0.93) | B (0.08) | N | **−3.17** |
| E471K | Pathogenic (0.95) | B (0.1) | N | −0.68 |
| T568M | Pathogenic (0.93) | B (0.07) | N | −0.64 |

[a]E-SNPs&GO classification and associated probability (pathogenicity probability >0.5)
[b]E-pRSA: E = Exposed, B=Buried; among brackets is the predicted relative accessible surface area value
[c]ISPRED-SEQ computes protein–protein interaction (PPI). N = No PPI, IS = interaction site
[d]INPS: DDG value associated with the variant in kcal/mol. In bold, DDG values associated with protein destabilization [27]
[e]Binding site and site important for enzyme activity, respectively

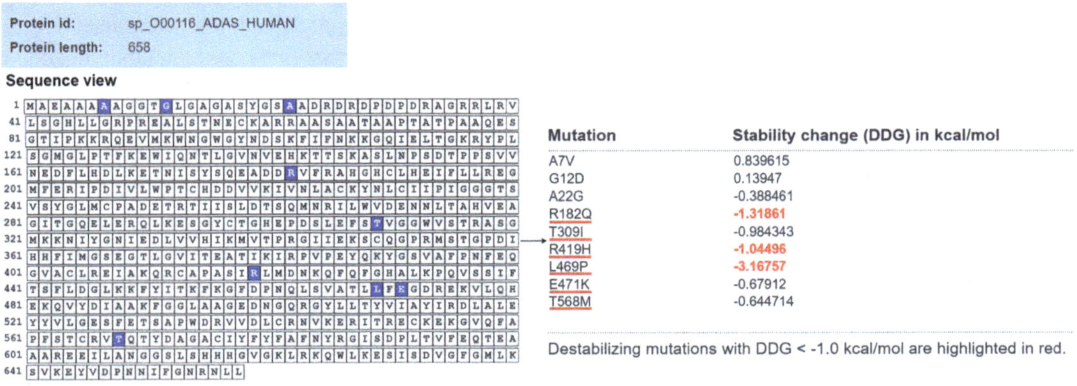


**Fig. 4** INPS output. INPS [21] is a predictor for annotating the effect of non-synonymous variations on the protein stability from its sequence. It is based on SVM regression, and it is trained to predict the thermodynamic free energy change upon single-point variations in protein sequences. It appears that three of the six pathogenic mutations (underlined in red) are affecting protein stability with a |DDG value| > 1 kcal/mol [27]

downloaded from the website, we can then map the pathogenic variants and realize to what extent their location in the protein structure is consistent with the predicted features of Table 1 (Fig. 5, [22]).

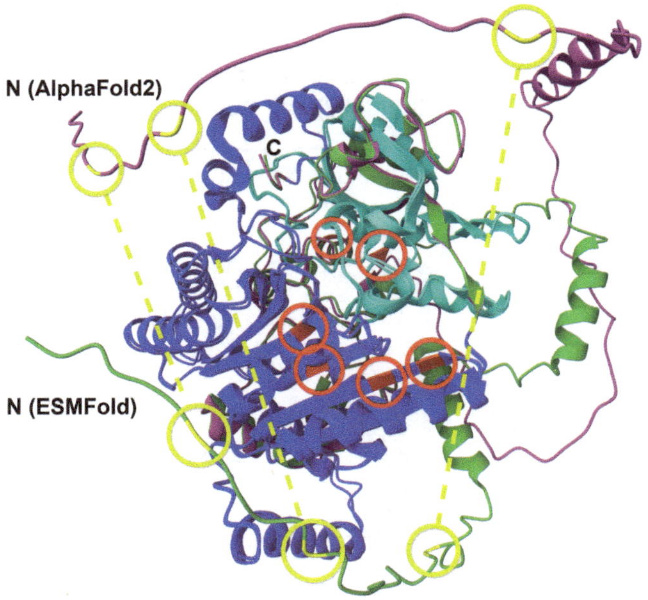


**Fig. 5** Visualization of the three-dimensional structural models of ADAS_HUMAN (UniProt ID: O00116) predicted with ESMFold (in green) and AlphaFold2 (in purple). The pairwise models are downloaded from Alpha&ESMhFolds (https:/alpha-esmhfolds.biocomp.unibo.it/viewer/O00116). The two models have a good overlap (TM-score 0.88). The light blue region indicates the FAD binding domain of the protein (Pfam PF01565, covering residues 206–345) and the dark blue region indicates the FAD-linked oxidase, C-terminal domain (Pfam PF02913, covering residues 385–654). The Pfam domains overlap in the two models, which have however no overlapping regions at the N-terminus. The six pathogenic variations characterized in this case study are marked in red: one is outside the two domains (R182Q), one (annotated as FAD binding domain) is in the light blue domain (T309I) and 4 are located in the same beta-sheet of the dark blue domain (R419H, L469P, E471K, and T568M). The three benign variations, to be found at N terminal of the protein, are located in the diverging portion of the two models, corresponding to the Transit peptide (1–58). They are marked in yellow in the figure and a dashed line links the corresponding positions in the two models. E-pRSA predicts the six pathogenic variations as buried; by adopting either model, this can be confirmed by running the DSSP program [31, 32] at https:/pdb-redo.eu/dssp. The pathogenic variants have RSA values ranging from 1% to 5%. The benign variants (predicted by E-pRSA Exposed) are solvent-accessible (RSA values ranging from 86% to 99%). For molecular visualizations, we adopt the PDBe Mol* viewer JavaScript plugin (https:/github.com/molstar/pdbe-molstar, version 3.2.0)

### 3.6 Analyzing Biological Processes

Finally, we can try to frame the protein into the Reactome, which includes a hierarchical graphical organization of known biological processes and pathways containing detailed information on components and their relations (https://curator.reactome.org/).

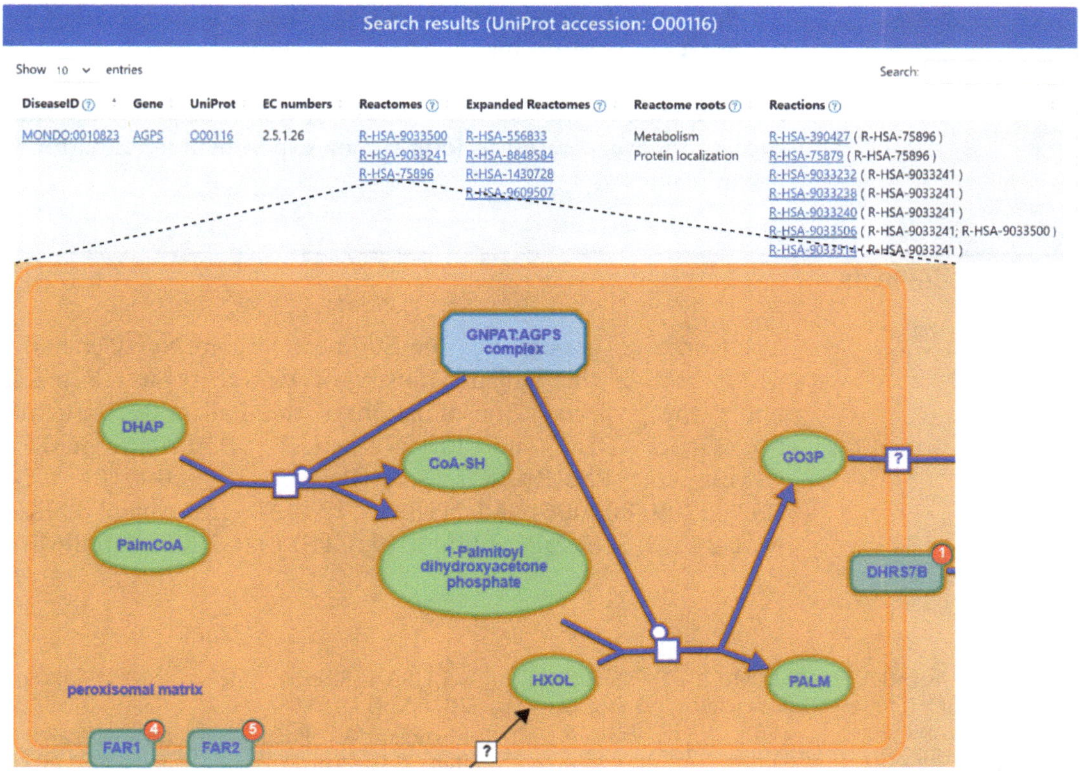


**Fig. 6** DAR output. Our *AGPS* gene is involved in three Reactome pathways: TYSND1 cleaves peroxisomal proteins (R-HSA-9033500), Peroxisomal protein import (R-HSA-9033241), and Plasmalogen biosynthesis (R-HSA-75896). This last one is highlighted in the insert. Here AGSP is in complex with GNPAT, Dihydroxyacetone phosphate acyltransferase, an enzyme that catalysis the first step in the biosynthesis of plasmalogens, and whose variants are associated with Rhizomelic chondrodysplasia punctata type 2

Reactome describes human biological processes, and by mapping the disease-associated enzyme, a Reactome-disease association can be established [30]. This procedure allows the integration of the protein in a more complex scenario of interactions with other proteins whose functioning can also affect the overall network of biological processes. By tagging "database" in Bioinformatic Sweeties, we find DAR, which provides the mapping of human disease-associated enzymes into Reactome and allows the characterization of disease groups and their interactions [23] (Fig. 6).

## 4 Conclusions

We implemented a suite of different state-of-the-art predictors based on protein embedding and shallow and/or deep learning methods. Taking advantage of the suite we derived different physicochemical properties related to pathogenic variants of a selected protein, and how the protein can be characterized in the context of the Reactome pathways for better exploitation of the genotype-to-phenotype relationship. The predictors, starting from E-SNPs&GO for predicting whether a variant is pathogenic or not, are state-of-the-art. This, however, does not guarantee their success in every case, as long as an experimental validation is not available.

## Acknowledgments

The work was supported by the European Union- NextGenerationEU through the Italian Ministry of University and Research under the projects "Consolidation of the Italian Infrastructure for Omics Data and Bioinformatics" (ElixirxNextGenIT)" (Investment PNRRM4C2-I3.1, Project IR_0000010, CUP B53C22001800006) and "HEAL ITALIA" (Investment PNRR-M4C2-I1.3, ProjectPE_00000019, CUP J33C22002920006).

### References

1. Ofer D, Brandes N, Linial M (2021) The language of proteins: NLP, machine learning & protein sequences. CSBJ 19:1750–1758. https://doi.org/10.1016/j.csbj.2021.03.022
2. Bepler T, Berger B (2021) Learning the protein language: evolution, structure, and function. Cell Syst 16:654–669. https://doi.org/10.1016/j.cels.2021.05.017
3. Kandathil SM, Lau AM, Jones DT (2023) Machine learning methods for predicting protein structure from single sequences. Curr Opin Struct Biol 81:102627. https://doi.org/10.1016/j.sbi.2023.102627
4. Rost B, Sander C (1993) Improved prediction of protein secondary structure by use of sequence profiles and neural networks. Proc Natl Acad Sci USA 90:7558–7562. https://doi.org/10.1073/pnas.90.16.7558
5. Rost B, Sander C (1994) Conservation and prediction of solvent accessibility in protein families. Proteins 20:216–226. https://doi.org/10.1002/prot.340200303
6. Lesk AM (2016) Introduction to protein science, 3rd edn. Oxford University Press, Oxford
7. Meier J, Rao R, Verkuil R, Lui J, Sercu T, Rives A (2021) Language models enable zero-shot prediction of the effects of mutations on protein function. Adv Neural Inf Process Syst 34: 29287–29303
8. Rives A, Meier J, Sercu T, Goyal S, Lin Z, Liu J et al (2021) Biological structure and function emerge from scaling unsupervised learning to 250 million protein sequences. Proc Natl Acad Sci USA 118. https://doi.org/10.1073/pnas.2016239118
9. Elnaggar A, Heinzinger M, Dallago C, Rehawi G, Wang Y, Jones L et al (2022) ProtTrans: toward understanding the language of life through self-supervised learning. IEEE Trans Pattern Anal Mach Intell 44:7112–7127. https://doi.org/10.1109/TPAMI.2021.3095381
10. Lin Z, Akin H, Rao R, Hie B, Zhu Z, Lu W et al (2023) Evolutionary-scale prediction of atomic-level protein structure with a language model. Science 379:1123–1130. https://doi.org/10.1126/science.ade2574
11. Hayes T, Rao R, Akin H, Sofroniew NJ, Oktay D, Lin Z et al (2024) Simulating 500 million years of evolution with a language model. biorxiv https://doi.org/10.1101/2024.07.01.600583
12. Hie BL, Yang KK, Kim PS (2022) Evolutionary velocity with protein language models predicts evolutionary dynamics of diverse proteins. Cell Syst 13:274–285. https://doi.org/10.1016/j.cels.2022.03.004
13. Jha K, Saha S, Singh H (2022) Prediction of protein–protein interaction using graph neural networks. Sci Rep 12:8360. https://doi.org/10.1038/s41598-022-12201-9
14. Manfredi M, Savojardo C, Martelli PL, Casadio R (2023) ISPRED-SEQ: deep neural networks and embeddings for predicting interaction sites in protein sequences. J Mol Biol 435:167963.

https://doi.org/10.1016/j.jmb.2023.167963

15. Hamamsy T, Morton JT, Blackwell R, Berenberg D, Carriero N, Gligorijevic V et al (2023) Protein remote homology detection and structural alignment using deep learning. Nat Biotechnol 42:975–985. https://doi.org/10.1038/s41587-023-01917-2
16. Manfredi M, Savojardo C, Martelli PL, Casadio R (2024) E-pRSA: embeddings improve the prediction of residue relative solvent accessibility in protein sequence. J Mol Biol 436:168494. https://doi.org/10.1016/j.jmb.2024.168494
17. Pantolini L, Studer G, Pereira J, Durairaj J, Tauriello G, Schwede T (2024) Embedding-based alignment: combining protein language models with dynamic programming alignment to detect structural similarities in the twilight-zone. Bioinformatics 40. https://doi.org/10.1093/bioinformatics/btad786
18. Vazzana G, Savojardo C, Martelli PL, Casadio R (2024) Testing the capability of embedding-based alignments on the GST superfamily classification: the role of protein length. Molecules 29:4616. https://doi.org/10.3390/molecules29194616
19. Babbi G, Manfredi M, Bertolini E, Savojardo C, Martelli PL, Casadio R (2024) Bioinformatic sweeties: a unified portal for characterizing human proteins and their variants. Top Ital Sci J 1. https://doi.org/10.62684/HDYO1899
20. Manfredi M, Savojardo C, Martelli PL, Casadio R (2022) E-SNPs&GO: embedding of protein sequence and function improves the annotation of human pathogenic variants. Bioinformatics 38:5168–5174. https://doi.org/10.1093/bioinformatics/btac678
21. Fariselli P, Martelli PL, Savojardo C, Casadio R (2015) INPS: predicting the impact of non-synonymous variations on protein stability from sequence. Bioinformatics 31:2816–2821. https://doi.org/10.1093/bioinformatics/btv291
22. Manfredi M, Savojardo C, Iardukhin G, Salomoni D, Costantini A, Martelli PL, Casadio R (2024) Alpha&ESMhFolds: a web server for comparing AlphaFold2 and ESMFold models of the human reference proteome. J Mol Biol 436:168593. https://doi.org/10.1016/j.jmb.2024.168593
23. Savojardo C, Baldazzi D, Babbi G, Martelli PL, Casadio R (2022) Mapping human disease-associated enzymes into reactome allows characterization of disease groups and their interactions. Sci Rep 12:17963. https://doi.org/10.1038/s41598-022-22818-5
24. Edera AA, Milone DH, Stegmayer G (2022) Anc2vec: embedding gene ontology terms by preserving ancestors relationships. Brief Bioinform 23. https://doi.org/10.1093/bib/bbac003
25. Bishop CM (2006) Pattern recognition and machine learning. Springer, New York
26. Shrake A, Rupley JA (1973) Environment and exposure to solvent of protein atoms. Lysozyme and insulin. J Mol Biol 79:351–371. https://doi.org/10.1016/0022-2836(73)90011-9
27. Casadio R, Savojardo C, Fariselli P, Capriotti E, Martelli PL (2022) Turning failures into applications: the problem of protein ΔΔG prediction. In: Data mining techniques for the life sciences. Springer, pp 169–185
28. Savojardo C, Manfredi M, Martelli PL, Casadio R (2024) DDGemb: predicting protein stability change upon single- and multi-point variations with embeddings and deep learning. biorxiv https://doi.org/10.1101/2024.09.05.611455
29. Zhang Y, Skolnick J (2004) Scoring function for automated assessment of protein structure template quality. Proteins 57:702–710. https://doi.org/10.1002/prot.20264
30. Milacic M, Beavers D, Conley P, Gong C, Gillespie M, Griss J et al (2024) The Reactome pathway knowledgebase 2024. Nucl Acids Res 52:D672–D678. https://doi.org/10.1093/nar/gkad1025
31. Kabsch W, Sander C (1983) Dictionary of protein secondary structure: pattern recognition of hydrogen-bonded and geometrical features. Biopolymers 22:2577–2637. https://doi.org/10.1002/bip.360221211
32. Joosten RP, te Beek TAH, Krieger E, Hekkelman ML, Hooft RWW, Schneider R et al (2010) A series of PDB related databases for everyday needs. Nucleic Acids Res 39:D411–D419. https://doi.org/10.1093/nar/gkq1105

# Chapter 13

# A Survey of Biological Function Prediction Methods with Focus on Natural Language Processing (NLP) and Large Language Models (LLM)

**Dana Mary Varghese, T. Athulya, Vikash K. Mohani, and Shandar Ahmad**


## Abstract

Protein function prediction from sequence, structure, gene expression profiles, and published literature are needed to understand all biological processes. Natural language processing of biological text and large language model (LLM)-based encoding of sequence and structure opens powerful paths to rapid function annotation and novel training models. In this survey, we take a look at the available models for function prediction, especially the NLP- and LLM-based models. The survey highlights the major advances made and the ground that still needs to be covered to automate the process of function prediction from two major sources namely protein sequences and published research documents.


**Key words** Function prediction, Natural language processing, Large language models, Protein function, Gene ontologies

## 1 Introduction

Protein function is most confidently determined through experimental methods like inhibition of gene expression [1], gene knockout [2], expression of fusion proteins [3], chromatography [4], protein-protein interaction [3, 5], biochemical and enzymatic assay [6], spectroscopy [7], and crystallography [5]. However, direct determination or even a systematic curation of function from literature for each protein is costly and perhaps not needed in view of the ability of computational methods to transfer annotations from one protein to another with various degrees of accuracy. Many successful methods of protein function prediction have been developed, including those based on homology relationship [8, 9], gene expression or co-expression-based guilt-by-association law [10, 11], protein-protein interactions or structural, computed or predicted features [8, 12–18].

Dukka B. KC (ed.), *Large Language Models (LLMs) in Protein Bioinformatics*, Methods in Molecular Biology, vol. 2941, https://doi.org/10.1007/978-1-0716-4623-6_13,

Problem of function prediction can be defined in a focused manner of predicting the ontology terms, i.e., predicting a specific molecular function, cellular composition or biological process. The least accurate models among them are the ones trying to predict the cellular component or the localization of the proteins [19–21].

On the other hand, the prediction of protein-protein and other biomolecular interactions itself, may be regarded as a kind of function prediction. Protein function prediction in both broad and narrow contexts has been carried out using feature sets individually or through an integration of different levels of information with various degrees of success [22]. For example, amino acid sequence alignment [23], structural information [24], gene expression data in the form of co-expressed genes at proteomic or RNA expression level, protein-protein interaction [25] networks, domains or motif contents [26, 27], subcellular localization [28] and biophysical properties of amino acids have resulted in improved prediction of protein function as seen in critical assessment of functional annotation (CAFA2 and CAFA3) community challenges [14, 29, 30].

Apart from the predictive features of proteins described above, a path to function annotation is by direct or indirect evidence already discovered but not systematically documented, i.e., literature search. Faster mining of the existing knowledge of protein functions has become possible with dedicated databases and efficient text mining methods [31, 32] and genomic context-based predictions [33]. However, when functional annotation cannot be directly inferred from literature, function transfer relying on amino acid sequence, gene expression environment, three-dimensional structure and protein-protein interaction networks, as outlined above is indispensable. The most obvious among them, are the methods of function prediction based on homology relationships. Sequence similarity frequently indicates evolutionary relationship and functional similarity. However, this association is not absolute. Proteins have linear and structural motifs [34], that indicate their implicit functional capabilities. The existence of these motifs has been used to predict protein function, e.g., by a Basic Local Alignment Search Tool (BLAST) search of a database of already annotated proteins [35] or CATHEDRAL [36] and pair-wise annotation transfer. Unfortunately, as many as 30% of proteins cannot be computationally annotated using the above features as they are not closely related to well-annotated proteins [37, 38]. This highlights the need for alternative methods of sequence-based function prediction beyond homology relationships. For this purpose, protein sequences have been represented as k-mer frequencies and evolutionary profiles or characterized by smaller patterns contained in them. Single amino acid composition or K-mer counts of peptide subsequences are some of the basic so-called alignment-free methods of sequence comparison, applied to function prediction [39–41]. K-mer counts have been widely used but the success of this method is limited as protein function is encoded in a more complex manner than a simple 20 letter amino acid language or even longer k-mers at the sequence level.

Despite considerable success with prediction methods based on k-mers or single amino acid compositions, the real amino acid language of protein function has remained unclear. A major breakthrough to represent this latent language has recently been provided by the advent of large language-based models [42, 43]. Proteins are made up of sequences of letters from a predefined set of alphabets, the natural amino acids. These sequences store a major part of all the information about the protein's structure and function [42, 44]. They come together to form secondary structural elements and domains which are equivalent to words and sentences in the natural language [45–47]. If we could decode the amino acid language of proteins that describes their function, it would be possible to directly annotate proteins from their sequence.

In Table 1, we have collected a representative set of function prediction methods using different feature sets and computational models. We summarize here that the protein function annotation methods can be grouped into four categories as follows:

1. Classical models for direct prediction by sequence using homology, evolutionary profiles, or its derived features such as k-mer compositions [48–50].
2. Indirect predictions from interaction networks, gene expression patterns, and related properties [11, 51–56].
3. Mining of published literature using NLP techniques [57, 58] and methods based on language using large language model (LLM) techniques [59].

Being among the early groups of protein function prediction, a number of reviews and books are available for covering variants of methods that fall in (1) and (2). However, review literature is scarce on NLP- and LLM-based methods mentioned in (3). We, therefore, review the methods in all three groups but specifically focus on the last group of two methods in this chapter. Protein function annotation methods based on NLP and LLM share the language properties of literature or amino acid sequences, but are indeed quite different from one another and need independent treatment, as discussed in the later sections. Indeed several computational models, specifically dedicated to interactions have relied on both LLMs and NLP techniques, some of which will be covered in this survey.

**Table 1**
**We have tricollected the benchmark methods in protein function annotation trying to group them into three main groups: (1) automated methods that do not use machine learning and those methods that try to create an ensemble of existing methods. (2) Methods that use machine learning models like SVM, CNN, GAT, and others. (3) Automated function prediction methods that use embedding of protein sequences or other feature sets using transformer models**

| Method | Features | Input/query | Model | Performance | Publication | Availability |
|---|---|---|---|---|---|---|
| ProFunc | Sequence motifs, superfamily, gene neighbors, folds, clefts, conservation,binding sites | 3D structure | SURFNET algorithm, HMM, secondary structure matching | NA | [126] | http://www.ebi.ac.uk/thornton-srv/databases/ProFunc |
| I-TASSER | Sequence and structure profile comparison | Protein sequence | Structure-based function annotation | NA | [127] | https://zhanggroup.org/I-TASSER/ |
| InterProScan 5 | Searches against InterPro member databases for domains or patterns | Protein sequence | NA | NA | [128] | ftp://ftp.ebi.ac.uk/pub/software/unix/iprscan/5/ http://code.google.com/p/interproscan/ |
| SIFTER | Phylogeny-based sequence features | Protein accession, GO ID, function name, species name or ID, | Evolutionary relationship (not machine learning) | Precision—72% Recall—30% | [129] | http://sifter.berkeley.edu/ |
| GeneMANIA | Co-expression, protein-protein and genetic interactions, protein domains, co-localization, pathways, predictions via ortholog | Gene list | Network weighted method | ~0.7 (precision) (exact value not reported) | [130] | https://genemania.org/ |

| | | | | | | |
|---|---|---|---|---|---|---|
| NETGO | Protein sequences and their network information with GO annotation | NA | Rank framework | 0.557 | [131] | NA |
| I-TASSER-MTD | Sequence and structure profile comparison | Protein sequence | Structure-based function annotation for multidomain | NA | [132] | https://zhanggroup.org/I-TASSER-MTD/ |
| DEPICTER 2 | Ensemble of flDPnn disorder predictor and disorder function predictors (DFLpred, ANCHOR2, MoRFs, DisoRDPbind, DisoLipPred) | Protein sequence | NA | NA | [133] | http://biomine.cs.vcu.edu/servers/DEPICTER2/ |
| CATH/Gene3D | Functional annotations using sequence-based and structure-based methods (DeepFRI,PROST, Foldseek scans, STRING/IntAct | Protein sequence | Ensemble of servers | NA | [134] | https://cathdb.info/search/by_sequence |
| DeepGraphGO | Protein network features and InterProScan binary feature vector | Protein sequence | GCN | 0.543 (AUPR) Fmax (0.623) | [135] | https://github.com/yourh/DeepGraphGO |
| CombFunc | Protein sequence (BLAST/PSI-BLAST annotation transfer, domain information), gene expression, protein-protein interaction | Protein sequence | SVM | 0.714 (precision) and 0.41 (recall) | [136] | NA |
| Mashup | Topological context of multiple interaction networks (physical interaction, genetic | Networks | SVM | 0.59 (AUPR) | [137] | NA |

(continued)

**Table 1**
**(continued)**

| Method | Features | Input/query | Model | Performance | Publication | Availability |
|---|---|---|---|---|---|---|
| | interaction, a co-expression, and functional association) using RWR | | | | | |
| FFPRED3 | Secondary structure, transmembrane helices, intrinsically disordered regions, signal peptides, and other motifs | NA | SVM | Per GO term performance with highest precision of 1 | [138] | NA |
| deepNF | Network-based features (neighborhood, fusion, cooccurrence, coexpression, experimental, database) | Networks | MDA, SVM | ~60–72% m-PR | [139] | https://github.com/VGligorijevic/deepNF |
| DeepText2GO | Deep semantic text representation(text-based), sequence homology, families, domains, and motifs (sequence-based) | NA | CNN with 16 1D convolutional layers | 0.647 (Fmax) | [111] | NA |
| GOLabeler | GO term frequency, sequence alignment, amino acid trigram, domains and motifs, and biophysical properties<br>Automated function prediction<br>Proteins with annotations | Protein sequence | BLAST-KNN, Logistic regression | 0.546 | [140] | NA |

| | | | | | | |
|---|---|---|---|---|---|---|
| ECPRED | Subsequences, sequence similarities, and amino acid physicochemical features | Protein sequence | SPMap, BLAST-kNN and Pepstats-SVM | 0.95 (macro precision) | [141] | https://github.com/cansyl/ECPred<br>http://cansyl.metu.edu.tr/ECPred.html |
| PANNZER 2 | Sequence similarity and enrichment statistics | Protein sequence | KNN | ~0.84 (Fmx exact value not reported) | [142] | http://ekhidna2.biocenter.helsinki.fi/sanspanz/ |
| STRING2GO | Network embedding using mashup and Node2vec | PPI network | DMNN and SVM | ~0.80 F1 score (exact value not reported) | [143] | https://github.com/psipred/STRING2GO |
| DEEPred | Functional annotations of proteins, structural properties, protein-protein interaction | Protein sequence | Multitask feed forward deep neural network | 0.67 | [144] | https://github.com/cansyl/DEEPred |
| Graph2GO | Interaction networks (sequence similarity and protein-protein) and protein features (amino acid sequence, subcellular location, and protein domains) | Protein sequence | VGAE, DNN classifier | 0.528 (m-AUPR)<br>0.620 (F-max) | [145] | https://github.com/yanzhanglab/Graph2GO |
| NetQUILT | Sequence identity and IsoRank from BLAST, network, and annotation from STRING-db | NA | DNN | ~0.68 (m-AUPR) and ~ 0.82 (Fmax) (exact values not reported) | [146] | https://github.com/nowittynamesleft/NetQuilt |
| TALE & Tale+ | Transformer-based sequence and GO label embedding<br>TALE+(+sequence similarity) | Protein sequence | CNN | 0.604 | [147] | https://github.com/Shen-Lab/TALE |
| GOPREDSIM | Embedding-based annotation transfer | Protein sequence | KNN | 52 ± 2%(Fmax) | [148] | https://github.com/Rostlab/goPredSim |

(continued)

**Table 1**
**(continued)**

| Method | Features | Input/query | Model | Performance | Publication | Availability |
|---|---|---|---|---|---|---|
| DEEPGOWeb | Sequence features (sequence similarity, sequence motifs), PPI network features | Protein sequence | CNN | 0.487 | [149] | https://deepgo.cbrc.kaust.edu.sa/deepgo/ |
| PredictProtein | Evolutionary information, secondary structure, solvent accessibility, transmembrane segments, disordered regions, protein flexibility, and disulfide bridges (embedding-based), effect of variants or mutation, subcellular localization, binding sites | Protein sequence | NA | NA | [150] | https://predictprotein.org/ |
| DeepFRI | Sequence features extracted from protein language model and protein structures | Protein structure and sequence | GCN | ~0.434 (term-centric AUPR) and 0.55 (protein-centric Fmax) (exact values not reported) | [151] | https://beta.deepfri.flatironinstitute.org/ |
| GATGO | 1D features (SA/SS/PSSM), inter-residue contact graphs with node-level feature embeddings | Protein sequence | GAT | 0.662 (AUPRC) Fmax (0.637) | [18] | NA |
| PANDA2 | Topology of the GO DAG and features from transformer protein language | Protein sequence | GNN | 0.564 (AUPR) 0.598 (Fmax) | [152] | http://dna.cs.miami.edu/PANDA2/ |

| | | | | | | |
|---|---|---|---|---|---|---|
| DeepGATGO | Embeddings of the sequence and label data | Protein sequence | GAT and contrastive learning | 0.568 | [113] | NA |
| Struct2GO | SeqVec sequence embeddings, protein structure features using GCN | Protein structure and sequence | Different convolutional, pooling layers available | 0.796 (AUPR) and 0.701 (Fmax) | [153] | |

Abbreviations: *DEPICTER2* DisorderEd PredictIon CenTER, *PANNZER2* Protein ANNotation with Z-scoRE, *VGAE* variational graph auto encoder, *RWR* random walks with restart, *GAT* graph attention network, *GCN* graph convolutional network, *MDA* multimodal deep autoencoder, *DMNN* Deep maxout neural networks

## 2 Classical Methods of Function Annotation

### 2.1 Prediction from Sequence, Structure and Evolutionary Profile

The fact that sequence determines the structure and structure determines the function of a protein has led to various methodologies in which computer-encoded protein sequences are the primary features for predictive models [60, 61]. A classical method of annotating proteins is the transfer of annotation of functions from proteins with high homology in their sequence or structure [62]. High sequence similarity between proteins or genes generally can suggest an evolutionary relationship, which often implies functional similarity. Function prediction based on sequence similarity can be performed using different tools like FASTA, BLAST, PSI-BLAST, HMMER, SSEARCH which try to capture global and local regions of similarity between sequences. Functional annotation from homology searches is typically derived from sequences that meet threshold scores, such as an *E*-value of 0.01 or 0.001 and a Smith-Waterman alignment score of 200, on a general scale [63, 64].

Apart from the whole protein sequence comparisons, other properties can be utilized. For example, many proteins possess basic functional units called sequential, and structural motifs representing the functional capability of that protein. Function annotation of domains, motifs, and protein families is available comprehensively in several databases like SCOP [65], CATH [66], Gene3D [27], InterPro [67], SMART [68], Pfam [69, 70], PRINTS [71], and PROSITE [72] which have been used to transfer function by local sequence or structure comparison through a database. Some proteins possess structural motifs which are fingerprints of globular domains while some groups have linear motifs that are locally unstructured regions with a few specificity-determining residues and a tendency to transition from disordered to ordered states upon binding [34]. These sequence regions are usually evolutionarily conserved because of their functional importance and hence the presence of these motifs has been used to predict protein function.

However, an approach based on motif search or any sequence comparison method has limitations. First of all, such an approach does not account for the varying levels of sequence conservation across different protein families. Secondly, many homologs exist that perform different functions facilitated by very small changes in the sequence or structure. Inadequate or erroneous detection of multiple sequence alignment also poses a problem in function transfer via homology search [73]. For example, it has been stated that a lack of enough representative members of some protein families makes it difficult to extract HMM profiles leading to a lot of unannotated sequences in Pfam [74]. Consequently, many genes in most genomes remain unannotated and cannot be used for homology-based function search because none of them is annotated in the first place [29]. The use of structural features in function annotation has been shown to play an important role in determining function, which suffered from the lack of structures for a large number of proteins until recently. Arguably, with the advances in structure prediction using AlphaFold2 [75], ESMfold [76], and RoseTTAFold [77] the availability of high-quality structures is poised to solve this problem to an extent.

### 2.2 Moonlighting and Multifunctionality

The eukaryotic cell has membrane-bound and membrane-less organelles. Different biological processes can occur simultaneously because proteins relocate to these locations to function. The compartmentalization of cells allows the maintenance of different microenvironments and different concentrations of interacting partners that lead to proteins' capability to perform different functions. This enhances the capability of a protein to perform different functions that increase the complexity of the protein function prediction problem.

A protein domain is a conserved region of a protein sequence or structure that can operate and evolve separately from the other parts of the chain. Many proteins are multifunctional having multiple domains which during evolution undergo domain shuffling. Each domain might perform its function independently or in cooperation with the other domains. Thus, remote homologs might share the same function requiring a domain-level comparison than at the global sequence level. Most of the domain-related functional annotation was in free text and the mapping to gene ontology came Pfam2GO and Inter-Pro2GO. Thus protein function prediction is a "Mult-Instance Multi-Label"(MIML) task [78]. The presence of multifunctional proteins especially moonlighting proteins which perform multiple functions without processing different domains to perform these different functions by partnering with different ligands, under different cellular conditions, localizations, or using different binding sites makes protein function prediction more of a micro-environment based problem. They follow varied mechanisms and the characteristics that enable a protein to moonlight have not been fully understood which adds chaos to the automation of function prediction. Homologs of moonlighting proteins do not necessarily perform the same moonlighting function, making the comprehensive prediction of function even more challenging.

## 3 Indirect Function Prediction Using Protein-Protein Interactions and Gene Expression Patterns

Several methods are available that can predict the interactions of a protein from its sequence or structure. Protein interactions can also be predicted via homology through inheritance from similar proteins. STRING [79] transfers association from orthologous proteins in the COG database [80]. Other tools that attempt to predict protein-protein interaction include DeepBank [81]. The protein interaction-based function prediction is affected by the noise and lack of complete interaction data [82]. The lack of conserved features or specific properties in relation to some types of functions makes it difficult to automate their functional identification. This kind of ambiguity is seen in the case of protein-protein interactions, intrinsically disordered proteins, etc.

## 4 Function Prediction Using NLP and LLM Techniques

### *4.1 General Principles of NLP and LLM*

Natural language is a field of computer science that tries to formalize the natural human spoken language to enable computers to understand the meaning, analyze the structure, and generate human-like language [83, 84]. Modeling natural language with computational techniques has been approached in several ways, which can be enumerated as (1) rule-based, (2) statistical methods, and (3) neural network–based methods, which form the foundation of modern techniques such as transformers and large language models (LLMs) [83, 85].

Historically, NLP research dates back to the 1940s after World War II in an effort to create machines that can translate one language to another. Early approaches in NLP focused on automating the analysis of linguistic structures and developing foundational technologies such as document summarization, sentiment analysis, content categorization, machine translation, and transliteration [86–88]. The concept of machine translation (MT) was proposed by Warner Weaver in 1947 [89, 90]. A rule-based system was the initial approach. The grammar of language has been divided into multiple levels like phonetics (speech sounds), phonology (sound patterns), morphology (words), syntax (phrases and sentences), semantics (literal meaning of phrases and sentences), pragmatics (meaning in context), and discourse (meaning in extra-linguistic context) [91]. It was proposed that the main difference between different spoken languages was the vocabulary and the order of words so the tools looked up in a dictionary and arranged the words based on the syntax of the target language. These techniques use predefined patterns and language-specific rules which were applied to tasks like pattern recognition, tokenization, part-of-speech (POS) tagging, named entity recognition (NER), morphological analysis, and rule-based sentiment analysis [85]. This compounded the lexical ambiguity in language and scalability with the difficulty of maintaining and transferring manmade rules to newer domains.

A key turning point in the field of NLP was achieved through the works of Naom Chomsky who introduced the concept that grammar was a mathematical system with precise and finite rules using which sentences are generated [92]. The focus was shifted from a rationalist approach to empirical or corpus-based method which was automated by statistical methods and later by machine learning methods. One of the key applications of statistical NLP models was language modeling which involves predicting the probability of a word given the context. Statistical language models (SLMs) are word prediction models based on statistical learning [93]. In this model, probabilities are assigned to the word sequence to predict the next word based on the previous data sequence using methods such as Markov assumption. These techniques help to capture the probabilistic nature of language, leading to N-gram models, hidden Markov models for part of speech tagging, bag of words models, Topic Modeling using methods like Latent Dirichlet Allocation (LDA) [94], word embedding methods like Word2Vec, Global vectors (GloVe), Term Frequency Inverse Document Frequency (TF-IDF). The main challenge of such methods lies in dimensionality reduction.

The next and perhaps even bigger revolution in NLP came with the application of neural networks namely, the convolutional neural networks (CNNs) [95], recurrent neural networks (RNNs) [96], and lengthy short-time memory (LSTM) networks, allowing machines to capture sequential dependencies in language. Bengio et al., introduced neural network–based language models [97]. The use of multilayer perceptron (MLP) and RNN in neural language models became popular to understand the relationships in the language.

LLMs have advanced significantly from statistical language models (SLMs). The developmental phase included statistical language models, neural language models, and large language models. Neural language models emerged to improve upon the limitations of traditional SLMs. This shift marked the beginning of deep learning methods in language modeling, ultimately leading to the development of large language models [98]. The advancement of transformers and pretraining models enhanced the performance of language modeling tasks extending to a generative nature. Vithanage et al., 2018 have reviewed the contextualized word representation models [99]. In that direction, BERT [100] is a language representation model that uses a transformer-based neural network. The multilayer bidirectional transformer encoder is the state-of-the-art feature of the BERT. In the implementation of BERT, there are two phases namely pretraining and fine-tuning. During the pretraining phase, the BERT model is trained on unannotated data by means of different pretraining tasks. In the fine-tuning phase, after initializing the pretrained parameters on the BERT model, all of these parameters are finetuned using the annotated data for downstream tasks. Later on various LLMs were introduced by extending the pretrained model mainly scaling up in terms of the size of the model and parameters of the model

[98]. The state-of-the-art performance of LLMs relies on in-context learning, instruction following, and reasoning. LLMs significantly employs the role of understanding the natural language and generating the text data although it is based on transformers and a large number of parameters, trained on huge amounts of data. During the embedding phase, LLMs extracts the syntax and meaning from the language followed by a feedforward layer which has fully connected layers. Recurrent layer understands the relationships from the input text or between the words. The scope of LLMs is eminent because of its attention mechanism. In the self-attention phase, the model can focus on some relevant texts and it is helpful to detect the dependencies between the words.

The foundational technology for these LLMs lies in pretrained models (PTMs). In 2018, PTMs utilizing self-supervised learning over large-scale text corpora gained significant attention [101], fostering a two-stage learning paradigm of pretraining followed by fine-tuning. Notable examples of such models include ELMo [102], BERT [100], and GPT-3 [103]. These pretrained models have entered an era of large-scale parameters, commonly known as foundation models. Such models have recently achieved unprecedented success across various fields; for example, DALL.E 2 [104] and GPT-4 have made strides in computer vision and natural language generation (NLG). In the biological domain, models like Enformer [105], which predicts gene expression from DNA sequences, scGPT [106], a foundational model for single-cell biology, and scVI [107], a tool for single-cell data integration, cell annotation, and function prediction, have demonstrated the versatile applications of LLMs in single-cell biology and other biological fields. Large language models (LLMs) like ChatGPT, Gemini [108], and LLaMA [109], developed by OpenAI, Google, and Meta AI, have demonstrated remarkable capabilities in language understanding and generation, especially in conversational contexts.

### 4.2 NLP Application to Function Annotation

Initially, functional annotation from literature was carried out in an ad hoc fashion using free text with a whole spectrum of terminologies and synonyms. Automated prediction of protein function for large-scale datasets will be possible if there is a universal, controlled, and well-established vocabulary across the species. At the protein level itself, there is conflict in annotation with a lot of protein name variations, synonyms of protein names, non-specific descriptions of function with the textual description of function often varying with different investigators. To reduce the inconsistencies in terminology, Gene ontology consortium (GO)(Gene ontology consortium, 2004) and Enzyme Commission (EC) number have tried to organize the functional annotation data in a structured manner with minimum ambiguity but with a lot of manual reviewing. These processes are time-consuming and there is a huge gap between the annotations and published literature. Experimental annotation is only available for a restricted number of proteins. The function annotation of a vast number of proteins is achieved through ongoing extrapolation, which means that errors are anticipated. Even in manual curation, there can be a conflict in the assignment by different curators. Tools like Unifunc [110] and DeepText2GO [111] which use semantic text representation along with other protein information like homology, families, domain, and motif, that try to unify functional information from multiple sources to increase confidence in functional annotation, do automated sequence analysis with automated annotation using text mining either from gene ontology annotation or literature.

One of the main complexities in using automated processes or NLP is that the rules are very complex to define the manual annotation process, that has been followed for decades. Yet another complexity lies in the fact that it is difficult to assign the correct ontology, especially in the case of new functions where new terms need to be assigned.

NLP-based models for literature mining can be applied for the identification of a protein's capability to perform a certain function in a binary form. Literature mining in these cases may be able to capture an already-reported function from published literature repositories like PubMed, which was not added to the functional annotation of the protein in standard databases.

## 4.3 Large Language Model (LLM) Applications to Function Prediction

### 4.3.1 General Principles

An alternative to utilizing evolutionary information or sequence and structural attributes for function identification is the application of deep learning to derive insights from the ever-expanding protein sequence data, which can be perceived as the language of proteins, similar to the natural language processing of textual data. These protein language models can learn the evolutionary, structural, and functional aspects of the proteins via self-supervised learning. These methods can be trained on the billions of protein sequences found in the databases and then used for answering different questions with fine-tuning on the specific dataset. The main advantage of these methods is the fact that because of the capability of transfer learning the computationally costly training needs to be performed only once during training after which the embeddings can be used for further tasks [112, 113].

Earlier, a supervised NLP technique that trains input sequences alongside their labels has been utilized for detecting structural similarity and predicting stability [42, 114]. The introduction of transformers in language modeling has facilitated the emergence of unsupervised training across various tasks. Many of the current models utilize a BERT-like architecture with denoising autoencoding, wherein the training process includes masking tokens and attempting to reconstruct the original text. The primary function of these is encoding, although this can be expanded to develop generative models. The representation learning for proteins has been reviewed in detail by Ofer et al. [47].

Naturally developed proteins are usually made up of repeated modular parts with small differences that can be rearranged and put together in a hierarchical way, just like natural language. Most of these language models have been developed for natural language processing and cannot be directly applied to protein sequences. Proteins are not completely made of pre-defined blocks like words or sentences, have highly variable lengths, and involve very distant interactions due to folding, making LLM-based function prediction more complex [115].

#### 4.3.2 LLMs for Proteins

Several large language models (LLMs) trained on protein data have been developed and UniProtKB also provides per-protein and per-residue embeddings generated from ProtT5 model (a sample use case is shown in Fig. 1 and detailed tutorial is being available at https://github.com/Sciwhylab/protein-embedding-and-classification.git) for several model organism (https://ftp.uniprot.org/pub/databases/uniprot/current_release/knowledgebase/embeddings/). A detailed survey of various other embeddings available for protein function prediction methods is presented in Table 2, with the key characteristics of each of them. In most cases, LLM-based encodings and in some examples combinations of multiple embeddings have been shown to provide high predictive performance [116–121]. These embeddings are particularly helpful in high throughput benchmarks [122]. A recent review compared the performance of several protein language models in several tasks and found evolutionary scale modeling (ESM) and ProtTans both bidirectional encoder representations from transformers (BERT) for models that outperform other protein representation methods [123]. The embedding-based annotation transfer (EAT) has been shown to be very efficient in attempts to predict the molecular function, binding sites, secondary structure, disorder, and transmembrane regions. Structure prediction models (RoseTTAFold [77], ALPHAFold 2 [75]) can be used to add structural

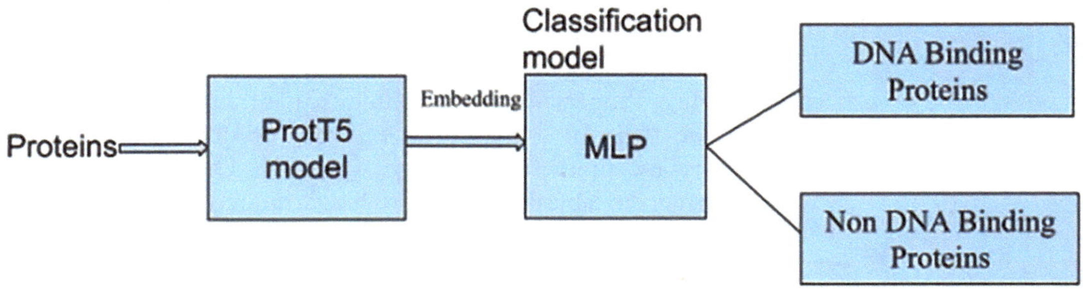


**Fig. 1** General workflow of utilizing LLM-based embeddings in predicting DNA-binding proteins

**Table 2**
**A summary of different embeddings for proteins**

| Protein LLMs | Data source | Implementation specification | Prediction task | Reference |
|---|---|---|---|---|
| ProtGPT2 | Uniclust30 database version- 2018_08<br>pdb version-2021_04<br>Uniref50 | Autoregressive transformer model with 738 million parameters | Generates de novo sequences with properties | [45] |
| ProGen2 | Uniref90,<br>BFD30 | Scaled upto 6.4B parameters,<br>Google's TPU-V3 hardware,<br>Adam optimizer | Capturing distribution of observed proteins,<br>Generate novel sequences,<br>Fitness prediction | [154] |
| ProtLLM | PubMed,<br>Uniprot,<br>STRING | Protein-encoder,<br>Cross-modal connectors,<br>LLM,<br>Dynamic protein mounting mechanism | Protein-as-word language modeling, interleaved protein-text dataset generation-InterPT | [155] |
| ProLLaMA | UniRef50,<br>InterPro | LoRA rank:128<br>AdamW optimizer<br>Cosine annealing scheduler<br>Learning rate:0.05<br>LoRA rank for instruction tuning:64<br>Protein property prediction task—Accuracy: Nearly 100% | Controllable protein generation and protein property prediction | [46] |
| protT5 | UniRef50,BFD100 | 420 M parameters,<br>Encoder-decoder-based architecture,<br>Trained two models with 3B and 11B parameters,<br>Ada factor optimization | Extracting biophysical features of amino acids,<br>Capturing protein function in conserved motifs,<br>Per-residue secondary structure prediction | [124] |
| prostT5 | PDB | Fine-tuning the existing ProtT5 model | Protein structure prediction,<br>Inverse folding,<br>Generating structure sequence | [156] |

(continued)

**Table 2**
**(continued)**

| Protein LLMs | Data source | Implementation specification | Prediction task | Reference |
|---|---|---|---|---|
| ESM2 | UniRef50 | Used attention mechanism,<br>15 billion parameters trained,<br>Executed completely on GPU | End-to-end atomic level structure prediction,<br>Efficient prediction of protein structure | [76] |
| ProteinBERT | UniProtKB/UniRef90 | Deep neural network architecture,<br>~16 M trainable parameters | Improves protein modeling,<br>Protein prediction,<br>Capturing the local and global representations of the protein | [115] |
| ProGen | Uniparc, UniprotKB, SWISS-PROT, TrEMBL, pfam, NCBI taxonomic information | Implemented in TensorFlow,<br>Cloud TPU v3 pod for one million iterations,<br>1.2 billion trainable parameters | Protein generation | [157] |

information to these prediction models to increase the accuracy of function prediction. Several studies have used pretrained protein models, which include ESMFold [76], ProtTrans [124], UniRep [114], ProteinBert [115], and ProtNLM [125].

#### 4.3.3 Tokenization

Since proteins do not have well-defined vocabulary, word-level tokenization is not an option. The simplest and most common method of tokenization used for proteins is an individual amino acid or character-level tokenization. However, in character-level tokenization though the number of tokens will be much less, more tokens must be used to form a sentence which again increases the memory requirement. Bag of words (BoW) approach where the amino acid count is utilized has been used in many bioinformatics approaches. Other sequence representations like the content or occurrence of 2D structure or representation of amino acid properties like hydrophobicity/polarity can be used with the BoW. Amino acid level N-gram/k-mer representations can also be used but could lead to a sparsity in the feature space. K-mer symmetry can be used meaning "ALG" and "GLA" can be treated as the same feature to reduce the dimensionality [47].

Altogether LLMs for function prediction have provided a strong alternative to sequence-based annotation by homology detection or template-based search and continue to show an improving performance. It is hoped function prediction of natural and synthetic proteins will become possible from sequence in the future just like the structure determination has been revolutionized by AlphaFold-like programs.

## 5 Conclusion

Function prediction from sequence, structure, and other contextual information has remained a highly active area of bioinformatics research. Natural language and LLM-based methods have not only enabled faster mining of published literature but also paved the way for representing amino acid sequences in latent lower-dimensional space, which allows their use in machine learning models. Different types of embedding, tokenization, transfer learning, and training methods distinguish LLM models, and improvements are being made in each direction. It is expected that in the future most function prediction can be made directly from sequence with a comprehensive use of sequence and structure-based encoding, just like similar success stories of structure prediction.

## References


1. Caplen NJ (2002) A new approach to the inhibition of gene expression. Trends Biotechnol 20(2):49–51
2. Hall B, Limaye A, Kulkarni AB (2009) Chapter 19: Overview: generation of gene knockout mice. Curr Protoc Cell Biol:Unit 19.12 19.12.1–Unit 19.12 19.1217
3. Alberts B et al (2002) Analyzing protein structure and function, in molecular biology of the cell, 4th edn. Garland Science
4. Walls D et al (2017) Protein chromatography. Springer
5. Zhou M, Li Q, Wang R (2016) Current experimental methods for characterizing protein-protein interactions. ChemMedChem 11(8):738–756
6. Gerlt JA et al (2011) The enzyme function initiative. Biochemistry 50(46):9950–9962
7. Saima M et al (2022) Elucidation of Protein Function Using Raman Spectroscopy, in Analytical Techniques for the Elucidation of Protein Function, pp 69–100
8. Rost B et al (2003) Automatic prediction of protein function. Cell Mol Life Sci 60(12): 2637–2650
9. Sinha S, Eisenhaber B, Lynn AM (2018) Predicting protein function using homology-based methods. In: Shanker A (ed) Bioinformatics: sequences, structures, phylogeny. Springer, Singapore, pp 289–305
10. Aravind L (2000) Guilt by association: contextual information in genome analysis. Genome Res 10(8):1074–1077
11. Quackenbush J (2003) Genomics. Microarrays–guilt by association. Science 302(5643):240–241
12. Gherardini PF, Helmer-Citterich M (2008) Structure-based function prediction: approaches and applications. Brief Funct Genomic Proteomic 7(4):291–302
13. Lee D, Redfern O, Orengo C (2007) Predicting protein function from sequence and structure. Nat Rev Mol Cell Biol 8(12):995–1005
14. Shehu A, Barbará D, Molloy K (2016) A survey of computational methods for protein function prediction. In: Wong K-C (ed) Big data analytics in genomics. Springer, Cham, pp 225–298
15. Pakhrin SC et al (2021) DeepNGlyPred: a deep neural network-based approach for human N-linked glycosylation site prediction. Molecules 26(23)
16. Ismail H et al (2022) FEPS: a tool for feature extraction from protein sequence. Methods Mol Biol 2499:65–104
17. Pakhrin SC et al (2022) Deep learning-based advances in protein posttranslational modification site and protein cleavage prediction. Methods Mol Biol 2499:285–322
18. Lai B, Xu J (2022) Accurate protein function prediction via graph attention networks with predicted structure information. Brief Bioinform 23(1)
19. Boadu F, Lee A, Cheng J (2024) Deep learning methods for protein function prediction. Proteomics:e2300471
20. Bonetta R, Valentino G (2020) Machine learning techniques for protein function prediction. Proteins 88(3):397–413
21. Zhang F et al (2019) DeepFunc: a deep learning framework for accurate prediction of protein functions from protein sequences and interactions. Proteomics 19(12): e1900019
22. Boadu F, Cheng J (2024) Improving protein function prediction by learning and integrating representations of protein sequences and function labels. Bioinformatics Advances 4(1)
23. Clark WT, Radivojac P (2011) Analysis of protein function and its prediction from amino acid sequence. Proteins 79(7): 2086–2096
24. Dukka BK (2013) Structure-based methods for computational protein functional site prediction. Comput Struct Biotechnol J 8: e201308005
25. Tuncbag N, Gursoy A, Keskin O (2011) Prediction of protein-protein interactions: unifying evolution and structure at protein interfaces. Phys Biol 8(3):035006
26. Ibtehaz N, Kagaya Y, Kihara D (2023) Domain-PFP allows protein function prediction using function-aware domain embedding representations. Commun Biol 6(1):1103
27. Rentzsch R, Orengo CA (2013) Protein function prediction using domain families. BMC Bioinformatics 14(3):S5
28. Casadio R, Martelli PL, Pierleoni A (2008) The prediction of protein subcellular localization from sequence: a shortcut to functional genome annotation. Brief Funct Genomic Proteomic 7(1):63–73
29. Jeffery CJ (2023) Current successes and remaining challenges in protein function prediction. Front Bioinform 3:1222182

30. Zhou N et al (2019) The CAFA challenge reports improved protein function prediction and new functional annotations for hundreds of genes through experimental screens. Genome Biol 20(1):244
31. Funk CS et al (2015) Evaluating a variety of text-mined features for automatic protein function prediction with GOstruct. J Biomed Semantics 6(1):9
32. Papanikolaou N et al (2015) Protein-protein interaction predictions using text mining methods. Methods 74:47–53
33. Gabaldón T, Huynen MA (2004) Prediction of protein function and pathways in the genome era. Cell Mol Life Sci 61(7–8): 930–944
34. Neduva V, Russell RB (2005) Linear motifs: evolutionary interaction switches. FEBS Lett 579(15):3342–3345
35. Altschul SF et al (1997) Gapped BLAST and PSI-BLAST: a new generation of protein database search programs. Nucleic Acids Res 25(17):3389–3402
36. Redfern OC et al (2007) CATHEDRAL: a fast and effective algorithm to predict folds and domain boundaries from multidomain protein structures. PLoS Comput Biol 3(11):e232
37. Das S, Orengo CA (2016) Protein function annotation using protein domain family resources. Methods 93:24–34
38. Price MN et al (2018) Mutant phenotypes for thousands of bacterial genes of unknown function. Nature 557(7706):503–509
39. Guo Y et al (2021) Prediction of hormone-binding proteins based on K-mer feature representation and naive Bayes. Front Genet 12: 797641
40. Siebert KJ (2003) Modeling protein functional properties from amino acid composition. J Agric Food Chem 51(26):7792–7797
41. Vinga S, Almeida J (2003) Alignment-free sequence comparison-a review. Bioinformatics 19(4):513–523
42. Bepler T, Berger B (2021) Learning the protein language: evolution, structure, and function. Cell Syst 12(6):654–669.e3
43. Vu MH et al (2023) Linguistically inspired roadmap for building biologically reliable protein language models. Nature Machine Intelligence 5(5):485–496
44. Yang KK, Wu Z, Arnold FH (2019) Machine-learning-guided directed evolution for protein engineering. Nat Methods 16(8): 687–694
45. Ferruz N, Schmidt S, Höcker B (2022) ProtGPT2 is a deep unsupervised language model for protein design. Nat Commun 13(1):4348
46. Lv L, et al (2024) ProLLaMA: a protein language model for multi-task protein language processing
47. Ofer D, Brandes N, Linial M (2021) The language of proteins: NLP, machine learning & protein sequences. Comput Struct Biotechnol J 19:1750–1758
48. Kaundal R, Saini R, Zhao PX (2010) Combining machine learning and homology-based approaches to accurately predict subcellular localization in Arabidopsis. Plant Physiol 154(1):36–54
49. Valencia A (2005) Automatic annotation of protein function. Curr Opin Struct Biol 15(3):267–274
50. Kc DB (2017) Recent advances in sequence-based protein structure prediction. Brief Bioinform 18(6):1021–1032
51. Amitai G et al (2004) Network analysis of protein structures identifies functional residues. J Mol Biol 344(4):1135–1146
52. Bernardes JS, Pedreira CE (2013) A review of protein function prediction under machine learning perspective. Recent Pat Biotechnol 7(2):122–141
53. Deng M et al (2002) Prediction of protein function using protein–protein interaction data. In: Proceedings / IEEE computer society bioinformatics conference, vol 1. IEEE Computer Society Bioinformatics Conference, pp 197–206
54. Marcotte EM et al (1999) Detecting protein function and protein-protein interactions from genome sequences. Science 285(5428): 751–753
55. Sivashankari S, Shanmughavel P (2006) Functional annotation of hypothetical proteins – a review. Bioinformation 1(8):335–338
56. Vazquez A et al (2003) Global protein function prediction from protein-protein interaction networks. Nat Biotechnol 21(6): 697–700
57. Kulmanov M (2020) Predicting gene functions and phenotypes by combining deep learning and ontologies
58. Verspoor KM (2014) Roles for text mining in protein function prediction. Methods Mol Biol 1159:95–108
59. Wang Z et al (2022) LM-GVP: an extensible sequence and structure informed deep learning framework for protein property prediction. Sci Rep 12(1):6832

60. Anfinsen CB (1973) Principles that govern the folding of protein chains. Science 181(4096):223–230
61. Redfern OC, Dessailly B, Orengo CA (2008) Exploring the structure and function paradigm. Curr Opin Struct Biol 18(3):394–402
62. Yang J et al (2015) The I-TASSER suite: protein structure and function prediction. Nat Methods 12(1):7–8
63. Pearson WR (1996) Effective protein sequence comparison. Methods Enzymol 266:227–258
64. Pearson WR (2000) Flexible sequence similarity searching with the FASTA3 program package. Methods Mol Biol 132:185–219
65. Murzin AG et al (1995) SCOP: a structural classification of proteins database for the investigation of sequences and structures. J Mol Biol 247(4):536–540
66. Orengo CA et al (1997) CATH–a hierarchic classification of protein domain structures. Structure 5(8):1093–1108
67. Apweiler R et al (2001) The InterPro database, an integrated documentation resource for protein families, domains and functional sites. Nucleic Acids Res 29(1):37–40
68. Schultz J et al (1998) SMART, a simple modular architecture research tool: identification of signaling domains. Proc Natl Acad Sci USA 95(11):5857–5864
69. Mistry J et al (2021) Pfam: the protein families database in 2021. Nucleic Acids Res 49(D1):D412–d419
70. Sonnhammer EL, Eddy SR, Durbin R (1997) Pfam: a comprehensive database of protein domain families based on seed alignments. Proteins 28(3):405–420
71. Attwood TK et al (1994) PRINTS–a database of protein motif fingerprints. Nucleic Acids Res 22(17):3590–3596
72. Bairoch A (1992) Prosite: a dictionary of sites and patterns in proteins. Nucleic Acids Res 20 (suppl):2013–2018
73. Bordin N et al (2023) Novel machine learning approaches revolutionize protein knowledge. Trends Biochem Sci 48(4):345–359
74. Vitale R et al (2024) Evaluating large language models for annotating proteins. Brief Bioinform 25(3)
75. Jumper J et al (2021) Highly accurate protein structure prediction with AlphaFold. Nature 596(7873):583–589
76. Lin Z, et al (2022) Language models of protein sequences at the scale of evolution enable accurate structure prediction
77. Baek M et al (2021) Accurate prediction of protein structures and interactions using a three-track neural network. Science 373(6557):871–876
78. Wu JS, Huang SJ, Zhou ZH (2014) Genome-wide protein function prediction through multi-instance multi-label learning. IEEE/ACM Trans Comput Biol Bioinform 11(5): 891–902
79. Franceschini A et al (2013) STRING v9.1: protein-protein interaction networks, with increased coverage and integration. Nucleic Acids Res 41(Database issue):D808–D815
80. Tatusov RL et al (2000) The COG database: a tool for genome-scale analysis of protein functions and evolution. Nucleic Acids Res 28(1): 33–36
81. Renaud N et al (2021) DeepRank: a deep learning framework for data mining 3D protein-protein interfaces. Nat Commun 12(1):7068
82. Peng X et al (2017) Protein-protein interactions: detection, reliability assessment and applications. Brief Bioinform 18(5):798–819
83. Joseph S et al (2016) Natural Language Process Rev 6:207–210
84. Juang B-H, Furui S (2000) Automatic recognition and understanding of spoken language—a first step toward natural human-machine communication. Proc IEEE 88: 1142–1165
85. Lior G, Meysam G, Asha S (2024) Mastering NLP from foundations to LLMs: apply advanced rule-based techniques to LLMs and solve real-world business problems using Python. Packt Publishing, p 1
86. Feng Z (2023) Past and present of natural language processing. In: Feng Z (ed) Formal analysis for natural language processing: a handbook. Springer, Singapore, pp 3–48
87. Montejo-Ráez A, S.M. (2022) Jiménez-Zafra current approaches and applications in natural language processing. Appl Sci 12. https://doi.org/10.3390/app12104859
88. Sen O et al (2022) Bangla natural language processing: a comprehensive analysis of classical, machine learning, and deep learning-based methods. IEEE Access 10:38999–39044
89. Weaver W (1952) Translation. Massachusetts Institute of Technology
90. Garg A, Agarwal M (2018) Machine translation: a literature review. arXiv preprint arXiv:1901.01122
91. Alothman A, Alsalman A (2020) Arabic morphological analysis techniques. Int J Adv Comput Sci Appl:11

92. Chomsky N (1957) Logical structures in language. Am Doc 8(4):284–291
93. Rosenfeld R, Rosenfeld R (2000) Two decades of statistical language modeling: where do we go from here? Proceedings of the IEEE. Proc IEEE 88(8):1270–1278
94. Blei D, Ng A, Jordan M (2001) Latent Dirichlet allocation, vol 3, pp 601–608
95. Wu J (2017) Introduction to convolutional neural networks. National key lab for novel software technology. Nanjing University. China. 5(23):495
96. Sherstinsky A (2020) Fundamentals of recurrent neural network (RNN) and long short-term memory (LSTM) network. Physica D: Nonlinear Phenomena 404:132306
97. Bengio Y, Ducharme R, Vincent P (2000) A neural probabilistic language model, vol 3, pp 932–938
98. Zhao WX et al (2023) A survey of large language models. ArXiv:abs/2303.18223
99. Vithanage D et al (2024) Contextual word embedding for biomedical knowledge extraction: a rapid review and case study. J Healthcare Informatics Res 8(1):158–179
100. Devlin J et al (2019) BERT: Pre-training of deep bidirectional transformers for language understanding. In: North American chapter of the association for computational linguistics
101. Qiu X et al (2020) Pre-trained models for natural language processing: a survey. SCIENCE CHINA Technol Sci 63:1872–1897
102. Sarzynska-Wawer J et al (2021) Detecting formal thought disorder by deep contextualized word representations. Psychiatry Res 304:114135
103. Brown TB et al (2020) Language models are few-shot learners. ArXiv:abs/2005.14165
104. Ramesh A et al (2022) Hierarchical text-conditional image generation with CLIP Latents. ArXiv:abs/2204.06125
105. Avsec Ž et al (2021) Effective gene expression prediction from sequence by integrating long-range interactions. Nat Methods 18(10):1196–1203
106. Cui H et al (2024) scGPT: toward building a foundation model for single-cell multi-omics using generative AI. Nat Methods 21(8):1470–1480
107. Gayoso A et al (2022) A python library for probabilistic analysis of single-cell omics data. Nat Biotechnol 40(2):163–166
108. Team G et al (2023) Gemini: a family of highly capable multimodal models. arXiv preprint arXiv:2312.11805
109. Touvron H et al (2023) *Llama: Open and efficient foundation language models.* arXiv preprint arXiv::2302.13971
110. Queirós P et al (2021) Unification of functional annotation descriptions using text mining. Biol Chem 402(8):983–990
111. You R, Huang X, Zhu S (2018) DeepText2GO: improving large-scale protein function prediction with deep semantic text representation. Methods 145:82–90
112. Chaudhari M et al (2021) DTL-DephosSite: deep transfer learning based approach to predict Dephosphorylation sites. Front Cell Dev Biol 9:662983
113. Li Z, Jiang C, Li J (2023) Deepgatgo: a hierarchical pretraining-based graph-attention model for automatic protein function prediction. arXiv preprint arXiv:2307.13004
114. Alley EC et al (2019) Unified rational protein engineering with sequence-based deep representation learning. Nat Methods 16(12): 1315–1322
115. Brandes N et al (2022) ProteinBERT: a universal deep-learning model of protein sequence and function. Bioinformatics 38(8):2102–2110
116. Pokharel S et al (2022) Improving protein succinylation sites prediction using embeddings from protein language model. Sci Rep 12(1):16933
117. Pakhrin SC et al (2023) LMNglyPred: prediction of human N-linked glycosylation sites using embeddings from a pre-trained protein language model. Glycobiology 33(5): 411–422
118. Pokharel S et al (2023) Integrating Embeddings from multiple protein language models to improve protein O-GlcNAc site prediction. Int J Mol Sci 24(21)
119. Pratyush P et al (2023) pLMSNOSite: an ensemble-based approach for predicting protein S-nitrosylation sites by integrating supervised word embedding and embedding from pre-trained protein language model. BMC Bioinformatics 24(1):41
120. Pratyush P et al (2025) LMPTMSite: a platform for PTM site prediction in proteins leveraging transformer-based protein language models. Methods Mol Biol 2867: 261–297
121. Kulmanov M et al (2024) Protein function prediction as approximate semantic entailment. Nature Machine Intelle 6(2):220–228
122. Shaojun W et al (2023) NetGO 3.0: protein language model improves large-scale functional annotations. Genomics Proteomics Bioinformatics 21(2):349–358

123. Fenoy E, Edera AA, Stegmayer G (2022) Transfer learning in proteins: evaluating novel protein learned representations for bioinformatics tasks. Brief Bioinform 23(4)
124. Elnaggar A et al (2022) ProtTrans: toward understanding the language of life through self-supervised learning. IEEE Trans Pattern Anal Mach Intell 44(10):7112–7127
125. Gane, A., et al., ProtNLM: model-based natural language protein annotation.. Preprint, 2022
126. Laskowski RA, Watson JD, Thornton JM (2005) ProFunc: a server for predicting protein function from 3D structure. Nucleic Acids Res 33(Web Server issue):W89–W93
127. Roy A, Kucukural A, Zhang Y (2010) I-TASSER: a unified platform for automated protein structure and function prediction. Nat Protoc 5(4):725–738
128. Jones P et al (2014) InterProScan 5: genome-scale protein function classification. Bioinformatics 30(9):1236–1240
129. Sahraeian SM, Luo KR, Brenner SE (2015) SIFTER search: a web server for accurate phylogeny-based protein function prediction. Nucleic Acids Res 43(W1):W141–W147
130. Franz M et al (2018) GeneMANIA update 2018. Nucleic Acids Res 46(W1):W60–w64
131. You R et al (2019) NetGO: improving large-scale protein function prediction with massive network information. Nucleic Acids Res 47 (W1):W379–W387
132. Zhou X et al (2022) I-TASSER-MTD: a deep-learning-based platform for multi-domain protein structure and function prediction. Nat Protoc 17(10):2326–2353
133. Basu S, Gsponer J, Kurgan L (2023) DEPICTER2: a comprehensive webserver for intrinsic disorder and disorder function prediction. Nucleic Acids Res 51(W1):W141–w147
134. Waman VP et al (2024) CATH 2024: CATH-AlphaFlow doubles the number of structures in CATH and Reveals nearly 200 new folds. J Mol Biol 436(17):168551
135. You R et al (2021) DeepGraphGO: graph neural network for large-scale, multispecies protein function prediction. Bioinformatics 37(Supplement_1):i262–i271
136. Wass MN, Barton G, Sternberg MJ (2012) CombFunc: predicting protein function using heterogeneous data sources. Nucleic Acids Res 40(Web Server issue):W466–W470
137. Cho H, Berger B, Peng J (2016) Compact integration of multi-network topology for functional analysis of genes. Cell Systems 3(6):540–548.e5
138. Cozzetto D et al (2016) FFPred 3: feature-based function prediction for all gene ontology domains. Sci Rep 6(1):31865
139. Gligorijevic V, Barot M, Bonneau R (2018) deepNF: deep network fusion for protein function prediction. Bioinformatics 34(22): 3873–3881
140. You R et al (2018) GOLabeler: improving sequence-based large-scale protein function prediction by learning to rank. Bioinformatics 34(14):2465–2473
141. Dalkiran A et al (2018) ECPred: a tool for the prediction of the enzymatic functions of protein sequences based on the EC nomenclature. BMC Bioinformatics 19(1):334
142. Törönen P, Medlar A, Holm L (2018) PANNZER2: a rapid functional annotation web server. Nucleic Acids Res 46(W1):W84–w88
143. Wan C et al (2019) Using deep maxout neural networks to improve the accuracy of function prediction from protein interaction networks. PLoS One 14(7):e0209958
144. Sureyya Rifaioglu A et al (2019) DEEPred: automated protein function prediction with multi-task feed-forward deep neural networks. Sci Rep 9(1):7344
145. Fan K, Guan Y, Zhang Y (2020) Graph2GO: a multi-modal attributed network embedding method for inferring protein functions. GigaScience 9(8)
146. Barot M et al (2021) NetQuilt: deep multispecies network-based protein function prediction using homology-informed network similarity. Bioinformatics 37(16):2414–2422
147. Cao Y, Shen Y (2021) TALE: transformer-based protein function annotation with joint sequence–label embedding. Bioinformatics 37(18):2825–2833
148. Littmann M et al (2021) Embeddings from deep learning transfer GO annotations beyond homology. Sci Rep 11(1):1160
149. Kulmanov M, Zhapa-Camacho F, Hoehndorf R (2021) DeepGOWeb: fast and accurate protein function prediction on the (semantic) web. Nucleic Acids Res 49(W1):W140–W146
150. Bernhofer M et al (2021) PredictProtein – predicting protein structure and function for 29 years. Nucleic Acids Res 49(W1):W535–w540
151. Gligorijević V et al (2021) Structure-based protein function prediction using graph convolutional networks. Nat Commun 12(1): 3168
152. Zhao C, Liu T, Wang Z (2022) PANDA2: protein function prediction using graph

neural networks. NAR Genomics Bioinformatics 4(1)
153. Jiao P et al (2023) Struct2GO: protein function prediction based on graph pooling algorithm and AlphaFold2 structure information. Bioinformatics 39(10)
154. Nijkamp E et al (2023) ProGen2: exploring the boundaries of protein language models. Cell Syst 14(11):968–978.e3
155. Zhuo L et al (2024) Protllm: an interleaved protein-language llm with protein-as-word pre-training. arXiv preprint arXiv:2403.07920
156. Heinzinger M et al (2024) Bilingual language model for protein sequence and structure. NAR Genomics Bioinfo 6(4)
157. Madani A, et al (2020) ProGen: Language modeling for protein generation

# Chapter 14

# PLMSearch and PLMAlign: Protein Language Model (PLM)-Based Homologous Protein Sequence Search and Alignment

**Wei Liu, Ziye Wang, Ronghui You, Chenghan Xie, Hong Wei, Yi Xiong, Jianyi Yang, and Shanfeng Zhu**

## Abstract

Homologous protein search and sequence alignment are fundamental tasks in computational biology. While structure search methods can provide valuable insights, inferring distant evolutionary relationships with sequences alone remains challenging. Here, we introduce PLMSearch and PLMAlign, two protein language model (PLM)-based tools for searching and aligning remote homologous sequences. PLMSearch is a novel method for homologous protein search with only sequences as input. It leverages deep representations from a pretrained protein language model and trains the similarity prediction model with a large number of real structure similarities, enabling it to detect remote homologies embedded behind the sequences. PLMAlign is a protein sequence alignment tool designed to accurately align the identified homologous pairs. We offer two distinct Web servers for PLMSearch and PLMAlign, freely available at https://dmiip.sjtu.edu.cn/PLMSearch and https://dmiip.sjtu.edu.cn/PLMAlign, respectively. We demonstrate how to operate the Web servers and interpret the resulting outputs. Furthermore, we offer a large range of free and open resources, including open-source code, datasets, and containers, all aimed at advancing scientific research. We also demonstrate how these resources can be leveraged to run PLMSearch and PLMAlign either locally or in the cloud.

**Key words** Protein language model, Homologous protein search, Protein sequence alignment, Large-scale vector database search, Protein structure similarity prediction

## 1 Introduction

Homologous protein search is a critical and fundamental component in computational biology, serving as a cornerstone for nearly all sequence-related research, such as protein function prediction [1–11], protein–protein interaction prediction [12], and protein–phenotype association prediction [13]. The goal of homologous protein search is to identify homologous proteins for a given query from a target dataset (typically a large-scale standardized database

Dukka B. KC (ed.), *Large Language Models (LLMs) in Protein Bioinformatics*, Methods in Molecular Biology, vol. 2941, https://doi.org/10.1007/978-1-0716-4623-6_14,

like UniProt [14]), ranking those with a higher likelihood of homology at the top. Homologous protein search methods can be broadly categorized into sequence search and structure search, depending on the type of input data.

Due to the low cost and extensive availability of sequence data, sequence similarity-based methods are the most commonly used for homologous protein search. Examples include MMseqs2 [15], BLASTp [16], and Diamond [17]. Despite their success in homology detection, identifying distant evolutionary relationships based solely on sequence similarity remains challenging [18]. When sequence identity falls below 0.3, profile HMM-based methods such as HMMER [19], HHsearch [20], and HHblits [21, 22] tend to be more effective for homologous protein search.

In cases involving highly distant evolutionary relationships, sequences may have diverged so significantly that detecting their relatedness becomes increasingly difficult. Given that protein structures tend to evolve more slowly than sequences, detecting similarities through 3D structural superposition offers higher sensitivity [23–27]. Nonetheless, sequence search methods remain faster and more convenient in the vast majority of cases.

Meanwhile, protein language models (PLMs), such as ESMs [28–31] and ProtTrans [32], which are trained on hundreds of millions of unlabeled protein sequences using self-supervised tasks like masked amino acid prediction, have shown impressive performance across various downstream tasks [33]. These include structure-related tasks like secondary structure prediction and contact prediction [34]. Recently, DEDAL [35], DeepBLAST [36], and the latest pLM-BLAST [37] have demonstrated that integrating deep representations of protein sequences with alignment algorithms such as Smith-Waterman (SW) [38] or Needleman-Wunsch (NW) [39] can lead to more accurate pairwise sequence alignment and homology detection. However, protein language models are not fully utilized for the large-scale protein sequence search.

To enhance sensitivity while preserving the universality and efficiency of sequence search, we introduce PLMSearch (Fig. 1a–c [40], reproduced from PLMSearch: Protein language model powers accurate and fast sequence search for remote homology 2024 with permission from Nature Communications). In addition, we present PLMAlign (Fig. 1d), a sequence alignment tool designed to precisely align the homologous pairs identified by PLMSearch, both leveraging protein language models. We outline the architecture of our methods and provide detailed guidance on how to use the Web server and interpret the results. Finally, we explain how to utilize the extensive open-source resources we offer to run PLMSearch and PLMAlign either locally or in the cloud, enabling the reproduction of our experimental findings.

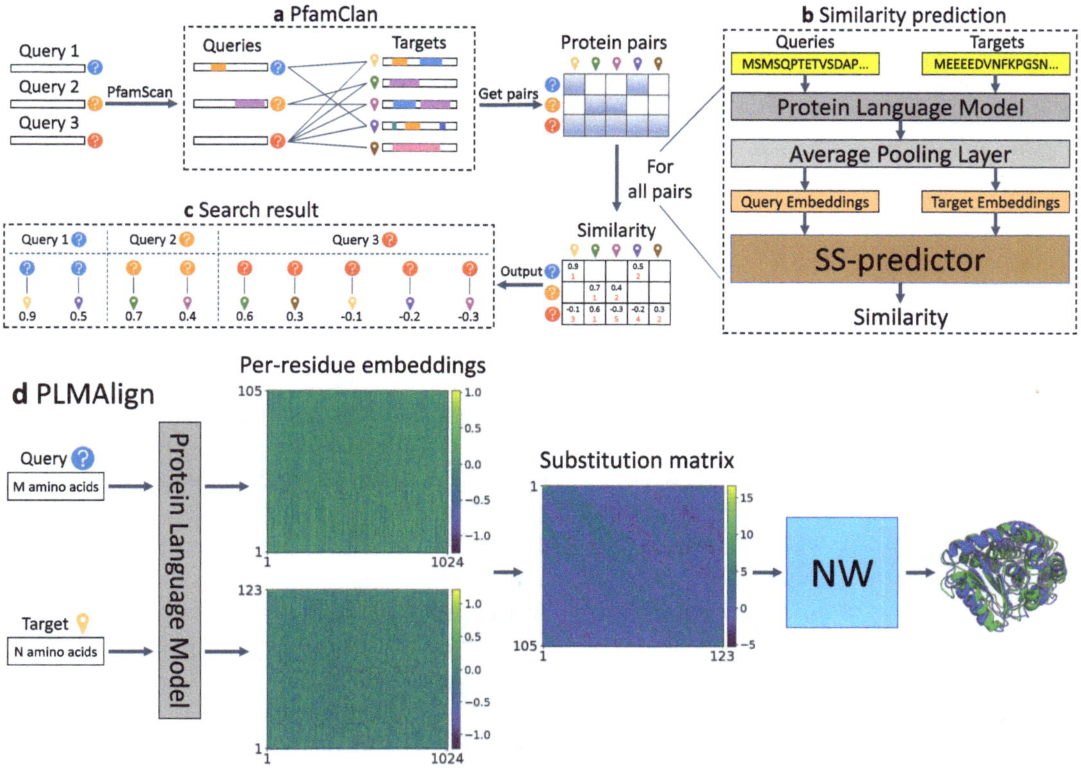


Fig. 1 Overview of PLMSearch and PLMAlign

## 2 Materials and Method

### 2.1 Architecture

PLMSearch primarily consists of three key steps (Fig. 1a–c): (1) PfamClan filters out protein pairs that share the same Pfam clan domain [41]. (2) The SS-predictor (Structural Similarity predictor) estimates the similarity between all query-target pairs using embeddings generated by the protein language model. PLMSearch will not lose much sensitivity without structures as input, because it uses the protein language model to capture remote homology information from deep sequence embeddings. In addition, the SS-predictor used in this step uses the structural similarity (TM-score) as the ground truth for training. This allows PLMSearch to acquire reliable similarity even without structures as input. (3) PLMSearch then ranks the pre-filtered pairs by their predicted similarity and outputs the search results for each query protein.

It is important to note that PLMSearch can only predict the similarity between protein pairs but is unable to provide alignment suggestions. To address this, we introduce PLMAlign, a sequence alignment method based on protein language models (Fig. 1d). In the final integrated pipeline, PLMSearch + PLMAlign, PLMAlign generates sequence alignments and alignment scores for the protein pairs filtered by PLMSearch.

### 2.2 Predictive Quality and Runtime

The original manuscript [40] presents our method's evaluation results on the all-versus-all search test using the SCOPe40-test dataset. PLMSearch performs favorably across most metrics, particularly at the superfamily and fold levels, which have less significant similarity between protein sequences and are more challenging to detect. Specifically, PLMSearch outperforms MMseqs2 in AUROC by factors of 3, 16, and 219 at the family level (improving from 0.318 to 0.928), superfamily level (from 0.050 to 0.826), and fold level (from 0.002 to 0.438), respectively. The primary driver of this improvement is PLMSearch's robustness, which reduces the rank of the first false positive (FP), significantly increasing the number of true positives (TPs) up to the first FP (outperforming MMseqs2 by 38 times, from 2.74 to 104.78). In the integrated workflow of PLMSearch + PLMAlign, PLMAlign aligns protein pairs that PLMSearch predicts to have a similarity score above 0.3. The alignment score is subsequently used for reranking, resulting in improved precision, particularly in cases involving novel proteins that lack identifiable Pfam domains.

The original manuscript [40] also demonstrates that by utilizing the SS-predictor to estimate similarity instead of computing structural similarity (TM-score) across all protein pairs, both SS-predictor (10 s) and PLMSearch (4 s) rank among the fastest methods available, achieving speeds over four orders of magnitude faster than TM-align (11,303 s).

Overall, PLMSearch consistently ranks among the top methods, offering an optimal balance between accuracy and speed. Specifically, like MMseqs2, PLMSearch can process millions of query-target protein pairs within seconds, yet it achieves a more than threefold increase in sensitivity, rivaling state-of-the-art structure search methods. Additionally, the integrated pipeline of PLMSearch + PLMAlign not only addresses the limitation of PLMSearch's inability to provide alignments but also avoids the inefficiency of direct use of PLMAlign without pre-filtering, thereby ensuring the efficiency of the pipeline.

It is noteworthy that PLMSearch can achieve comparable efficiency on our publicly accessible Web server, which relies solely on CPU (64 * Intel(R) Xeon(R) CPU E5–2682 v4 @ 2.50 GHz with 512 GB RAM). Searching a query protein against Swiss-Prot (568 K proteins) and UniRef50 (53.6 M proteins), followed by employing PLMAlign to align the query with the Top-10 targets [14], takes approximately 0.2 and 2.3 min, respectively (Fig. 2). This efficiency is achieved as PLMSearch pre-generates and preloads the embeddings for all target proteins. This strategy significantly reduces time by eliminating the need for repeated forward propagations of the protein language model, which is parameter-

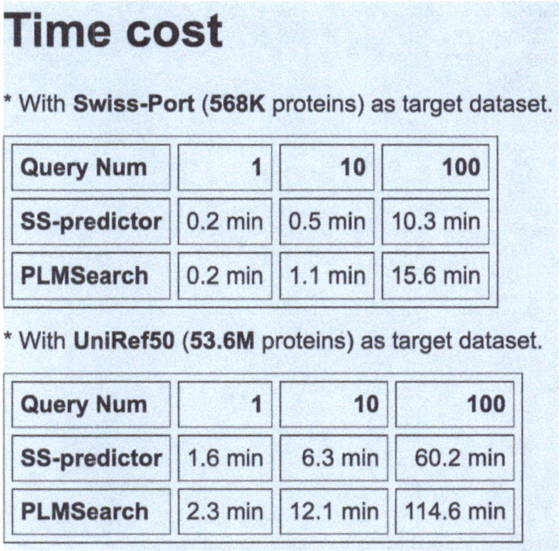

Time cost

* With **Swiss-Port (568K** proteins) as target dataset.

| **Query Num** | **1** | **10** | **100** |
|---|---|---|---|
| **SS-predictor** | 0.2 min | 0.5 min | 10.3 min |
| **PLMSearch** | 0.2 min | 1.1 min | 15.6 min |

* With **UniRef50 (53.6M** proteins) as target dataset.

| **Query Num** | **1** | **10** | **100** |
|---|---|---|---|
| **SS-predictor** | 1.6 min | 6.3 min | 60.2 min |
| **PLMSearch** | 2.3 min | 12.1 min | 114.6 min |

**Fig. 2** Time cost of PLMSearch Web server

heavy, and minimizes the time spent loading embeddings from disk to RAM. Consequently, only a single forward pass through the SS-predictor network is required to predict the similarity for millions of query-target pairs.

## 3 Web Server

The Web server interfaces for PLMSearch and PLMAlign are highly user-friendly. PLMSearch is freely accessible at https://dmiip.sjtu.edu.cn/PLMSearch, and PLMAlign is available at https://dmiip.sjtu.edu.cn/PLMAlign. Users simply require a modern Web browser (such as Firefox, Internet Explorer, Microsoft Edge, Mac Safari, or Chrome) and an internet connection to access the Web servers.

### *3.1 Submission to PLMSearch*

Users can submit queries and check results through the following five steps (Fig. 3):

1. **Input a protein sequence(s).** Users can either type the sequences directly into the provided text area or upload them from a file. If both methods are used, the server will prioritize the sequence(s) in the uploaded file. Only the FASTA format is acceptable, and all sequences must be in the standard single-letter amino acid code ( ACDEFGHIKLMNPQRSTVWYVBZX* ).
2. **Choose a target dataset.** For the target dataset, users have the option to select from Swiss-Prot (568 K proteins), PDB (680 K proteins), UniRef50 (53.6 M proteins), or the query dataset itself (Self).

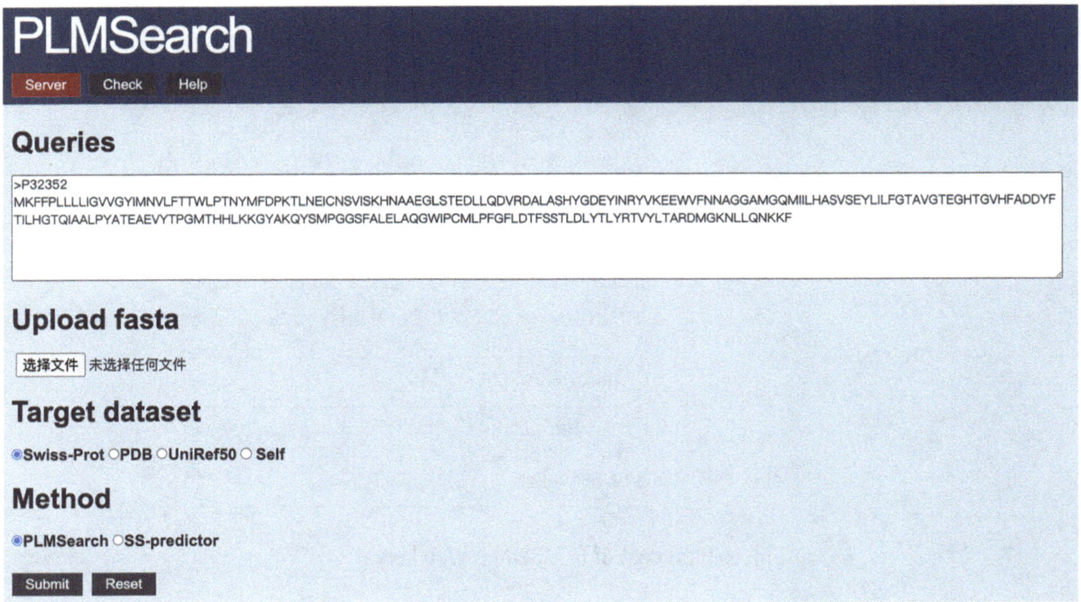


**Fig. 3** PLMSearch portal page

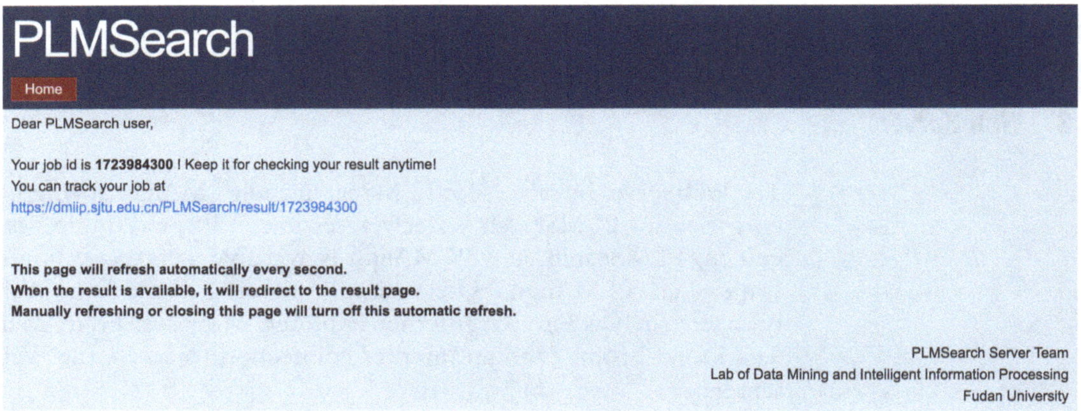


**Fig. 4** PLMSearch guidance page after submissions

3. **Choose a method.** Users can choose between PLMSearch or the SS-predictor. Brief descriptions are as follows:
   - **PLMSearch.** This method uses pairs pre-filtered by Pfam-Clan, rather than searching all protein pairs from scratch.
   - **SS-predictor.** This calculates the similarity across all proteins, ranking pairs based on this similarity.
4. **Submission.** After making their selections, users can press the "Submit" button to process the prediction. The server then provides a Job ID and a Web link to the results (Fig. 4). For example, using the submitted protein "P32352," the Job ID is

**Fig. 5** PLMSearch check page

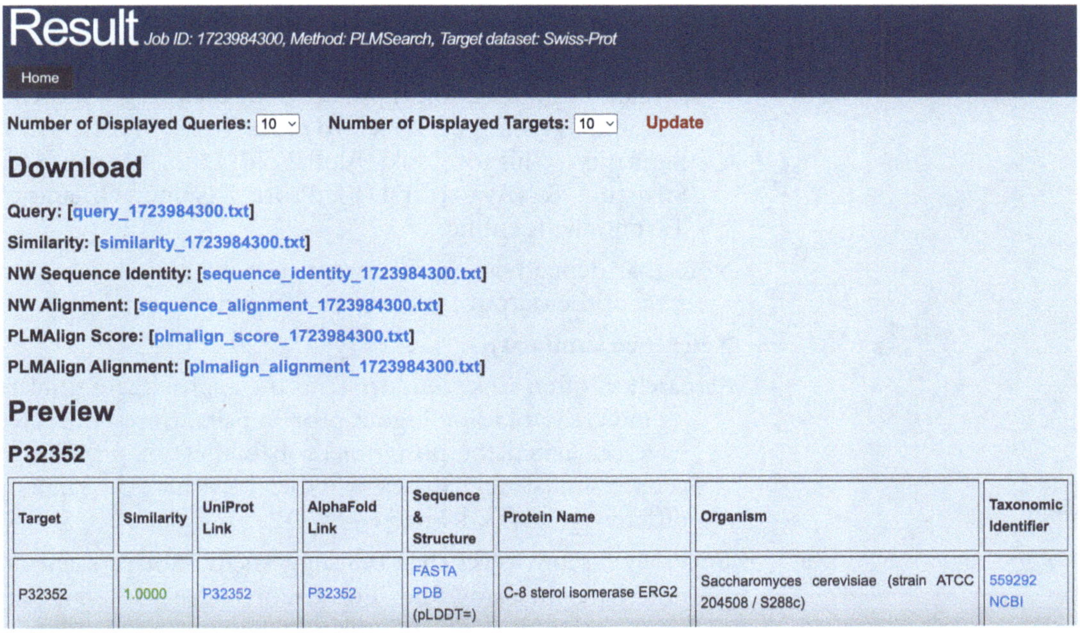


**Fig. 6** PLMSearch result page

"1723984300"users can also use the "Reset" button to clear all inputs.

5. **Checking Results.** Users can access their search results at any time by entering the previously provided Job ID on the "Check" page (Fig. 5). Alternatively, users can directly enter the Web link provided by the server into their browser to access the results.

### 3.2 Results Generated by PLMSearch

Users can "Download" or "Preview" the search results on the results page (Fig. 6).

1. **Download**
   - **Query**: All the query proteins in the Fasta format.
   - **Similarity**: The query-target pairs and their predicted similarity.

- **NW sequence identity**: Sequence identity from Needleman-Wunsch algorithm.
- **NW alignment**: Sequence alignment from Needleman-Wunsch algorithm.
- **PLMAlign score**: Score from PLMAlign.
- **PLMAlign alignment**: Sequence alignment from PLMAlign.

2. **Preview**

   Each table is presented for a query protein.

   Each row contains a target protein, arranged in descending order of similarity.

   Some relevant information is provided for each target protein. Column by column from left to right: Target (Protein ID), Similarity, UniProt Link, AlphaFold Link, Sequence & Structure & (Avg. pLDDT), Protein Name, Organism, Taxonomic Identifier.

   Note that depending on the target data set you choose, the style of the output table will be different.

3. **Reference similarity**

   Researchers often seek to understand the approximate similarity indicative of homologous protein pairs. In response, we have calculated the posterior probabilities for proteins at certain similarities to assess whether they share the same or different folds, as defined by SCOP.

   Generally, a score lower than 0.3 suggests that the proteins do not share the same fold. For a more detailed discussion, refer to the "Reference Similarity" Supplement Section in the original manuscript [40].

### 3.3 Submission to PLMAlign

Users can submit query and target proteins and check results through the following three steps (Fig. 7):

1. **Input query and target protein sequence(s)**

   The sequences can either be typed directly into the text area, or uploaded from a file. If both, the server will only consider the sequence(s) in the uploaded file.

   Only the FASTA format is acceptable. All sequences have to be amino acids specified in a single-letter code ( ACDEF GHIKLMNPQRSTVWYVBZX* ).

2. **Submit**

   Press the "Submit" button to process the alignment. A Job ID and a Web link to the results will be provided.

   Use the "Reset" button to clear all the inputs.

**Fig. 7** PLMSearch portal page

3. **Checking Results**

   Similar to the PLMSearch Web server, users can access their search results through either a Job ID or the provided Web link.

### *3.4 Results Generated by PLMAlign*

Users can "Download" or "Preview" the alignment results on the results page (Fig. 8).

1. **Download**
   - **Query**: All the query proteins in the Fasta format.
   - **Target**: All the target proteins in the Fasta format.
   - **Score**: The query-target pairs and their score.
   - **Alignment**: Sequence alignment results of query-target pairs.
   - **Sorted score**: Consistent with "Score," but the protein pairs are sorted from high to low scores.

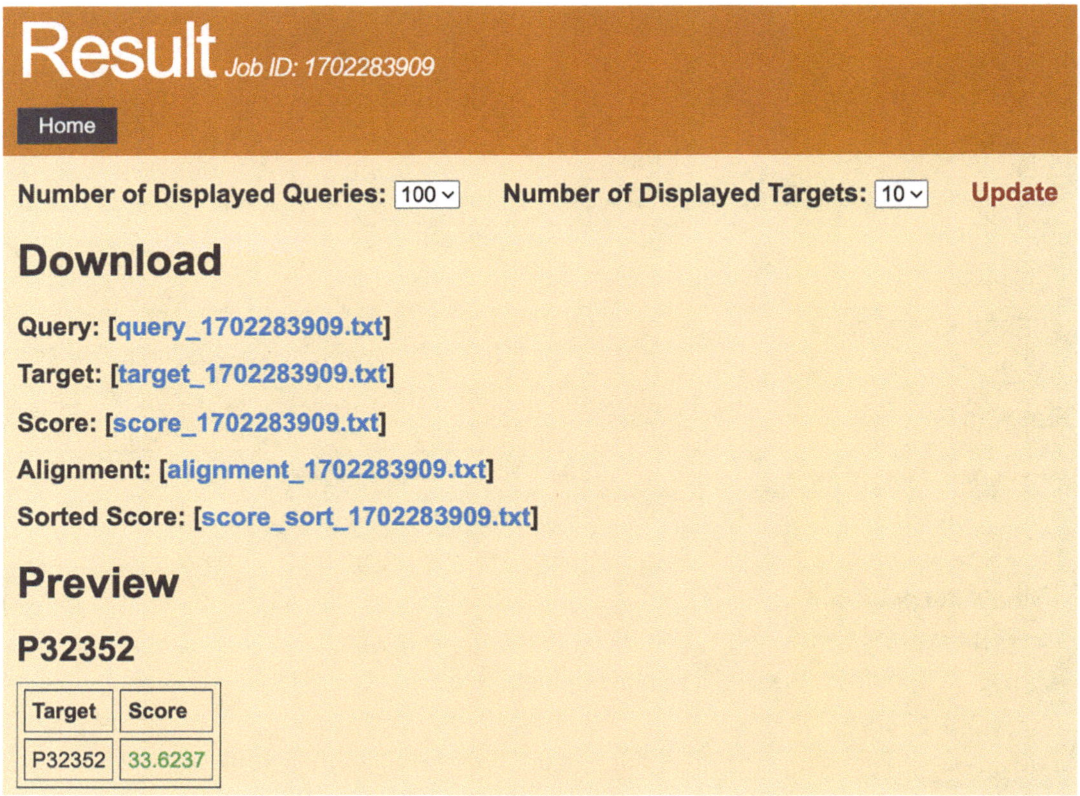


**Fig. 8** PLMAlign result page

2. **Preview**

   Each table is presented for a query protein.

   Each row contains a target protein, arranged in descending order of score.

3. **Reference similarity**

   For a more detailed discussion, refer to the "Reference Similarity" Supplement Section in the original manuscript [40].

## 4 Open-Source Resources

### 4.1 Code and Visual Reproduction

Our structural visualizations employ PyMOL v.2.4.0, accessible at https://github.com/schrodinger/pymol-open-source. For data visualization related to PLMSearch, we utilize Python (version 3.8.16), Seaborn (version 0.12.2), and matplotlib-base (version 3.6.2).

1. **PLMSearch**

   PLMSearch is freely available on GitHub at https://github.com/maovshao/PLMSearch. Specifically, researchers can set up and utilize the PLMSearch pipeline locally with a Jupyter Notebook, available at https://github.com/maovshao/PLMSearch/blob/main/pipeline.ipynb. Additionally, researchers can replicate PLMSearch experiments and regenerate figures in both the main text and supplementary materials using another Jupyter Notebook, found at https://github.com/maovshao/PLMSearch/blob/main/main.ipynb.

2. **PLMAlign**

PLMAlign is also freely available to the public on GitHub at https://github.com/maovshao/PLMAlign. Researchers can deploy and use the PLMAlign pipeline locally with a Jupyter Notebook available at https://github.com/maovshao/PLMAlign/blob/main/pipeline.ipynb. Additionally, experiments on malidup, malisam, and scope40 can be replicated using three separate Jupyter Notebooks at https://github.com/maovshao/PLMAlign/blob/main/malidup.ipynb, https://github.com/maovshao/PLMAlign/blob/main/malisam.ipynb, and https://github.com/maovshao/PLMAlign/blob/main/scope40.ipynb, respectively.

3. **Dataset**

Furthermore, we have published our experimental data for PLMSearch and PLMAlign on Zenodo at https://zenodo.org/records/11480660 and the target datasets used on our Web server at https://zenodo.org/records/11393990.

### 4.2 Containers

The Code Ocean platform standardizes the creation, tracking, and sharing of computational experiments in scientific research. It offers open, exportable, and readily reproducible containers—container technologies specifically designed for scientific investigations. Each container represents an independent computational experiment, encompassing the computing environment, code, data, version history, and results, facilitating the reproducibility of experiments contained within for researchers at any later date.

The results of PLMSearch can be directly replicated using a container published on Code Ocean, accessible at https://doi.org/10.24433/CO.8325548.v1. This allows researchers to replicate findings without the need to download source code or set up a local environment.

### 4.3 Limitation

1. **Query**: The maximum supported number of query proteins for each submission of PLMSearch and PLMAlign is 100. If the number of proteins in your job exceeds, just divide them into separate jobs and submit them.
2. **Similarity**: Only protein pairs with a similarity higher than 0.3 will be retained (0.5 for UniRef50). The maximum value of the total number of query-target pairs is 10,000. In other words, the server will return the Top-K highest similarity for each

protein in n queries, and $K = 10{,}000/n$ (If one query's actual search results are less than K, it will not be affected).

3. **NW & PLMAlign**: For each query, always return Top-10 alignment results.

   Use our source code to implement more searches (github.com/maovshao/PLMSearch) and alignments (github.com/maovshao/PLMAlign).

## Acknowledgments

This work has been supported by the National Natural Science Foundation of China (Grant No. 62272105, T2225007, 62172274), the Shanghai Municipal Science and Technology Major Project (Grant No. 2018SHZDZX01), 111 Project (Grant No. B18015), the ZJ Lab, the Shanghai Research Center for Brain Science and Brain-inspired Intelligence Technology, and Beijing Academy of Artificial Intelligence (BAAI).

### References

1. Yao S, You R, Wang S, Xiong Y, Huang X, Zhu S (2021) NetGO 2.0: improving large-scale protein function prediction with mass ive sequence, text, domain, family and network information. Nucleic Acids Res 49(W1): W469–W475
2. You R, Yao S, Xiong Y, Huang X, Sun F, Mamitsuka H, Zhu S (2019) NetGO: improving large-scale protein function prediction with massive network information. Nucleic Acids Res 47(W1):W379–W387
3. You R, Yao S, Mamitsuka H, Zhu S (2021) DeepGraphGO: graph neural network for large-scale, multispecies protein function prediction. Bioinformatics 37:i262–i271
4. Torres M, Yang H, Romero AE (2021) Paccanaro a protein function prediction for newly sequenced organisms. Nat Mach Intell 3(12): 1050–1060
5. Unsal S, Atas H, Albayrak M, Turhan K, Acar AC, Doğan T (2022) Learning functional properties of proteins with language models. Nat Mach Intell 4(3):227–245
6. Wang S, You R, Liu Y, Xiong Y, Zhu S (2023) NetGO 3.0: protein language model improves large-scale functional annotations. Genomics Proteomics Bioinformatics 21(2):349–358
7. You R, Zhang Z, Xiong Y, Sun F, Mamitsuka H, Zhu S (2018) GOLabeler: improving sequence-based large-scale protein function prediction by learning to rank. Bioinformatics 34(14):2465–2473. https://doi.org/10.1093/bioinformatics/bty130
8. Yan H, Wang S, Liu H, Mamitsuka H, Zhu S (2024) GORetriever: reranking protein-description-based GO candidates by literature-driven deep information retrieval for protein function annotation. Bioinformatics 40 (Supplement_2):ii53–ii61. https://doi.org/10.1093/bioinformatics/btae401
9. Radivojac P, Clark WT, Oron TR, Schnoes AM, Wittkop T, Sokolov A, Graim K, Funk C, Verspoor K, Ben-Hur A, Pandey G, Yunes JM, Talwalkar AS, Repo S, Souza ML, Piovesan D, Casadio R, Wang Z, Cheng J, Fang H, Gough J, Koskinen P, Törönen P, Nokso-Koivisto J, Holm L, Cozzetto D, Buchan DWA, Bryson K, Jones DT, Limaye B, Inamdar H, Datta A, Manjari SK, Joshi R, Chitale M, Kihara D, Lisewski AM, Erdin S, Venner E, Lichtarge O, Rentzsch R, Yang H, Romero AE, Bhat P, Paccanaro A, Hamp T, Kaßner R, Seemayer S, Vicedo E, Schaefer C, Achten D, Auer F, Boehm A, Braun T, Hecht M, Heron M, Hönigschmid P, Hopf TA, Kaufmann S, Kiening M, Krompass D, Landerer C, Mahlich Y, Roos M, Björne J, Salakoski T, Wong A, Shatkay H, Gatzmann F, Sommer I, Wass MN, Sternberg MJE, Škunca N, Şupek F, Bošnjak M, Panov P, Džeroski S, Šmuc T, Kourmpetis YAI et al (2013) A large-scale evaluation of computational protein function prediction. Nat Methods 10(3):221–227. https://doi.org/10.1038/nmeth.2340

10. Jiang Y, Oron TR, Clark WT, Bankapur AR, D'Andrea D, Lepore R, Funk CS, Kahanda I, Verspoor KM, Ben-Hur A, Koo DCE, Penfold-Brown D, Shasha D, Youngs N, Bonneau R, Lin A, Sahraeian SME, Martelli PL, Profiti G, Casadio R, Cao R, Zhong Z, Cheng J, Altenhoff A, Skunca N, Dessimoz C, Dogan T, Hakala K, Kaewphan S, Mehryary F, Salakoski T, Ginter F, Fang H, Smithers B, Oates M, Gough J, Törönen P, Koskinen P, Holm L, Chen C-T, Hsu W-L, Bryson K, Cozzetto D, Minneci F, Jones DT, Chapman S, Bkc D, Khan IK, Kihara D, Ofer D, Rappoport N, Stern A, Cibrian-Uhalte E, Denny P, Foulger RE, Hieta R, Legge D, Lovering RC, Magrane M, Melidoni AN, Mutowo-Meullenet P, Pichler K, Shypitsyna A, Li B, Zakeri P, ElShal S, Tranchevent L-C, Das S, Dawson NL, Lee D, Lees JG, Sillitoe I, Bhat P, Nepusz T, Romero AE, Sasidharan R, Yang H, Paccanaro A, Gillis J, Sedeño-Cortés AE, Pavlidis P, Feng S, Cejuela JM, Goldberg T, Hamp T, Richter L, Salamov A, Gabaldon T, Marcet-Houben M, Supek F, Gong Q, Ning W, Zhou Y, Tian W, Falda M, Fontana P, Lavezzo E, Toppo S, Ferrari C, Giollo M, Piovesan D, Tosatto SCE, del Pozo A, Fernández JM, Maietta P, Valencia A, Tress ML, Benso A, Di Carlo S, Politano G, Savino A, Rehman HU, Re M, Mesiti M, Valentini G, Bargsten JW, van Dijk ADJ, Gemovic B, Glisic S, Perovic V, Veljkovic V, Veljkovic N, Almeida-e-Silva DC, Vencio RZN, Sharan M, Vogel J, Kansakar L, Zhang S, Vucetic S, Wang Z, Sternberg MJE, Wass MN, Huntley RP, Martin MJ, O'Donovan C, Robinson PN, Moreau Y, Tramontano A, Babbitt PC, Brenner SE, Linial M, Orengo CA, Rost B, Greene CS, Mooney SD, Friedberg I, Radivojac P (2016) An expanded evaluation of protein function prediction methods shows an improvement in accuracy. Genome Biol 17(1):184. https://doi.org/10.1186/s13059-016-1037-6

11. Zhou N, Jiang Y, Bergquist TR, Lee AJ, Kacsoh BZ, Crocker AW, Lewis KA, Georghiou G, Nguyen HN, Hamid MN, Davis L, Dogan T, Atalay V, Rifaioglu AS, Dalkıran A, Cetin Atalay R, Zhang C, Hurto RL, Freddolino PL, Zhang Y, Bhat P, Supek F, Fernández JM, Gemovic B, Perovic VR, Davidović RS, Sumonja N, Veljkovic N, Asgari E, Mofrad MRK, Profiti G, Savojardo C, Martelli PL, Casadio R, Boecker F, Schoof H, Kahanda I, Thurlby N, McHardy AC, Renaux A, Saidi R, Gough J, Freitas AA, Antczak M, Fabris F, Wass MN, Hou J, Cheng J, Wang Z, Romero AE, Paccanaro A, Yang H, Goldberg T, Zhao C, Holm L, Törönen P, Medlar AJ, Zosa E, Borukhov I, Novikov I, Wilkins A, Lichtarge O, Chi P-H, Tseng W-C, Linial M, Rose PW, Dessimoz C, Vidulin V, Dzeroski S, Sillitoe I, Das S, Lees JG, Jones DT, Wan C, Cozzetto D, Fa R, Torres M, Warwick Vesztrocy A, Rodriguez JM, Tress ML, Frasca M, Notaro M, Grossi G, Petrini A, Re M, Valentini G, Mesiti M, Roche DB, Reeb J, Ritchie DW, Aridhi S, Alborzi SZ, Devignes M-D, Koo DCE, Bonneau R, Gligorijević V, Barot M, Fang H, Toppo S, Lavezzo E, Falda M, Berselli M, Tosatto SCE, Carraro M, Piovesan D, Ur Rehman H, Mao Q, Zhang S, Vucetic S, Black GS, Jo D, Suh E, Dayton JB, Larsen DJ, Omdahl AR, McGuffin LJ, Brackenridge DA, Babbitt PC, Yunes JM, Fontana P, Zhang F, Zhu S, You R, Zhang Z, Dai S, Yao S, Tian W, Cao R, Chandler C, Amezola M, Johnson D, Chang J-M, Liao W-H, Liu Y-W, Pascarelli S, Frank Y, Hoehndorf R, Kulmanov M, Boudellioua I, Politano G, Di Carlo S, Benso A, Hakala K, Ginter F, Mehryary F, Kaewphan S, Björne J, Moen H, Tolvanen MEE, Salakoski T, Kihara D, Jain A, Šmuc T, Altenhoff A, Ben-Hur A, Rost B, Brenner SE, Orengo CA, Jeffery CJ, Bosco G, Hogan DA, Martin MJ, O'Donovan C, Mooney SD, Greene CS, Radivojac P, Friedberg I (2019) The CAFA challenge reports improved protein function prediction and new functional annotations for hundreds of genes through experimental screens. Genome Biol 20(1):244. https://doi.org/10.1186/s13059-019-1835-8

12. Hu L, Wang X, Huang Y-A, Hu P, You Z-H (2021) A survey on computational models for predicting protein–protein interactions. Brief Bioinform 22(5):bbab036

13. Liu L, Huang X, Mamitsuka H, Zhu S (2020) HPOLabeler: improving prediction of human protein–phenotype associations by learning to rank. Bioinformatics 36(14):4180–4188

14. UniProt Consortium, T (2018) UniProt: the universal protein knowledgebase. Nucleic Acids Res 46(5):2699–2699

15. Steinegger M, Söding J (2017) MMseqs2 enables sensitive protein sequence searching for the analysis of massive data sets. Nat Biotechnol 35(11):1026–1028

16. Altschul SF, Gish W, Miller W, Myers EW, Lipman DJ (1990) Basic local alignment search tool. J Mol Biol 215(3):403–410

17. Buchfink B, Reuter K, Drost H-G (2021) Sensitive protein alignments at tree-of-life scale using DIAMOND. Nat Methods 18(4): 366–368

18. Mahlich Y, Steinegger M, Rost B, Bromberg Y (2018) HFSP: high speed homology-driven function annotation of proteins. Bioinformatics 34(13):i304–i312
19. Eddy SR (2011) Accelerated profile HMM searches. PLoS Comput Biol 7(10):e1002195
20. Söding J (2005) Protein homology detection by HMM–HMM comparison. Bioinformatics 21(7):951–960
21. Remmert M, Biegert A, Hauser A, Söding J (2012) HHblits: lightning-fast iterative protein sequence searching by HMM-HM M alignment. Nat Methods 9(2):173–175
22. Steinegger M, Meier M, Mirdita M, Vöhringer H, Haunsberger SJ, Söding J (2019) HH-suite3 for fast remote homology detection and deep protein annotation. BMC Bioinformatics 20(1):473
23. Illergård K, Ardell DH, Elofsson A (2009) Structure is three to ten times more conserved than sequence – a study of structural response in protein cores. Proteins 77(3):499–508
24. Jumper J, Evans R, Pritzel A, Green T, Figurnov M, Ronneberger O, Tunyasuvunakool K, Bates R et al (2021) Highly accurate protein structure prediction with AlphaFold. Nature 596(7873):583–589
25. Baek M, DiMaio F, Anishchenko I, Dauparas J, Ovchinnikov S, Lee GR, Wang J, Cong Q, Kinch LN, Schaeffer RD, Millán C, Park H, Adams C, Glassman CR, DeGiovanni A, Pereira JH, Rodrigues AV, Dijk AA, Ebrecht AC, Opperman DJ, Sagmeister T, Buhlheller C, Pavkov-Keller T, Rathinaswamy MK, Dalwadi U, Yip CK, Burke JE, Garcia KC, Grishin NV, Adams PD, Read RJ, David B (2021) Accurate prediction of protein structures and interactions using a three-track neural network. Science 373(6557):871–876
26. Varadi M, Anyango S, Deshpande M, Nair S, Natassia C, Yordanova G, Yuan D, Stroe O, Wood G, Laydon A, Žídek A, Green T, Tunyasuvunakool K, Petersen S, Jumper J, Clancy E, Green R, Vora A, Lutfi M, Figurnov M, Cowie A, Hobbs N, Kohli P, Kleywegt G, Birney E, Hassabis D, Velankar S (2022) AlphaFold protein structure database: massively expanding the structural coverage of protein-sequence space with high-accuracy models. Nucleic Acids Res 50(D1):D439–D444
27. Van Kempen M, Kim SS, Tumescheit C, Mirdita M, Lee J, Gilchrist CL, Söding J, Steinegger M (2024) Fast and accurate protein structure search with Foldseek. Nat Biotechnol 42(2):243–246
28. Rives A, Meier J, Sercu T, Goyal S, Lin Z, Liu J, Guo D, Ott M, Zitnick CL, Ma J, Rob F (2021) Biological structure and function emerge from scaling unsupervised learning to 250 million protein sequences. Proc Natl Acad Sci 118(15):e2016239118
29. Meier J, Rao R, Verkuil R, Liu J, Sercu T, Rives A (2021) Language models enable zero-shot prediction of the effects of mutations on protein function. Adv Neural Inf Proces Syst 34: 29287–29303
30. Lin Z, Akin H, Rao R, Hie B, Zhu Z, Lu W, Smetanin N, Verkuil R, Kabeli O, Shmueli Y, Santos Costa A, Fazel-Zarandi M, Sercu T, Candido S (2023) Evolutionary-scale prediction of atomic-level protein structure with a language model. Science 379(6637): 1123–1130
31. Hayes T, Rao R, Akin H, Sofroniew NJ, Oktay D, Lin Z, Verkuil R, Tran VQ, Deaton J, Wiggert M, Badkundri R, Shafkat I, Gong J, Derry A, Molina RS, Thomas N, Khan Y, Mishra C, Kim C, Bartie LJ, Nemeth M, Hsu PD, Sercu T, Candido S, Rives A (2024) Simulating 500 million years of evolution with a language model. bioRxiv:2024.2007.2001.600583. https://doi.org/10.1101/2024.07.01.600583
32. Elnaggar A, Heinzinger M, Dallago C, Rehawi G, Wang Y, Jones L, Gibbs T, Feher T, Angerer C, Steinegger M, Bhowmik D, Rost B (2021) ProtTrans: toward understanding the language of life through self-supervised learning. IEEE Trans Pattern Anal Mach Intell 44(10):7112–7127
33. Hu M, Yuan F, Yang K, Ju F, Su J, Wang H, Yang F, Ding Q (2022) Exploring evolution-aware &-free protein language models as protein function predictors. Adv Neural Inf Proces Syst 35:38873–38884
34. Rao R, Bhattacharya N, Thomas N, Duan Y, Chen X, Canny J, Abbeel P, Song YS (2019) Evaluating protein transfer learning with TAPE. Adv Neural Inf Process Syst 32:9689–9701
35. Llinares-López F, Berthet Q, Blondel M, Teboul O, Vert J-P (2023) Deep embedding and alignment of protein sequences. Nat Methods 20(1):104–111
36. Hamamsy T, Morton JT, Blackwell R, Berenberg D, Carriero N, Gligorijevic V, Strauss CE, Leman JK, Cho K, Bonneau R (2024) Protein remote homology detection and structural alignment using deep learning. Nat Biotechnol 42(6):975–985
37. Kaminski K, Ludwiczak J, Pawlicki K, Alva V, Dunin-Horkawicz S (2023) pLM-BLAST: distant homology detection based on direct

comparison of sequence representations from protein language models. Bioinformatics 39(10):btad579

38. Smith TF, Waterman MS (1981) Identification of common molecular subsequences. J Mol Biol 147(1):195–197
39. Needleman SB, Wunsch CD (1970) A general method applicable to the search for similarities in the amin o acid sequence of two proteins. J Mol Biol 48(3):443–453
40. Liu W, Wang Z, You R, Xie C, Wei H, Xiong Y, Yang J, Zhu S (2024) PLMSearch: protein language model powers accurate and fast sequence se arch for remote homology. Nat Commun 15(1):2775
41. Mistry J, Chuguransky S, Williams L, Qureshi M, Salazar GA, Sonnhammer ELL, Tosatto SCE, Paladin L, Raj S, Richardson LJ, Finn RD, Bateman A (2021) Pfam: The protein families database in 2021. Nucleic Acids Res 49(D1):D412–D419

# Chapter 15

# Large Context, Deeper Insights: Harnessing Large Language Models for Advancing Protein–Protein Interaction Analysis

**Kaicheng U, Sophia Meixuan Zhang, Suresh Pokharel, Pawel Pratyush, Farah Qaderi, Dongfang Liu, Junhan Zhao, Dukka B. KC, and Siwei Chen**

## Abstract

Protein–protein interactions (PPIs) are involved in nearly all biological processes. Understanding and analysis of PPI is key to revealing biological networks and identifying new therapeutic targets. Various computational approaches have been proposed as an alternative to the experimental investigation of PPIs. More recently, with the advent of Large Language Models (LLMs), a plethora of approaches using LLMs have been developed, enabling efficient analysis of interaction networks and binding sites directly from protein sequences. These models capture intricate biological patterns, offering scalability and adaptability across diverse datasets. However, challenges remain, including computational costs, data imbalance, and the integration of multimodal information. Advancements in addressing these limitations are set to further enhance the potential of LLMs in protein–protein interaction analysis, driving deeper insights and broader applications in biological research.

**Key words** Protein language model, Protein–protein interaction (PPI), Large language models (LLMs), Sequence-based models, PPI prediction

## 1 Introduction

Protein–protein interactions (PPIs) are the functional orchestration of cellular mechanisms involved in nearly all biological processes. Understanding PPIs and their specific interaction interfaces is key to revealing biological networks and identifying new therapeutic targets, especially in diseases driven by dysregulated PPIs [1–5]. Experimental investigation of PPIs is often labor-intensive and costly with traditional methods like protein arrays, Yeast Two-Hybrid (Y2H) systems, or newer technologies such as mass spectrometry and protein-fragment complementation assays

The authors Kaicheng U and Sophia Meixuan Zhang have contributed equally to this work. J. Zhao, D. KC and S. Chen served as senior authors.

Dukka B. KC (ed.), *Large Language Models (LLMs) in Protein Bioinformatics*, Methods in Molecular Biology, vol. 2941,
https://doi.org/10.1007/978-1-0716-4623-6_15,

(PCAs) [6–9]. In contrast, in silico bioinformatics approaches offer promising and efficient alternatives [10–13].

Deep learning has paved new avenues for exploring PPIs at an unprecedented scale and depth. Previous studies have utilized various deep neural networks, including convolutional, graph, and language models, to extract protein structural features for predicting PPIs [14–16]. The field has observed the application of large language models (LLMs), originally developed for natural language processing (NLP). Pretrained on large-scale protein sequence datasets, these models have shown an unmatched ability to capture intricate biological patterns and predict protein structures and functions [17–19]. This chapter outlines key biologically relevant applications in PPI research where LLMs hold transformative potential and highlights recent exploratory advancements in LLM-based approaches, from predicting PPI networks to protein interaction sites and beyond. In the beginning, we describe LLM-based approaches for predicting protein–protein interactions followed by LLM-based approaches for predicting protein–protein interaction sites. In both sections, we initially summarize some existing computational approaches followed by LLM-based approaches, summary and trends, and then limitations and future directions. Finally, we briefly describe other potential applications of LLMs for protein–protein interaction analysis.

## 2 Predicting Protein–Protein Interactions: Unveiling New Connections

Approximately 80% of proteins interact with others to perform essential cellular functions, and disruption of such interactions can lead to disease states such as cancer, neurodegenerative disorders, and infectious diseases [20–22]. Understanding PPIs is fundamental for interpreting cellular mechanisms, identifying therapeutic targets, and characterizing the roles of proteins with limited functional annotation [23]. Experimental techniques for determining PPIs, meanwhile, are time-consuming, resource-intensive, and are challenged by high rates of false positive and false negative predictions [24, 25]. These limitations highlight the necessity for computational approaches that can predict PPIs efficiently and accurately, leveraging the wealth of genomic, and proteomic data now available [22, 26].

### *2.1 Computational Approaches for Protein–Protein Interaction Prediction*

Early computational methods for PPI prediction relied on machine learning (ML) algorithms such as support vector machines (SVMs), random forests, and decision trees [22, 25, 26]. These approaches utilized manually engineered features derived from protein sequences, such as amino acid composition, physicochemical properties, and evolutionary information like position-specific scoring matrices (PSSMs) [25, 27]. While these methods expanded the field's understanding of PPI prediction, they were constrained by their dependence on feature engineering and often lacked scalability and generalizability across different datasets [25, 26]. The advent of deep learning introduced models capable of automatic feature extraction, significantly improving prediction performance [22]. Convolutional neural networks (CNNs) and recurrent neural networks (RNNs) were employed to capture local and global sequence patterns, respectively. For example, models like DeepPPI incorporated statistical descriptors alongside deep neural networks to refine feature representation [28]. Additionally, graph-based models, such as graph convolutional networks (GCNs) and graph neural networks (GNNs), modeled proteins as nodes and their interactions as edges, effectively capturing the structural and topological characteristics of interaction networks [27]. Despite these advancements, traditional deep learning models faced methodological limitations, including overfitting in scenarios with constrained training data and a lack of generalizability across species, often due to biases introduced by sequence similarity [29].

### 2.2 LLM-Based Approaches for Protein–Protein Interaction Prediction

In this section, we briefly review recent advances in PPI prediction using LLMs. LLMs pretrained on extensive protein sequence databases have transformed PPI prediction. These models, including ESM-2 [20], ProtBERT [21, 30], ProtT5 [30], and SeqVec [31], treat protein sequences analogously to sentences in natural language, capturing contextual and evolutionary information without relying on multiple sequence alignments [23–27, 32, 33]. Transformer architectures enable these models to identify long-range dependencies within protein sequences, which are critical for understanding structural and functional properties [23, 24, 32, 33]. One example is the work by Jha et al. [24], who developed a framework integrating SeqVec [31] embeddings based on ELMo language models [34] with Graph-BERT [24, 35], a graph-based transformer model. By representing protein pairs as nodes within a graph and using linkless subgraphs, the model addresses limitations like suspended animation and over-smoothing encountered in traditional GCNs [24]. It achieved 99.10% accuracy on the human PPI dataset, demonstrated strong cross-species generalization, and effectively managed imbalanced datasets. Performance metrics, including sensitivity and specificity, consistently outperformed earlier approaches across multiple benchmarks.

Similarly, xCAPT5 [23] uses a hybrid classifier that integrates ProtT5-XL-UniRef50 [30] embeddings with multi-kernel CNNs within a Siamese architecture. ProtT5 provides rich amino acid embeddings capturing contextual, evolutionary, and functional information. The multi-kernel CNNs capture both local and global sequence patterns, refining these features through pooling and concatenation. By integrating XGBoost [36] for enhanced classification, xCAPT5 achieved state-of-the-art performance across various datasets, with accuracy rates of 97.27% on the Martin dataset [37] and over 99% on the Guo and Pan datasets [38, 39]. The model outperformed existing models, demonstrating robustness across species and maintaining high recall and F1 scores even under stringent sequence similarity thresholds. However, the model's recall-oriented design may compromise precision in certain applications, and it requires extensive hyperparameter tuning.

The sequence-structure-function (S2F) multimodal pretraining model broadens the scope of LLM applications by integrating sequence, structural topology, and functional annotations to produce embeddings that outperform single-modality methods [26]. S2F employs a single-stream transformer for encoding and pretraining tasks like masked sequence modeling and sequence-structure-function alignment. This multimodal framework enhances PPI predictions, particularly in cases involving structural flexibility and disordered regions. The study reported that S2F embeddings outperformed existing methods such as ELMo [34], TAPE [40], and ProSE [41] across tasks, including cross-species PPIs, antibody–antigen interactions, and mutation-driven affinity predictions.

Moreover, Wang et al. introduced a biological language model for PPI prediction that only relies on protein sequence data [25]. Their method employs Bio2Vec, a feature representation technique based on "bio-word" segmentation and the Skip-Gram word model, to generate embeddings capturing local and global semantic features of protein sequences [42]. These embeddings are processed through CNN for prediction, achieving superior performance across datasets, including human, yeast, and *Helicobacter pylori*. This approach reinforces the value of analyzing protein sequences with language models.

In another application, TT3D combines sequence and structural information to predict PPIs genome-wide [43]. By combining one-hot-encoded 3Di sequences from Foldseek with embeddings from a protein language model [44], TT3D demonstrated higher precision and recall in cross-species PPI prediction compared to baseline methods. The model excels in generalizing predictions to nonhuman species, achieving significantly higher precision than sequence/structure homology transfer approaches. However, it relies on precomputed structural models; predictions for proteins without such data default to uninformative 3Di sequences.

Furthermore, ProLLM represents a novel approach by converting PPI prediction into a natural language processing task using Protein Chain of Thought (ProCoT) [33]. ProCoT simulates stepwise signal transduction in proteins, enhancing the model's understanding of signaling pathways. By replacing standard embeddings with those derived from ProtTrans [30] and fine-tuning the model on the Mol dataset, ProLLM outperformed GNN- and LLM-based baselines with high micro-F1 scores across multiple benchmark datasets. The model excels in reasoning indirect interactions and predicting complex multi-step relationships in signaling pathways. However, dependency on curated datasets means performance is tied to the quality of signaling pathway annotations, and larger LLMs exhibited instability compared to smaller, optimized backbones. LLM-based approaches described for PPI prediction described in this chapter are summarized in Table 1.

### 2.3 Summary and Trends for LLM-Based Approaches for PPI Prediction

The integration of LLMs into PPI prediction has enabled more nuanced and comprehensive analyses, facilitating advancements across various biological and biomedical applications. For instance, the Graph-BERT framework has been instrumental in mapping large-scale PPI networks [24], providing insights into complex interaction patterns that were previously difficult to discern. Its ability to generalize across species makes it a valuable tool for comparative genomics and evolutionary studies, allowing researchers to infer PPIs in less-studied organisms based on human data. xCAPT5 has proven effective in diverse PPI prediction scenarios, including intra-species and cross-species interactions [23]. Its robust performance under stringent sequence similarity thresholds suggests its utility in studying novel or highly divergent proteins, which is particularly relevant in the context of emerging pathogens or rare genetic variants. The integration of XGBoost further enhances its classification capabilities [36], making it suitable for applications requiring high precision and recall, such as identifying potential drug targets or understanding disease mechanisms. The S2F model has significant implications for drug discovery and therapeutic development [26]. By effectively predicting antibody–antigen interactions and mutation-driven affinity changes, it aids in the design of more effective vaccines and antibody-based therapies. Additionally, its ability to handle structurally flexible and intrinsically disordered regions expands its applicability to a wider range of proteins. Bio2Vec [25] offers a scalable and linguistically inspired approach to PPI prediction, making it suitable for high-throughput analyses and large-scale proteomic studies. TT3D enhances genome-wide PPI predictions by incorporating structural information [43], which is crucial for understanding the mechanistic basis of protein interactions. This method is particularly useful in structural biology and computational docking studies, where accurate PPI predictions can streamline the identification of viable protein complexes for experimental validation. ProLLM introduces a novel reasoning-based approach to PPI prediction, leveraging the concept of Protein Chain of Thought to model signaling pathways [33]. This capability is particularly valuable in systems biology and network medicine, where understanding the flow of signals through protein networks is essential for deciphering cellular processes and identifying intervention points for therapeutic strategies. Overall, the application of LLMs in PPI prediction has not only

**Table 1**
**Overview of recent large language model (LLM)-based approaches for predicting PPIs**

| Method | Method type | Key features | Applications/results |
|---|---|---|---|
| Graph-BERT with SeqVec embeddings [24] | GNN, LLM | Integrates SeqVec embeddings (based on ELMo language models) with graph-BERT<br>Represents protein pairs as nodes in a graph<br>Uses linkless subgraphs to improve representation | Achieved 99.10% accuracy on human PPI dataset<br>Demonstrated high sensitivity and specificity across multiple PPI datasets<br>Effectively handled imbalanced datasets |
| xCAPT5 [23] | DL, CNN, LLM | Uses ProtT5-XL-UniRef50 embeddings for rich contextual information<br>Employs multi-kernel CNNs within a Siamese architecture<br>Integrates XGBoost for classification | Achieved 97.27% accuracy on the Martin dataset<br>Over 99% accuracy on Guo and Pan datasets<br>Maintained high recall under stringent sequence similarity thresholds<br>Effective for intra-species and cross-species PPI prediction |
| S2F Model [26] | Transformer, LLM | Multimodal pretraining integrating sequence, structural topology, and functional annotations<br>Single-stream transformer encoding three modalities<br>Pretrained on multiple tasks | Improved cross-species PPI prediction<br>Enhanced antibody–antigen interaction prediction<br>Effective in mutation-driven affinity change prediction<br>Outperformed single-modality approaches in capturing complex PPIs |
| Bio2Vec [25] | DL, CNN | Generates protein sequence embeddings using bio-word segmentation and skip-gram model<br>Captures local and global semantic information<br>Utilizes CNN for prediction | Achieved 97.31% accuracy on human dataset<br>93.30% accuracy on yeast dataset<br>88.01% accuracy on *H.* pylori dataset<br>Demonstrated scalability and effectiveness across different species |
| ProtInteract [45] | DL, CNN | Uses Conv-TCN-AE for dimensionality reduction<br>Employs deep CNN for PPI prediction<br>Addresses imbalanced datasets with downsampling | Achieved 95.68% accuracy for Homo sapiens in two-class scenario<br>92.44% accuracy in three-class scenario<br>Faster training times compared to Conv-LSTM-AE baseline<br>Effective across different species datasets |

| | | | |
|---|---|---|---|
| TT3D [43] | DL | Combines one-hot-encoded 3Di sequences from Foldseek with protein language model embeddings<br>Integrates sequence and structural representations | Provided higher early precision in cross-species PPI prediction<br>Improved performance in identifying PPIs for downstream experiments<br>Effective in generalizing predictions to nonhuman species |
| ProLLM [33] | LLM, Transformer | Introduces protein chain of thought (ProCoT) for modeling signaling pathways<br>Uses ProtTrans embeddings capturing structural attributes<br>Fine-tuned with domain-specific data | Achieved micro-F1 scores:<br>91.05% on human dataset<br>78.09% on SHS27k<br>87.66% on SHS148k<br>89.21% on STRING dataset<br>Effective in reasoning over signaling pathways and indirect interactions |
| TUnA [32] | Transformer, LLM | Combines ESM-2 embeddings with dual transformer encoders<br>Integrates uncertainty estimation using SNGP<br>Quantifies prediction uncertainty | Achieved Matthews correlation coefficient (MCC) of 0.54 and precision of 0.57 on the Bernett dataset<br>Higher area under precision-recall curve (AUPR) on cross-species datasets compared to previous models<br>Effective in filtering high-confidence predictions |
| PEFT with LoRA [29] | PEFT, LLM | Uses LoRA to add low-rank matrices to attention layers<br>Updates less than 1% of model parameters<br>Reduces memory and computational costs | Achieved comparable performance to full fine-tuning with significantly less memory usage<br>Enabled fine-tuning large PLMs on standard GPUs<br>Demonstrated scalability to larger datasets |

This table summarizes various methods incorporating LLMs, detailing their methodological types, key features, and performance outcomes across different PPI datasets improved the accuracy and scalability of these models but also expanded their utility across a wide range of biological and biomedical research areas. These advancements facilitate deeper insights into cellular mechanisms, enhance drug discovery efforts, and contribute to the development of novel therapeutic interventions.

### 2.4 Limitations and Future Directions for LLM-Based PPI Prediction

Despite significant progress, LLM-based models for PPI prediction face several limitations. One major challenge is the high computational cost associated with training and deploying large models [23, 29, 32, 43]. Models like ProtT5-XL-UniRef50 [30] and ESM-2 [20], while powerful, require substantial computational resources, which may not be accessible to all research groups [23, 29, 32, 43]. Efforts to address this include the use of parameter-efficient fine-tuning (PEFT) methods such as low-rank adaptation (LoRA) [46], which significantly reduce memory and computational requirements by adding a small number of trainable parameters while achieving performance comparable to traditional fine-tuning [29, 43]. Sledzieski et al. showed that PEFT achieves state-of-the-art results on PPI prediction benchmarks, outperforming traditional fine-tuning and frozen embedding methods. Using embeddings from ESM2 with PEFT reduces GPU memory requirements, enabling deeper model adaptation within resource constraints [29, 43]. Another limitation is the reliance on high-quality training data. Models often perform well on datasets with abundant and accurate annotations but may struggle with proteins lacking sufficient sequence similarity or structural information [23]. Class imbalance within datasets further complicates performance, as models tend to prioritize recall over precision in imbalanced scenarios [23]. Moreover, models like TUnA incorporate uncertainty awareness to address the limitations of existing models but still face challenges related to computational cost and sequence length constraints [29, 32, 43].

Future research directions include enhancing model interpretability and integrating additional protein data modalities [47]. Integrating structural predictions from tools like AlphaFold2 [15] could enrich the training data and improve model generalization. Developing methods to better handle intrinsically disordered regions and transient interactions remains a critical area of focus, as these aspects are often underrepresented in current datasets [47]. Expanding the applicability of LLM-based models to genome-wide PPI predictions and exploring their integration with other computational approaches could provide more comprehensive interaction networks [33]. Additionally, further development of parameter-efficient methods tailored for protein language, such as QLoRA, which incorporates model quantization for resource efficiency, may unlock additional performance gains [48]. Scaling protein language models with larger and more diverse datasets, including those from metagenomes, could also enhance generalizability [33]. Efforts should also focus on optimizing transformer models by adapting specific layers to improve performance in memory-constrained environments and combining more informative embeddings with advanced architectures for PPI prediction [29].

LLM applications in PPI prediction have significantly improved accuracy, scalability, and the capacity to capture complex sequences and structural features. By leveraging large protein sequence datasets and sophisticated architectures, these models address many limitations of traditional computational approaches [45, 47]. While challenges remain, ongoing research and innovations continue to push the boundaries of what is possible in PPI prediction [23, 29, 43, 47]. The integration of LLMs holds immense potential for accelerating biological discoveries and advancing therapeutic development, contributing to a deeper understanding of cellular processes [47].

## 3 Mapping Interaction Interfaces: Predicting Protein–Protein Interaction Sites

PPIs form dynamic networks that govern a wide range of biological processes. Central to these interactions are the specific binding sites, known as PPI interfaces, which consist of distinct amino acid residues that determine the specificity, affinity, and kinetics of the interactions. These interfaces not only facilitate protein binding but also modulate the strength and stability of the resulting complexes, playing pivotal roles in the assembly and regulation of protein machinery within the cell.

Disruptions at PPI interfaces, whether due to genetic mutations, post-translational modifications, or environmental factors, can lead to aberrant protein interactions and dysregulated cellular activities. Such perturbations are implicated in a wide array of diseases, including cancer, neurodegenerative disorders, cardiovascular diseases, and developmental abnormalities. For instance, mutations that alter interface residues can impede critical signaling pathways or metabolic control mechanisms, leading to pathological conditions [49, 50]. Detailed molecular characterization of PPI interfaces is essential for understanding disease mechanisms and designing targeted therapeutic strategies.

### 3.1 Modern Computational Approaches for Protein–Protein Interaction Site Prediction

The computational prediction of PPI sites has evolved significantly, transitioning from traditional ML methods to sophisticated deep learning architectures. Early ML approaches relied heavily on engineered features derived from protein sequences and structures, such as amino acid composition, evolutionary conservation scores, physicochemical properties, and secondary structure elements. Statistical classifiers like support vector machines, random forests, and ensemble methods were employed to identify patterns indicative of interaction sites [49, 50]. For example, the Interactome INSIDER tool [50] employed the ECLAIR algorithm, an ensemble machine learning classifier, to predict unresolved PPI interfaces by integrating features like co-evolutionary metrics and docking-based scores [51]. By leveraging both sequence-based and structural features, ECLAIR achieved high predictive performance, allowing for the large-scale prediction of PPI interfaces across the human proteome and several model organisms. This approach deepened our understanding of how genomic variants may disrupt protein interactions, as evidenced by the enrichment of disease-associated mutations at predicted interface sites. Traditional ML methods, while valuable, faced limitations due to their reliance on handcrafted features and their inability to capture the complex, nonlinear relationships inherent in protein interactions. To address these challenges, researchers turned to deep learning, which offers the capability to automatically learn hierarchical representations from raw data.

CNNs have been widely applied due to their strength in capturing local spatial patterns. DeepRank, a CNN-based framework, transforms three-dimensional protein structures into voxelized grids encoding physicochemical properties [27]. By applying convolutional filters over these grids, DeepRank learns spatial features associated with PPI interfaces. The model was effective in classifying biological versus crystallographic interfaces and ranking docking models. However, CNNs face challenges in handling rotational and translational invariance inherent in protein structures, often necessitating extensive data augmentation to capture different orientations and conformations. To overcome these limitations, GNNs emerged as a powerful alternative. GNNs represent proteins as graphs, with residues as nodes and interactions as edges, naturally accommodating the complex topology of protein structures. DeepRank-GNN extends the DeepRank framework by employing GNNs to process protein–protein interfaces [52]. By applying message-passing algorithms, the model captures both local and global structural relationships, providing rotation and translation invariance without requiring data augmentation. DeepRank-GNN demonstrated strong performance in docking model scoring and differentiating between biological and crystallographic protein dimers, highlighting the efficacy of GNNs in PPI site prediction [52].

Another notable GNN-based approach is PIQLE, which employs a multi-head graph attention network to assess the quality of PPI interfaces [53]. PIQLE represents interfacial residues as nodes and incorporates multimeric interaction geometries, such as interatomic distances and orientations, as edge features. The attention mechanism allows the model to focus on critical interactions within complex multimeric interfaces, enhancing the estimation of interface quality. PIQLE has consistently outperformed existing methods like DProQA [54] and TRScore [55] in benchmarking on various datasets. Furthermore, methods that integrate evolutionary information have been shown to enhance prediction accuracy. PIQLE incorporates multiple sequence alignments (MSAs) to leverage evolutionary conservation, which allows the model to capture conserved patterns across homologous proteins, improving the detection of critical interface residues.

Ensemble neural networks have also been employed to combine the strengths of different architectures. PIPENN, an ensemble framework, integrates multiple neural network architectures including fully connected networks, CNNs, dilated convolutional networks, U-Nets, residual networks, and recurrent neural networks [56]. The ensemble approach leverages the strengths of each architecture to address the limitations of individual models, though it increases computational complexity and may require careful tuning to optimize the balance among the models. PIPENN utilized U-Net architectures, which are particularly effective in biomedical image segmentation and can capture both local and contextual information through their encoder-decoder structures. Residual networks (ResNets) [57] have been employed to facilitate the training of deeper networks by mitigating the vanishing gradient problem, allowing the models to learn more complex features. RNNs [58], including long short-term memory (LSTM) units, have been used to model sequential dependencies in protein sequences, which capture long-range interactions that may be critical for PPI site prediction.

Despite the advancements brought by deep learning, challenges remain in terms of computational demands and data requirements. Deep learning models often require large amounts of high-quality labeled data to train effectively, which can be a limitation where certain annotated datasets are scarce. Additionally, processing three-dimensional structures or large protein complexes can be computationally intensive. These limitations highlight the need for future LLMs that can effectively integrate sequence and structural information while being computationally efficient and scalable. By leveraging the vast amount of protein sequence data and advanced architectures capable of capturing complex dependencies, LLMs offer a promising avenue for improving PPI site prediction.

### 3.2 LLM-Based Approaches for PPI Site Prediction

Building upon the advancements in traditional computational methods, the integration of LLMs into bioinformatics has transformed PPI site prediction. Inspired by advancements in natural language processing (NLP), transformer architectures have been repurposed to analyze protein sequences, enabling the prediction of PPIs at the sequence level without reliance on structural data. Central to this shift is the conceptualization of protein sequences as linguistic constructs, where amino acids serve as "words" and the entire sequence forms a "sentence." This analogy allows for the utilization of transformer-based models to capture the intricate dependencies and contextual relationships among amino acids. Pretrained protein language models such as ESM (Evolutionary Scale Modeling) [59], ProtBERT [21], and ProtT5 [30] have been trained on large protein sequence datasets, generate embeddings that encapsulate local and global sequence features, serving as robust inputs for downstream tasks.

One of the pioneering efforts in this domain is the introduction of an ensemble deep learning model (EDLMPPI), which leverages the transformer-based protein language model ProtT5 [60]. By encoding amino acid sequences into distributed vector representations, ProtT5 captures the global contextual associations inherent in protein sequences. EDLMPPI further enhances its predictive capabilities by integrating Bidirectional Long Short-Term Memory (BiLSTM) networks and capsule networks within its architecture [61]. The BiLSTM networks are adept at capturing long-range dependencies in sequences, essential for modeling interactions that may occur between distant residues. Capsule networks, on the other hand, explore the intrinsic associations between features, allowing the model to understand hierarchical relationships within the data [62]. This ensemble approach not only improves the model's ability to capture complex patterns relevant to PPI sites but also addresses the challenge of data imbalance commonly encountered in PPI datasets. Moreover, the model offers explainability, providing insights into the protein binding site identification mechanisms and the understanding of biological processes.

In parallel, EnsemPPIS, an ensemble framework, combines transformer-based models with a model employing gated convolutional neural networks (GatCNNPPIS) to predict PPI sites using only protein sequence information [63]. The transformer component, TransformerPPIS, leverages the self-attention mechanism to capture global sequence features and model long-range dependencies. By utilizing ProtBERT [21], a pretrained protein language model, EnsemPPIS generates rich contextual embeddings for amino acid residues, effectively capturing the nuances of protein sequences learned from large-scale protein data. The ensemble nature of EnsemPPIS allows it to integrate the strengths of both models. While TransformerPPIS excels at capturing global features and long-range interactions, GatCNNPPIS focuses on local sequential features through gated convolutional operations. Notably, EnsemPPIS achieves high accuracy without relying on structural data, making it broadly applicable across the proteome, including proteins without known structures.

The combination of protein language models with GNNs represents another promising direction for PPI site prediction. Pair-EGRET, an edge-aggregated graph attention network, utilizes features extracted from ProtBERT to predict PPI sites [64]. Pair-EGRET represents proteins as k-nearest neighbor graphs based on their three-dimensional structures, capturing spatial relationships between residues. By combining ProtBERT embeddings with a graph neural network architecture [21], the model leverages both the sequential information captured by the language model and the structural information represented in the graph. A key innovation in Pair-EGRET is the use of a multi-headed cross-attention mechanism, which allows each protein to attend to relevant features of its interacting partner. This mechanism enhances the model's ability to accurately identify interfacial residues by focusing on important interactions between residues of different proteins. The integration of sequence and structural information, along with the attention mechanism, enables Pair-EGRET to outperform several state-of-the-art methods in predicting PPI sites. Another example is DeepRank-GNN-esm, which integrates ESM-2 embeddings into a graph neural network framework to predict interaction interfaces and score protein–protein docking models [27]. Unlike sequence-only methods, this approach constructs residue-level graphs from the 3D structures of protein complexes, treating residues within interaction interfaces as nodes. ESM-2 embeddings replace computationally expensive PSSM features, enabling the model to combine evolutionary, contextual, and structural information for improved accuracy in identifying biologically relevant interfaces and discriminating docking conformations from decoys.

Another significant contribution is MP-BERT (MindSpore Protein BERT), a bidirectional encoder representation model designed to process protein pairs directly [65]. By training MP-BERT on protein pairs, the model inherently understands the interactions between proteins. The pretrained MP-BERT is fine-tuned for specific tasks: MPB-PPI for predicting PPIs and MPB-PPISP for predicting PPI sites. MP-BERT demonstrates high accuracy in predicting PPIs across various organisms, achieving an aggregated accuracy of 92.65%. Importantly, MPB-PPISP predicts interaction site propensity using sequence data alone, achieving competitive performance compared to models that require additional structural features. This approach reduces the computational resources typically associated with methods that rely on evolutionary data such as multiple sequence alignments.

Advancements in inter-protein contact prediction using LLMs are exemplified by DRN-1D2D_Inter, a deep learning-based model that leverages dimensional hybrid residual networks and protein language model embeddings [66]. By employing embeddings from pretrained models like ESM-1b and ESM-MSA-1b, the model predicts residue–residue contacts between interacting proteins [59]. The dimensional hybrid residual networks combine one-dimensional and two-dimensional convolutional operations, effectively capturing both sequential and spatial features. Remarkably, DRN-1D2D_Inter outperforms state-of-the-art methods that rely on structural data, despite using only sequence information. Integrating the predicted contacts into protein–protein docking significantly improves the accuracy of protein complex structure prediction, opening the possibility of predicting protein complex structures without experimental structural data.

Addressing data challenges and enhancing model interpretability are critical aspects of advancing PPI site prediction. Data imbalance, where non-interacting residues vastly outnumber interacting ones, remains a pervasive issue in PPI site prediction. Ensemble deep learning architectures employed in models like EDLMPPI [60] and EnsemPPIS [63] effectively mitigate this problem by improving generalization and reducing overfitting. Techniques such as attention mechanisms and interpretable deep learning frameworks provide insights into model decision-making, fostering trust in predictions and guiding biological investigations. Graph-based approaches complement LLMs by offering alternative representations for processing protein data. MEG-PPIS utilizes multi-scale graph information and Equivariant Graph Neural Networks (EGNNs) for PPI site prediction [67]. MEG-PPIS focuses on leveraging structural information through graph representations and achieves improved speed and accuracy, outperforming other methods. Integrating the complementary strengths of graph-based and LLM methods could lead to hybrid models that combine rich contextual information from sequences and detailed structural relationships.

In conclusion, the integration of LLMs into PPI site prediction represents a significant leap forward in computational biology. Transformer-based architectures and pretrained embeddings allow these models to capture complex patterns and contextual relationships in protein sequences. Continued innovation in hybrid modeling, addressing data challenges, and improving interpretability will drive further progress in this domain, enabling deeper exploration of protein interactions and accelerating discoveries in biology and medicine. The tools described in this section are presented in Table 2.

### 3.3 Summary and Trends for LLM-Based Approaches for PPI Site Prediction

The development of LLM-based approaches in PPI site prediction represents a significant advancement in our understanding of cellular processes and the acceleration of drug discovery. One of the most impactful applications is the identification of novel therapeutic targets. Accurate prediction of PPI sites allows researchers to pinpoint critical residues involved in disease-related interactions, thereby facilitating the design of molecules that can modulate these interactions with high specificity. For instance, the EDLMPPI model has demonstrated high precision in predicting PPI binding sites, which has informed efforts to target protein interfaces implicated in cancer and neurodegenerative diseases [60]. The explainable nature of EDLMPPI's predictions enables the identification of key binding residues, which is instrumental in developing targeted therapies with minimized off-target effects. In immunology,

**Table 2**
**Comprehensive summary of contemporary computational techniques for predicting protein–protein interaction interfaces and sites**

| Method | Method type | Key features | Applications/results |
|---|---|---|---|
| *PPI interface prediction* | | | |
| Interactome INSIDER (ECLAIR) [50] | Traditional ML | Integrates genomic and structural PPI data<br>Predicts interfaces for 185,000+ interactions<br>Uses ECLAIR with sequence, structural, docking-based features | Predicted PPI interfaces for 185,000 interactions<br>Validated with >2000 mutagenesis experiments<br>Linked mutations to disease-associated PPI sites<br>Accessible via a web-based interface |
| Interactome Perturbation Framework [49] | Hybrid ML, Experimental | Combines computational predictions with yeast two-hybrid assays<br>Uses Interactome INSIDER for interface prediction<br>Integrates PolyPhen-2 for mutation impact | Achieved 80.8% accuracy in PPI prediction compared to Y2H assays<br>Identified that mutations in ASD probands disrupt PPIs more frequently<br>Enriched disruptive mutations in network hubs and haploinsufficient genes |
| DeepRank [68] | DL, 3D CNN | Converts 3D PPI structures into voxel grids<br>GAUSSIAN-based feature mapping<br>Modular architecture for various PPI tasks | Achieved 86% accuracy in classifying biological vs. crystallographic interfaces<br>Improved ranking of docking models based on scores<br>Enhanced classification accuracy for distinguishing interface types |
| DeepRank-GNN [52] | DL, GNN | Represents PPIs as residue-level graphs<br>Uses message passing in GNNs<br>Rotation and translation invariant<br>Handles varied interface sizes | Achieved AUC >0.95 on BM5 docking benchmark<br>82% accuracy in classifying biological vs. crystal interfaces<br>Reduced computational demands by up to 20× compared to CNN-based methods |
| DeepRank-GNN-esm [27] | DL, GNN, LLM | Replaces PSSMs with ESM-2 embeddings for evolutionary information<br>Integrates embeddings into residue-level graph for PPI modeling | Improved scoring of near-native models in protein docking tasks<br>Reduced computation time from hours to seconds compared to using PSSMs<br>Achieved 83% accuracy in distinguishing biological vs. crystallographic interfaces |
| PIQLE [53] | DL, GNN with GAT | Multi-head graph attention network<br>Integrates multimeric interaction geometries | Spearman correlation of 0.519 on Dockground dataset<br>Outperformed DProQA, TRScore, GNN-DOVE, DOVE with higher |

(continued)

**Table 2**
**(continued)**

| Method | Method type | Key features | Applications/results |
|---|---|---|---|
| | | and evolutionary info<br>Edge features include distances and orientations | success rates in docking model ranking<br>Improved estimation of interface quality in multimeric complexes |
| *PPI site prediction* | | | |
| PIPENN [56] | Ensemble DL | Combines CNN, U-net, ResNet, RNN architectures<br>Predicts PPI, protein–nucleotide, and protein–small molecule interfaces<br>Ensemble approach for accuracy | Achieved AUC of up to 0.95, MCC of 0.85, precision of 0.90, and recall of 0.88 on benchmark datasets<br>Superior performance across multiple interaction types compared to existing sequence-based models<br>Effective without needing structural data |
| EDLMPPI [61] | LLM, BiLSTM, Capsule Net | Uses ProtT5 for sequence encoding<br>Integrates BiLSTM and capsule networks<br>Captures long-range dependencies<br>Incorporates multisource biological features | Achieved up to 10% increase in average precision on benchmark datasets<br>Higher AUROC scores across Dset_448, Dset_72, and Dset_164 compared to ten alternative models<br>Effective in handling imbalanced data by using asymmetric bagging and ensemble predictions |
| Pair-EGRET [64] | LLM, GNN, GAT | Combines ProtBERT embeddings with graph attention networks<br>Represents proteins as k-NN graphs<br>Uses cross-attention for inter-protein residue interactions | Median AUROC of 0.888 on DBD5 dataset<br>AUROC of 0.9583 on MaSIF dataset<br>Outperformed AlphaFold Multimer in 16 out of 17 test antibody–antigen complexes<br>High precision in antibody–antigen interaction predictions |
| MEG-PPIS [67] | DL, EGNN | Utilizes multi-scale graph representations<br>Incorporates E (n) equivariant GNNs<br>Handles spatial transformations (rotation, reflection)<br>Fast runtime | Achieved AUROC of 0.92 and AUPRC of 0.85 on Test_315-28 dataset<br>Reduced prediction time to ~20% of AGAT-PPIS<br>Improved true positive rate by 15% and true negative rate by 10% compared to AGAT-PPIS |
| EnsemPPIS [63] | LLM, Transformer, GatCNN | Combines transformer and gated CNN models<br>Uses ProtBERT embeddings<br>Captures both global and local sequence features<br>Sequence-only input | Achieved AUROC of 0.93, MCC of 0.88, and AUPRC of 0.85 on DeepPPISP dataset<br>AUROC of 0.90, MCC of 0.84, and AUPRC of 0.80 on DELPHI dataset<br>AUROC of 0.91, MCC of 0.86, and AUPRC of 0.82 on GraphPPIS dataset |

(continued)

**Table 2**
**(continued)**

| Method | Method type | Key features | Applications/results |
|---|---|---|---|
| MP-BERT [65] | LLM, Transformer, Transfer Learning | Develops MP-BERT for protein pairs<br>Processes protein pairs directly<br>Fine-tuned for PPI and PPIS prediction<br>Sequence-only input | Aggregated accuracy of 92.65% in PPI prediction across multiple organisms<br>PPIS prediction achieved an F1 score of 0.85 on benchmark datasets<br>High generalization capabilities demonstrated across human, yeast, and other model organisms |
| DRN-1D2D_Inter [66] | LLM, ESM-1b, Dimensional Hybrid ResNet | Combines 1D and 2D convolutions<br>Uses embeddings from ESM-1b and ESM-MSA-1b<br>Predicts residue–residue contacts<br>Sequence-only input | Combines 1D and 2D convolutions<br>Uses embeddings from ESM-1b and ESM-MSA-1b<br>Predicts residue–residue contacts<br>Sequence-only input |

This table provides an overview of diverse methods, including their methodological classifications, distinguishing features, and key application results in the context of PPI interface and site prediction

models such as Pair-EGRET have excelled in predicting antibody–antigen interactions, supporting the design of antibodies with enhanced specificity and affinity, thereby improving therapeutic efficacy against pathogens [64].

Moreover, the integration of LLM-based predictions with structural biology techniques significantly enhances the pipeline for protein complex characterization. Si and Yan illustrated that incorporating predicted inter-protein contacts from their DRN-1D2D_Inter model into docking algorithms like HADDOCK markedly improved the accuracy of protein complex structure predictions [66]. This synergy accelerates the structural determination process by providing constraints that guide experimental methods such as cryo-electron microscopy and X-ray crystallography, thereby prioritizing targets for structural studies and conserving valuable time and resources. In biotechnology, understanding PPIs is essential for engineering proteins with novel functions. LLM-based models enable the rational design of proteins by predicting how mutations may affect interaction sites, thereby informing strategies for protein engineering. For example, MEG-PPIS, with its fast runtime and high accuracy, facilitates designing enzymes with improved catalytic properties, developing biosensors, and creating synthetic biological systems [67]. The ability to predict PPIs proteome-wide supports high-throughput screening and accelerates the development of bioengineered products.

The generalization capabilities of models like MP-BERT further expand the utility of LLM-based approaches by enabling cross-species analysis of PPIs [65]. This is particularly beneficial for studying zoonotic diseases and developing broad-spectrum therapeutics. By predicting PPIs in organisms with limited data, MP-BERT aids in exploring conserved interaction mechanisms and enhancing our understanding of biological diversity. Additionally, the interpretability of LLM-based models enhances their practical utility in biomedical research. EnsemPPIS, for instance, provides insights into the importance of specific residues and interactions, supporting hypothesis generation for experimental validation [63]. This transparency not only increases confidence in the predictions but also aids in uncovering novel biological mechanisms underlying protein interactions.

Overall, the integration of LLM-based methods into PPI site prediction workflows offers substantial improvements in both accuracy and applicability. By leveraging rich sequence information and advanced neural network architectures, these models advance biomedical research from elucidating fundamental biological processes to developing novel therapeutics. As these models continue to evolve, their applications are expected to expand further, driving significant breakthroughs in our understanding and manipulation of protein interactions for therapeutic and biotechnological advancements.

### 3.4 Limitations and Future Directions for LLM-Based PPI Site Prediction

Despite the significant advancements brought by LLM-based methods in PPI site prediction, several challenges and limitations remain that warrant further investigation. One major limitation is the computational demand associated with training and deploying large models, particularly those utilizing transformer architectures and processing long protein sequences. Models like EDLMPPI [60], EnsemPPIS [63], and DRN-1D2D_Inter [66] require substantial computational resources, which can hinder scalability and accessibility, especially for researchers with limited computational infrastructure.

Future work may focus on optimizing model architectures for efficiency, such as through model compression techniques, efficient transformer variants, or leveraging specialized hardware accelerators to reduce computational overhead. Recent studies have explored deep compression of pretrained transformer models through quantization and sparsity techniques, achieving significant reductions in model size while maintaining accuracy. For instance, Wang et al. demonstrated that transformer models could be quantized down to 4-bit precision and achieve 50% structural sparsity, resulting in a 16-fold compression without significant loss of performance [69]. Similarly, modular transformers can modularize transformer layers to allow flexible model compression ratios, providing a trade-off between performance and efficiency.

Adapting these compression techniques to protein sequence models could alleviate computational demands and make LLM-based PPI site prediction more accessible.

Another challenge lies in the reliance on high-quality sequence data. While LLMs have shown remarkable ability to learn from vast datasets, the presence of noisy or incomplete data can adversely affect model performance. Sources of noise in biological sequence data include sequencing errors, misannotations, and the presence of low-complexity regions or disordered segments that can confuse the models. For example, improper tokenization of protein sequences can lead to the loss of important contextual information. Tokenization of biological sequences significantly impacts the performance of transformer models [70], and applying alternative tokenization strategies not only increases accuracy but also reduces input length, mitigating the effects of noise and improving interpretability.

Moreover, imbalanced class distributions in PPI datasets pose significant challenges for model training and evaluation. The ratio of positive to negative samples in natural PPI datasets can be as low as 2%, leading to models that are biased toward the majority class and have poor performance on the minority class [71]. Addressing this issue requires advanced class balancing methods, such as under-sampling, over-sampling, or the use of cost-sensitive learning techniques [71]. For instance, Bhat and Patil highlighted the need to explore or devise advanced class balancing methods to enhance prediction performance in PPI site prediction [72]. Developing models that can effectively handle imbalanced data is essential for improving prediction accuracy.

The presence of homologous sequences in training datasets can also affect the generalization ability of PPI prediction models. Highly similar or identical sequences can lead to overestimation of model performance due to overfitting [73]. To mitigate this issue, constructing non-redundant datasets by removing protein sequence pairs with high identity thresholds is a common practice69. Ensuring diversity in training data helps models learn interaction-specific features rather than relying on sequence similarity alone.

Furthermore, the prediction of PPI sites in oligomeric proteins of higher order remains challenging. Most current methods focus on dimers or small complexes, and extending prediction capabilities to larger oligomeric assemblies requires developing efficient prediction systems that can handle the increased complexity [72]. Incorporating structural information, such as three-dimensional atomic coordinates or predicted distance maps, can provide valuable insights into PPI site prediction for these larger complexes69. Integrating structural predictions from tools like AlphaFold2 [15] could bridge the gap between sequence-based and structure-based methods, leveraging the strengths of both approaches. For instance, AF2Complex utilizes AlphaFold2 to predict both the presence of PPIs and the corresponding quaternary structures without the need for paired multiple sequence alignments [74]. Although methods exist that incorporate structural predictions into PPI site prediction, integrating these structural insights with LLM-based sequence models remains an open area of research.

Combining multimodal data, such as integrating sequence data with structural information and high-level semantic features, has shown promise in enhancing PPI site prediction models. For example, MIPPIS fuses multi-information from different perspectives by integrating a graph convolutional network (GCN) channel, a bidirectional long short-term memory (Bi-LSTM) channel [61], and a ProtT5 language model channel [75]. This approach effectively extracts and integrates features from amino acid spatial structures, primary sequences, and high-level semantic representations, leading to superior predictive performance. Future research could explore the integration of additional data types, such as protein localization and function information, to further boost prediction accuracy. Bhat and Patil emphasized the importance of incorporating the correlation information between different protein locations and investigating how the structural information of full-length proteins can add value to prediction models [72]. The integration of multimodal approaches, particularly with available imaging data, has been thoroughly explored, demonstrating promising potential for uncovering relationships between functional proteins and cell morphologies [76, 77]. Bhat and Patil emphasized the importance of incorporating the correlation information between different protein locations and investigating how the structural information of full-length proteins can add value to prediction models [72]. The integration of multimodal approaches, particularly with available imaging data, has been thoroughly explored, demonstrating promising potential for uncovering relationships between functional proteins and cell morphologies [76, 77].

In conclusion, while LLM-based methods have significantly advanced the field of PPI site prediction, ongoing efforts to address these limitations through methodological innovations and interdisciplinary collaboration will be essential. By continuing to refine these models, incorporating diverse data sources, and enhancing model robustness against data quality issues, the scientific community can deepen our understanding of protein interactions. These advancements will ultimately contribute to breakthroughs in biomedical research, therapeutic development, and our fundamental comprehension of cellular mechanisms.

## 4 Expanding Horizons: Applications of Large Language Models for Interactions

The versatility of LLMs in protein research extends beyond PPI site prediction, finding utility in diverse yet related domains. One such application is the extraction of PPI information from biomedical literature. With the exponential growth of published research, manually curating PPIs from text has become impractical. LLMs, particularly transformer-based models trained on biomedical corpora, have demonstrated effectiveness in automating the identification of PPIs within scientific texts [68]. By employing advanced natural language processing techniques, these models can parse complex sentences to extract relevant interaction data, thereby accelerating the compilation of PPI databases and facilitating downstream biological research.

Another promising application lies in the prediction of PPI hotspots—specific residues that are crucial for the stability and specificity of protein interactions. Traditional hotspot identification relies heavily on structural data and extensive mutagenesis experiments, which are time-consuming and resource-intensive. Recent approaches leverage protein language models to predict these critical residues directly from sequence information [78]. By capturing the evolutionary and contextual nuances encoded within protein sequences, LLMs can identify potential hotspots even in the absence of structural data, providing valuable insights for drug design and protein engineering efforts aimed at modulating PPIs.

LLMs have also been applied to predict protein–membrane interfaces, particularly in peripheral membrane proteins that transiently associate with lipid bilayers. Accurate identification of membrane-interacting residues is essential for understanding membrane-related functions and for the development of therapeutics targeting membrane-associated processes. Utilizing embeddings from protein language models, researchers have developed methods to predict membrane-binding sites with reduced computational cost compared to structure-based techniques [79]. These models can rapidly analyze protein sequences to identify potential membrane-interacting regions, aiding in the characterization of proteins with limited structural information.

Collectively, these applications underscore the expanding role of LLMs in addressing complex biological questions beyond traditional PPI site prediction. By harnessing the power of deep learning and vast protein sequence data, LLMs offer innovative solutions for extracting meaningful biological insights from both textual and sequence information. The continued exploration of these under-investigated areas holds significant promise for advancing our understanding of protein function, interactions, and their implications in health and disease.

## Acknowledgments

Siwei Chen is supported by P-Squared Philanthropies Early-stage Investigator Development Award. Junhan Zhao is supported by Harvard Spark Grant. DKC is supported by startup grant from RIT.

### References

1. Cai S et al (2024) Spatially resolved subcellular protein-protein interactomics in drug-perturbed lung-cancer cultures and tissues. Nat Biomed Eng. https://doi.org/10.1038/s41551-024-01271-x
2. Chun C et al (2024) Profiling protein-protein interactions to predict the efficacy of B-cell-lymphoma-2-homology-3 mimetics for acute myeloid leukaemia. Nat Biomed Eng 8:1379. https://doi.org/10.1038/s41551-024-01241-3
3. Kuzmanov U, Emili A (2013) Protein-protein interaction networks: probing disease mechanisms using model systems. Genome Med 5:37. https://doi.org/10.1186/gm441
4. Li M et al (2021) The protein-protein interaction ontology: for better representing and capturing the biological context of protein interaction. BMC Genomics 22:544. https://doi.org/10.1186/s12864-021-07827-4
5. Chen S et al (2022) A full-proteome, interaction-specific characterization of mutational hotspots across human cancers. Genome Res 32:135–149. https://doi.org/10.1101/gr.275437.121
6. Guo D, Hazbun TR, Xu XJ, Ng SL, Fields S, Kuo MH (2004) A tethered catalysis, two-hybrid system to identify protein-protein interactions requiring post-translational modifications. Nat Biotechnol 22:888–892. https://doi.org/10.1038/nbt985
7. Michnick SW, Ear PH, Manderson EN, Remy I, Stefan E (2007) Universal strategies in research and drug discovery based on protein-fragment complementation assays. Nat Rev Drug Discov 6:569–582. https://doi.org/10.1038/nrd2311
8. Schwikowski B, Uetz P, Fields S (2000) A network of protein-protein interactions in yeast. Nat Biotechnol 18:1257–1261. https://doi.org/10.1038/82360
9. von Mering C et al (2002) Comparative assessment of large-scale data sets of protein-protein interactions. Nature 417:399–403. https://doi.org/10.1038/nature750
10. Evans R et al (2022) Protein complex prediction with AlphaFold-Multimer. bioRxiv, 2021.2010.2004.463034. https://doi.org/10.1101/2021.10.04.463034
11. Abramson J et al (2024) Accurate structure prediction of biomolecular interactions with AlphaFold 3. Nature 630:493–500. https://doi.org/10.1038/s41586-024-07487-w
12. Bryant P, Pozzati G, Elofsson A (2022) Improved prediction of protein-protein interactions using AlphaFold2. Nat Commun 13:1265. https://doi.org/10.1038/s41467-022-28865-w
13. Zhang QC et al (2012) Structure-based prediction of protein-protein interactions on a genome-wide scale. Nature 490:556–560. https://doi.org/10.1038/nature11503
14. Gainza P et al (2023) De novo design of protein interactions with learned surface fingerprints. Nature 617:176–184. https://doi.org/10.1038/s41586-023-05993-x
15. Jumper J et al (2021) Highly accurate protein structure prediction with AlphaFold. Nature 596:583–589. https://doi.org/10.1038/s41586-021-03819-2
16. Xiong D et al (2024) A structurally informed human protein-protein interactome reveals proteome-wide perturbations caused by disease mutations. Nat Biotechnol. https://doi.org/10.1038/s41587-024-02428-4
17. Ferruz N, Schmidt S, Hocker B (2022) ProtGPT2 is a deep unsupervised language model for protein design. Nat Commun 13:4348. https://doi.org/10.1038/s41467-022-32007-7
18. Ruffolo JA, Madani A (2024) Designing proteins with language models. Nat Biotechnol 42:200–202. https://doi.org/10.1038/s41587-024-02123-4
19. Schmirler R, Heinzinger M, Rost B (2024) Fine-tuning protein language models boosts predictions across diverse tasks. Nat Commun 15:7407. https://doi.org/10.1038/s41467-024-51844-2
20. Lin Z et al (2023) Evolutionary-scale prediction of atomic-level protein structure with a

language model. Science 379:1123–1130. https://doi.org/10.1126/science.ade2574

21. Gao Q, Zhang C, Li M, Yu T (2024) Protein-protein interaction prediction model based on ProtBert-BiGRU-attention. J Comput Biol 31: 797–814. https://doi.org/10.1089/cmb.2023.0297
22. Pokharel S, Pratyush P, Heinzinger M, Newman RH, Kc DB (2022) Improving protein succinylation sites prediction using embeddings from protein language model. Sci Rep 12:16933. https://doi.org/10.1038/s41598-022-21366-2
23. Dang TH, Vu TA (2024) xCAPT5: protein-protein interaction prediction using deep and wide multi-kernel pooling convolutional neural networks with protein language model. BMC Bioinformatics 25:106. https://doi.org/10.1186/s12859-024-05725-6
24. Jha K, Karmakar S, Saha S (2023) Graph-BERT and language model-based framework for protein-protein interaction identification. Sci Rep 13:5663. https://doi.org/10.1038/s41598-023-31612-w
25. Wang Y, You ZH, Yang S, Li X, Jiang TH, Zhou X (2019) A high efficient biological language model for predicting protein(−)protein interactions. Cells 8. https://doi.org/10.3390/cells8020122
26. Xue Y et al (2021) Multimodal pre-training model for sequence-based prediction of protein-protein interaction. arXiv preprint arXiv: 2112.04814v1
27. Xu X, Bonvin A (2024) DeepRank-GNN-esm: a graph neural network for scoring protein-protein models using protein language model. Bioinform Adv 4:vbad191. https://doi.org/10.1093/bioadv/vbad191
28. Du X et al (2017) DeepPPI: boosting prediction of protein-protein interactions with deep neural networks. J Chem Inf Model 57:1499–1510. https://doi.org/10.1021/acs.jcim.7b00028
29. Sledzieski S, Kshirsagar M, Baek M, Dodhia R, Lavista Ferres J, Berger B (2024) Democratizing protein language models with parameter-efficient fine-tuning. Proc Natl Acad Sci USA 121:e2405840121. https://doi.org/10.1073/pnas.2405840121
30. Elnaggar A et al (2022) ProtTrans: toward understanding the language of life through self-supervised learning. IEEE Trans Pattern Anal Mach Intell 44:7112–7127. https://doi.org/10.1109/TPAMI.2021.3095381
31. Heinzinger M et al (2019) Modeling aspects of the language of life through transfer-learning protein sequences. BMC Bioinformatics 20: 723. https://doi.org/10.1186/s12859-019-3220-8
32. Ko YS, Parkinson J, Liu C, Wang W (2024) TUnA: an uncertainty-aware transformer model for sequence-based protein-protein interaction prediction. Brief Bioinform 25. https://doi.org/10.1093/bib/bbae359
33. Jin M et al (2024) ProLLM: protein chain-of-thoughts enhanced LLM for protein-protein interaction prediction. arXiv e-prints, arXiv--2405
34. Peters ME et al (2018) Deep contextualized word representations. In: Walker M, Ji H, Stent A (eds) Proceedings of the 2018 conference of the north American chapter of the association for computational linguistics: human language technologies, vol 1. Association for Computational Linguistics, pp 2227–2237
35. Zhang J et al (2020) Graph-Bert: only attention is needed for learning graph representations. arXiv preprint arXiv:2001.05140
36. Chen T, Guestrin C (2016) XGBoost: a scalable tree boosting system. In: Proceedings of the 22nd ACM SIGKDD international conference on knowledge discovery and data mining. Association for Computing Machinery, pp 785–794
37. Martin S, Roe D, Faulon JL (2005) Predicting protein-protein interactions using signature products. Bioinformatics 21:218–226. https://doi.org/10.1093/bioinformatics/bth483
38. Guo Y, Yu L, Wen Z, Li M (2008) Using support vector machine combined with auto covariance to predict protein-protein interactions from protein sequences. Nucleic Acids Res 36:3025–3030. https://doi.org/10.1093/nar/gkn159
39. Pan XY, Zhang YN, Shen HB (2010) Large-scale prediction of human protein-protein interactions from amino acid sequence based on latent topic features. J Proteome Res 9: 4992–5001. https://doi.org/10.1021/pr100618t
40. Rao R et al (2019) Evaluating protein transfer learning with TAPE. arXiv [cs.LG]
41. Wong BJH, Kong W, Peng H, Goh WWB (2023) PROSE: phenotype-specific network signatures from individual proteomic samples. Brief Bioinform 24. https://doi.org/10.1093/bib/bbad075
42. Mikolov T, Sutskever I, Chen K, Corrado G, Dean J (2013) Distributed representations of words and phrases and their compositionality. In: Proceedings of the 26th international conference on neural information

processing systems, vol 2. Curran Associates Inc, pp 3111–3119

43. Sledzieski S, Devkota K, Singh R, Cowen L, Berger B (2023) TT3D: leveraging precomputed protein 3D sequence models to predict protein-protein interactions. Bioinformatics 39. https://doi.org/10.1093/bioinformatics/btad663
44. van Kempen M et al (2024) Fast and accurate protein structure search with Foldseek. Nat Biotechnol 42:243–246. https://doi.org/10.1038/s41587-023-01773-0
45. Soleymani F, Paquet E, Viktor HL, Michalowski W, Spinello D (2023) ProtInteract: a deep learning framework for predicting protein-protein interactions. Comput Struct Biotechnol J 21:1324–1348. https://doi.org/10.1016/j.csbj.2023.01.028
46. Hu EJ et al (2021) LoRA: low-rank adaptation of large language models. arXiv Preprint arXiv:2106.09685
47. Zhang J, Durham J, Qian C (2024) Revolutionizing protein-protein interaction prediction with deep learning. Curr Opin Struct Biol 85: 102775. https://doi.org/10.1016/j.sbi.2024.102775
48. Dettmers T et al (2023) QLoRA: efficient finetuning of quantized LLMs. arXiv preprint arXiv:2305.14314
49. Chen S et al (2018) An interactome perturbation framework prioritizes damaging missense mutations for developmental disorders. Nat Genet 50:1032–1040. https://doi.org/10.1038/s41588-018-0130-z
50. Meyer MJ et al (2018) Interactome INSIDER: a structural interactome browser for genomic studies. Nat Methods 15:107–114. https://doi.org/10.1038/nmeth.4540
51. Giecold G, Marco E, Garcia SP, Trippa L, Yuan GC (2016) Robust lineage reconstruction from high-dimensional single-cell data. Nucleic Acids Res 44:e122. https://doi.org/10.1093/nar/gkw452
52. Reau M, Renaud N, Xue LC, Bonvin A (2023) DeepRank-GNN: a graph neural network framework to learn patterns in protein-protein interfaces. Bioinformatics 39. https://doi.org/10.1093/bioinformatics/btac759
53. Shuvo MH, Karim M, Roche R, Bhattacharya D (2023) PIQLE: protein-protein interface quality estimation by deep graph learning of multimeric interaction geometries. Bioinform Adv 3:vbad070. https://doi.org/10.1093/bioadv/vbad070
54. Chen X, Morehead A, Liu J, Cheng J (2022) DProQ: a gated-graph transformer for protein complex structure assessment. arXiv [cs.LG]
55. Behre P et al (2022) TRScore: a novel GPT-based readability scorer for ASR segmentation and punctuation model evaluation and selection. arXiv [cs.CL]
56. Stringer B, de Ferrante H, Abeln S, Heringa J, Feenstra KA, Haydarlou R (2022) PIPENN: protein interface prediction from sequence with an ensemble of neural nets. Bioinformatics 38:2111–2118. https://doi.org/10.1093/bioinformatics/btac071
57. He K, Zhang X, Ren S, Sun J (2016) Deep residual learning for image recognition. In: 2016 IEEE conference on computer vision and pattern recognition (CVPR), pp 770–778
58. Hopfield JJ (1982) Neural networks and physical systems with emergent collective computational abilities. Proc Natl Acad Sci USA 79: 2554–2558. https://doi.org/10.1073/pnas.79.8.2554
59. Hayes T et al (2024) Simulating 500 million years of evolution with a language model. bioRxiv, 2024.2007.2001.600583. https://doi.org/10.1101/2024.07.01.600583
60. Hou Z, Yang Y, Ma Z, Wong KC, Li X (2023) Learning the protein language of proteome-wide protein-protein binding sites via explainable ensemble deep learning. Commun Biol 6: 73. https://doi.org/10.1038/s42003-023-04462-5
61. Graves A, Schmidhuber J (2005) Framewise phoneme classification with bidirectional LSTM and other neural network architectures. Neural Netw 18:602–610. https://doi.org/10.1016/j.neunet.2005.06.042
62. Dang TV (2023) Capsule network with shortcut routing. arXiv preprint arXiv:2307.10212
63. Mou M et al (2023) A transformer-based ensemble framework for the prediction of protein-protein interaction sites. Research (Wash D C) 6:0240. https://doi.org/10.34133/research.0240
64. Alam R, Mahbub S, Bayzid MS (2024) Pair-EGRET: enhancing the prediction of protein-protein interaction sites through graph attention networks and protein language models. Bioinformatics 40. https://doi.org/10.1093/bioinformatics/btae588
65. Liu T et al (2023) Protein-protein interaction and site prediction using transfer learning. Brief Bioinform 24. https://doi.org/10.1093/bib/bbad376
66. Si Y, Yan C (2023) Improved inter-protein contact prediction using dimensional hybrid residual networks and protein language models. Brief Bioinform 24. https://doi.org/10.1093/bib/bbad039

67. Ding H et al (2024) MEG-PPIS: a fast protein-protein interaction site prediction method based on multi-scale graph information and equivariant graph neural network. Bioinformatics 40. https://doi.org/10.1093/bioinformatics/btae269
68. Renaud N et al (2021) DeepRank: a deep learning framework for data mining 3D protein-protein interfaces. Nat Commun 12: 7068. https://doi.org/10.1038/s41467-021-27396-0
69. Wang N et al (2024) Deep compression of pre-trained transformer models. In: Proceedings of the 36th international conference on neural information processing systems. Curran Associates Inc
70. Dotan E, Jaschek G, Pupko T, Belinkov Y (2024) Effect of tokenization on transformers for biological sequences. Bioinformatics 40. https://doi.org/10.1093/bioinformatics/btae196
71. Sun T, Zhou B, Lai L, Pei J (2017) Sequence-based prediction of protein protein interaction using a deep-learning algorithm. BMC Bioinformatics 18:277. https://doi.org/10.1186/s12859-017-1700-2
72. Bhat P, Patil N (2023) An exhaustive review of computational prediction techniques for PPI sites, protein locations, and protein functions. Netw Model Anal Health Inf Bioinform 12:31. https://doi.org/10.1007/s13721-023-00427-0
73. Chen M et al (2019) Multifaceted protein-protein interaction prediction based on Siamese residual RCNN. Bioinformatics 35: i305–i314. https://doi.org/10.1093/bioinformatics/btz328
74. Gao M, Nakajima An D, Parks JM, Skolnick J (2022) AF2Complex predicts direct physical interactions in multimeric proteins with deep learning. Nat Commun 13:1744. https://doi.org/10.1038/s41467-022-29394-2
75. Wang S, Dong K, Liang D, Zhang Y, Li X, Song T (2024) MIPPIS: protein-protein interaction site prediction network with multi-information fusion. BMC Bioinformatics 25: 345. https://doi.org/10.1186/s12859-024-05964-7
76. Wang X et al (2024) A pathology foundation model for cancer diagnosis and prognosis prediction. Nature 634:970–978. https://doi.org/10.1038/s41586-024-07894-z
77. Zhao J et al (2023) PhaseFIT: live-organoid phase-fluorescent image transformation via generative AI. Light Sci Appl 12:297. https://doi.org/10.1038/s41377-023-01296-y
78. Sargsyan K, Lim C (2024) Using protein language models for protein interaction hot spot prediction with limited data. BMC Bioinformatics 25:115. https://doi.org/10.1186/s12859-024-05737-2
79. Paranou D, Chatzigoulas A, Cournia Z (2024) Using deep learning and large protein language models to predict protein-membrane interfaces of peripheral membrane proteins. Bioinform Adv 4:vbae078. https://doi.org/10.1093/bioadv/vbae078

# Chapter 16

# Prediction of Protein–Peptide Binding Sites Using PepBCL

**Ruheng Wang, Kenta Nakai, and Leyi Wei**

## Abstract

Identifying the protein–peptide binding residues is fundamentally important to understanding the mechanisms of protein functions and drug discovery. Although several computational methods have been developed, they highly rely on third-party tools or information for feature design, easily resulting in low computational efficacy and suffering from low predictive performance. We describe how to use an end-to-end computational method PepBCL that is free with feature design for high-throughput prediction of protein–peptide binding sites. PepBCL outperforms the state-of-the-art methods under benchmarking comparison and achieves more robust performance based on protein sequences only. We can automatically extract and learn high-latent representations of protein sequences relevant to protein structure and functions by the introduction of a well pretrained protein large language model. We overview our method and discuss how to run the supported codes to reproduce our predictor.

**Key words** Protein, peptide binding residues, Protein large language model, Pretraining, Contrastive learning, PepBCL

## 1 Introduction

Protein–peptide interactions play a crucial role in understanding biological processes, including protein function and drug discovery [1, 2]. Identifying the binding sites where peptides interact with proteins can provide key insights into these interactions [3, 4]. Traditional computational approaches for predicting binding sites often rely on complex feature engineering and structural data. However, recent advancements in machine learning have paved the way for sequence-based methods that simplify this process. In this context, PepBCL [5] has emerged as a state-of-the-art tool designed to predict protein–peptide binding sites using deep learning techniques. This innovative approach leverages pretrained language models and contrastive learning to enhance prediction accuracy. By focusing on sequence embeddings, PepBCL minimizes the reliance on manually engineered features, offering a more automated and scalable solution.

Dukka B. KC (ed.), *Large Language Models (LLMs) in Protein Bioinformatics*, Methods in Molecular Biology, vol. 2941, https://doi.org/10.1007/978-1-0716-4623-6_16,

Historically, methods for predicting protein–peptide binding sites fall into two categories: structure-based and sequence-based approaches. Structure-based methods, such as PepSite [6] and PepNN-Struct [7], rely on the 3D conformation of proteins and peptides to identify potential binding regions. These methods often require detailed structural data, which can be difficult to obtain and computationally intensive to process. On the other hand, sequence-based methods, like PepBind [8] and SPRINT-Seq [9], predict binding sites directly from the amino acid sequence. Sequence-based approaches are generally faster and easier to apply since they do not require structural data. However, they often face challenges with prediction accuracy due to the limited information contained in the sequence alone.

To address these challenges, PepBCL combines the strengths of pretrained language models, which capture rich sequence representations, and contrastive learning, which improves performance on imbalanced datasets by learning more distinguishable features for binding and nonbinding sites. We overview our model architecture and provide details of our benchmark datasets. Finally, we show users how to reproduce our model based on our supported codes.

## 2 Materials and Method

### 2.1 Datasets

This study utilizes two widely used benchmark datasets, referred to as Dataset 1 and Dataset 2, to evaluate and compare our proposed method with existing approaches. Both datasets underwent similar pre-processing to ensure robust model training, including reducing sequence identity to 30% using "blastclust" in the BLAST package [10] to mitigate evaluation bias.

**Dataset 1** originally proposed for the SPRINT-Str [11] method, contains 1279 protein–peptide complexes. It is divided into a training set (TR1154) with 1154 proteins and an independent testing set (TE125) with 125 proteins. Specifically, TR1154 comprises 15,030 binding residues and 261,792 nonbinding residues, while TE125 contains 1719 binding residues and 29,151 nonbinding residues.

**Dataset 2** derived from Zhao et al.[8]'s work comprises the same 1279 protein–peptide complexes. It is split into a training set (TR640) with 640 complexes and an independent testing set (TE639) with the remaining complexes. TR640 includes 8259 binding residues and 149,103 nonbinding residues, while TE639 contains 8490 binding residues and 141,840 nonbinding residues.

Both datasets exhibit a significant class imbalance, with nonbinding residues substantially outnumbering binding residues. This imbalance is a common characteristic in protein–peptide binding site prediction tasks and presents a challenge for developing effective prediction models. To ensure a fair comparison with existing studies, we adhered to the same dataset-splitting procedure for model training and testing. This standardized approach enables consistent evaluation across different prediction methods in the field of protein–peptide interaction prediction.

### 2.2 Architecture

Recently, we developed the PepBCL method to predict protein–peptide binding residues in input protein sequences. Our method is based on an efficient four-module design; see Fig. 1. First, we represent the input protein using a sequence embedding module. Second, these embeddings are processed by a BERT-based encoder module. Third, a contrastive learning module optimizes the feature representations. Fourth, an output module generates the final predictions. Following we provide a more detailed explanation.

In the sequence embedding module, we translate the raw protein sequence into a digital sequence according to a defined vocabulary dictionary. Each amino acid is mapped to a digit value, with rare amino acids uniformly replaced by a specific digit. The encoded vector is then embedded by a pretrained embedding layer to

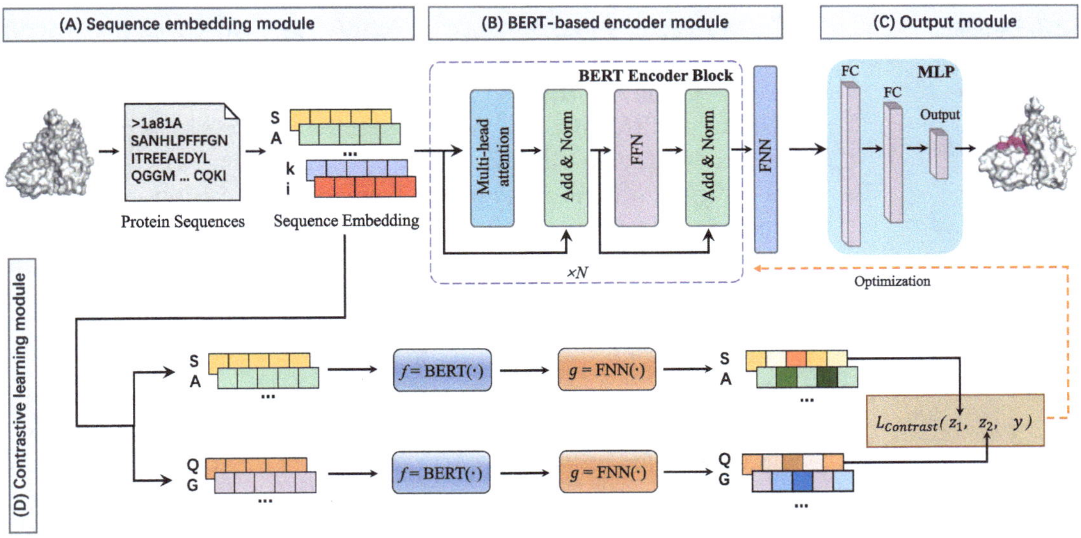


**Fig. 1** The architecture of the PepBCL method. The four modules are denoted by the corresponding letters (**a**), (**b**), (**c**), and (**d**). (**a**) The sequence embedding module translates raw protein sequences into numerical embeddings using the ProtBert-BFD model. (**b**) The BERT-based encoder module applies multi-head attention to capture contextual relationships between amino acids, generating high-dimensional feature representations. (**c**) The output module converts feature vectors to residue-level peptide-binding probabilities. (**d**) The contrastive learning module optimizes feature distinction between binding and nonbinding sites by minimizing a novel contrastive loss function

generate an improved initial embedding. In the BERT-based encoder module, we utilize a pretrained BERT model called ProtBert-BFD [12], which has been trained on 2.1 billion protein sequences [13]. This module applies multi-head attention to capture contextual relationships between amino acids, generating high-dimensional feature representations. These representations are then refined using fully connected layers to reduce dimensionality and enhance prediction accuracy. The contrastive learning module, a novel component of PepBCL, applies a contrastive loss function to improve feature distinction between binding and non-binding sites. This module is designed to address the inherent imbalance in binding site datasets by learning from both positive and negative sample pairs. In the output module, the residue representation vector generated by the previous modules is fed into a multi-layer perceptron (MLP) to convert the feature vector to a residue-level category output. This module generates a residue-level peptide-binding probability and determines whether each residue in the input sequence is peptide binding or not.

PepBCL's architecture is designed to provide accurate predictions while processing raw protein sequences without relying on complex feature engineering or structural data. By leveraging pre-trained language models and contrastive learning, PepBCL offers a more automated and scalable solution for predicting protein–peptide binding sites.

## 2.3 Model Performance Evaluation

### 2.3.1 Evaluation Metrics

To evaluate the performance of PepBCL, we employed five metrics commonly used in imbalanced classification tasks: Recall, Specificity, Precision, Matthews Correlation Coefficient (MCC), and Area Under the ROC Curve (AUC). Recall measures the proportion of correctly predicted binding residues, while Specificity assesses the accuracy of nonbinding residue predictions. Precision indicates the accuracy of residues predicted as binding. MCC provides a comprehensive assessment of prediction ability for both binding and non-binding residues. Additionally, AUC quantifies the overall performance of the model.

### 2.3.2 Model Performance

To assess the predictive power of PepBCL, the model was compared to eight existing methods, including both structure-based and sequence-based approaches. PepBCL demonstrated superior predictive performance across two independent testing sets (TE125 and TE639). On TE125, PepBCL achieved a Precision of 0.540, an AUC of 0.815, and an MCC of 0.385, outperforming PepBind by 7.1%, 2.2%, and 1.3%, respectively. Similarly, on TE639, PepBCL surpassed PepBind in AUC by 3.7% and outperformed state-of-the-art methods like PepNN-Seq and PepNN-Struct in MCC by 6.1% and 1.1%, respectively. The results showed that PepBCL outperformed most sequence-based methods in terms of precision, AUC, and MCC. Furthermore, PepBCL maintained high AUC values for over 87% of protein sequences, indicating its robust predictive capabilities across varying contexts. In addition to these metrics, PepBCL demonstrated significant improvements in handling imbalanced data. The contrastive learning module contributed to more accurate predictions by enhancing the separation between binding and nonbinding site features. Ablation studies, where the contrastive learning module was removed, showed a clear drop in performance, reinforcing the module's importance.

#### 2.3.3 Run Time

Using a modern GPU (Nvidia RTX 3090), we evaluated the computational efficiency of PepBCL compared to PepBind. As shown in Fig. 2, the runtime of PepBCL for a single protein (PDBID: 1a81A) is approximately 28 seconds, while processing ten proteins takes about 44 seconds. In contrast, PepBind requires around 600 seconds to process a single protein. This significant difference in processing time demonstrates the superior computational efficiency of PepBCL. Our method was implemented in Python using the PyTorch framework, allowing for optimal utilization of GPU resources. Here we also give some guide about how to run PepBCL on CPU (*see* **Note 1**).

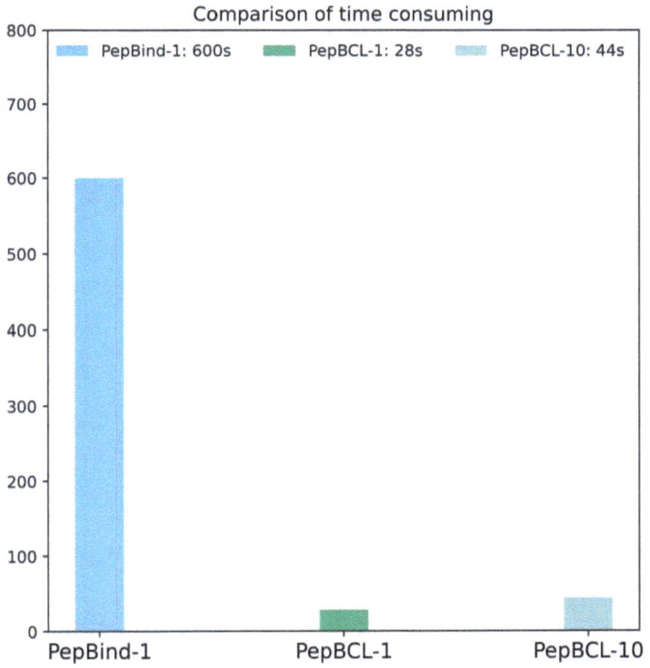


**Fig. 2** The comparison of time-consuming between our method and the PepBind model

To provide a comprehensive comparison, Fig. 2 illustrates three specific runtime scenarios: PepBind-1 (time for PepBind to process one sequence), PepBCL-1 (time for PepBCL to process one sequence), and PepBCL-10 (time for PepBCL to process ten sequences). For the PepBCL-10 test, we repeated the single sequence (1a81A) ten times to simulate a multi-protein prediction scenario. These results indicate that PepBCL can efficiently handle large-scale protein datasets, making it particularly suitable for high-throughput analyses of entire proteomes or genomes. The minimal increase in processing time from one to ten sequences, as visualized in Fig. 2, further highlights PepBCL's scalability and potential for rapid, large-scale predictions of protein–peptide binding sites. Discussions about some factors that influence running time are shown in Note 2.

## 3 Reproduction of PepBCL

To reproduce the PepBCL model or utilize it with custom data, adhere to the following protocol:

### 3.1 Environment Setup

1. Clone the repository:

```
git clone https://github.com/Ruheng-W/PepBCL
          cd PepBCL
```

2. Generate a new conda environment:

```
conda create -n PepBCL python=3.8
conda activate PepBCL
```

3. Install the required packages:

```
pip install -r requirements.txt
```

Should you encounter notifications of uninstalled packages post-execution of the above command, please manually install any missing dependencies.

### 3.2 Model Preparation

Establish a directory for the pretrained BERT model and acquire it:

```
mkdir -p prot_bert_bfd
wget https://huggingface.co/Rostlab/prot_bert_bfd/resolve/
main/pytorch_model.bin -O prot_bert_bfd/pytorch_model.bin
```

Ensure the pytorch_model.bin file is correctly situated within the prot_bert_bfd directory of your cloned PepBCL repository. This protocol facilitates the proper setup and preparation of the PepBCL model for subsequent utilization or customization.

### 3.3 Custom Training and Testing

To train and evaluate the model on custom datasets:

1. Locate the select_dataset() function within the train/protBert_main.py file. Modify the function to specify the paths to your custom training and testing datasets:

```
def select_dataset():
    path_train_data = '/path/to/your/training/dataset.tsv'
    path_test_data = '/path/to/your/testing/dataset.tsv'
   return path_train_data, path_test_data
```

Replace the placeholder paths with the actual file paths to your training and testing datasets, respectively.

1. If necessary, adjust other hyperparameters and configurations within the load_config() function to optimize the model for your specific use case. Refer to the inline comments for guidance on each parameter. Users could also improve Recall by customizing probability thresholds (*see* **Note 3**).
2. Execute the main program to initiate the training and testing process:

```
python train/protBert_main.py
```

This procedure allows for the customization of the PepBCL model to accommodate various datasets and experimental requirements. Ensure that your custom datasets adhere to the same format as the original training and testing files to maintain compatibility with the model architecture.

### 3.4 Reproducing Results on the Paper

To reproduce the results from the paper:

1. Download our saved model parameters of PepBCL from the following link: https://drive.google.com/drive/folders/1UTZxnR34UaUryKkXRaM0ts3ufjVQtCas

The directory contains two parameter files corresponding to two datasets (Dataset1 and Dataset2).

2. Create a directory for the model parameters and download the files:

```
mkdir -p model_parameters
# Download Dataset1 parameters
wget https://drive.google.com/uc?id=FILE_ID_FOR_DATASET1 -O
model_parameters/dataset1_parameters.pth
# Download Dataset2 parameters
wget https://drive.google.com/uc?id=FILE_ID_FOR_DATASET2 -O
model_parameters/dataset2_parameters.pth
```

Replace FILE_ID_FOR_DATASET1 and FILE_ID_FOR_DATASET2 with the actual file IDs from the Google Drive link. Alternatively, you can manually download the files from the webpage and place them in the model_parameters directory.

3. Create a directory for the model parameters and download the files:

```
import torch
import prot_bert

config = ... # Your model configuration
model = prot_bert.BERT(config)  # model instantiation
model_dict  =  torch.load('model_parameters/dataset1_parameters.pth')['model']  # load the state_dict from file path
model.load_state_dict(model_dict)   # load the state_dict to model
```

You can replace 'model_parameters/dataset1_parameters.pth' with 'model_parameters/dataset2_parameters.pth' to use the parameters for Dataset2.

### 3.5 Directory Structure

After completing these steps, your directory structure should look like this:

```
PepBCL/
├── prot_bert_bfd/
│   └── pytorch_model.bin
├── model_parameters/
│   ├── dataset1_parameters.pth
│   └── dataset2_parameters.pth
├── train/
│   └── protBert_main.py
└── ... (other files and folders)
```

### 3.6 Input file format

To facilitate the utilization of our model for custom training purposes, it is essential to adhere to a specific input file format. The input file should be structured as a tab-separated values (TSV) file, containing two columns: "seq" and "label."

The "seq" column represents the protein sequence, encoded as a string of single-letter amino acid codes. Each row corresponds to a single protein sequence. The "label" column contains the binary classification for each residue in the corresponding protein sequence, where "0" denotes a nonbinding residue and "1" denotes a binding residue.

An exemplary excerpt from the input file is as follows:

```
seq    label
MGSSEQELKAIVKDLGCGPY     00000000000000000000
ADVPAGVQLADKQTLVRNNG     00000000000000000001
IVEGQDAEVGLSPWQVMLFR     00000000000000001000
```

It is imperative that users preparing their own datasets for model training strictly adhere to this format to ensure compatibility with the PepBCL framework. The length of the "label" string must precisely match the length of the corresponding protein sequence in the "seq" column. This format enables the model to accurately associate each amino acid residue with its corresponding binding status during the training process.

## 4 Notes

1. Hardware Requirements and Running on CPU

   Due to the use of large protein language models, running computations on a GPU may require significant memory resources—approximately 20 GB. If users intend only to replicate the model and perform testing, they can execute inference on a CPU. For model training, it is advisable to utilize a GPU. To run the model on a CPU, users should remove code segments that allocate data and the model to CUDA devices. For example, change:

```
label = torch.tensor(label, dtype=torch.long).cuda()
```

   to:

```
label = torch.tensor(label, dtype=torch.long)
```

2. Impact of Sequence Length on Inference Time

   Owing to the transformer architecture, the model performs rapid inference on short sequences. However, increasing the sequence length can substantially extend the inference time. It is important to note that sequence length has a more pronounced effect on inference time than the number of sequences.

3. Customizing Probability Thresholds to Improve Recall

   Existing methods are often trained on benchmark datasets that are extremely imbalanced, leading to low recall values and the omission of many true protein–peptide binding sites in real-world applications. To mitigate this issue, different probability thresholds can be selected. Our findings indicate that decreasing the threshold significantly enhances recall, although it also impacts other evaluation metrics. Users wishing to set a custom threshold can modify the `model_eval` function in the **train/protBert_main.py** file. For instance, to set the threshold to 0.3, replace the following code:

```
# Probability of predicting positive classes [batch_size, seq_len]
pred_prob_sort = torch.max(pred_prob_all, 2)
# The maximum probability of prediction in each sample [batch_size]
pred_class = pred_prob_sort[1]

with:

threshold = 0.3
pred_class = (pred_prob_positive > threshold).long()
```

## References


1. Pawson T, Nash P (2003) Assembly of cell regulatory systems through protein interaction domains. science 300(5618):445–452
2. Rubinstein M, Niv MY (2009) Peptidic modulators of protein-protein interactions: progress and challenges in computational design. Biopolymers: Original Research on Biomolecules 91(7):505–513
3. Lee H, Heo L, Lee MS, Seok C (2015) GalaxyPepDock: a protein–peptide docking tool based on interaction similarity and energy optimization. Nucleic acids research 43(W1): W431–W435
4. Neduva V, Linding R, Su-Angrand I, Stark A, Masi F, Gibson TJ, Lewis J, Serrano L, Russell RB (2005) Systematic discovery of new recognition peptides mediating protein interaction networks. PLoS biology 3(12):e405
5. Wang R, Jin J, Zou Q, Nakai K, Wei L (2022) Predicting protein–peptide binding residues via interpretable deep learning. Bioinformatics 38(13):3351–3360. https://doi.org/10.1093/bioinformatics/btac352
6. Petsalaki E, Stark A, García-Urdiales E, Russell RB (2009) Accurate prediction of peptide binding sites on protein surfaces. PLoS computational biology 5(3):e1000335
7. Abdin O, Wen H, Kim PM (2020) Sequence and structure based deep learning models for the identification of peptide binding sites. Adv Neural Info ProcessSyst:33
8. Zhao Z, Peng Z, Yang J (2018) Improving sequence-based prediction of protein–peptide binding residues by introducing intrinsic disorder and a consensus method. J Chem Information Modeling 58(7):1459–1468. https://doi.org/10.1021/acs.jcim.8b00019
9. Taherzadeh G, Yang Y, Zhang T, Liew AWC, Zhou Y (2016) Sequence-based prediction of protein–peptide binding sites using support vector machine. J Comput Chem 37(13): 1223–1229
10. Altschul SF, Madden TL, Schäffer AA, Zhang J, Zhang Z, Miller W, Lipman DJ (1997) Gapped BLAST and PSI-BLAST: a new generation of protein database search programs. Nucleic Acids Res 25(17):3389–3402
11. Taherzadeh G, Zhou Y, Liew AW-C, Yang Y (2018) Structure-based prediction of protein–peptide binding regions using Random Forest. Bioinformatics 34(3):477–484
12. Elnaggar A, Heinzinger M, Dallago C, Rehawi G, Wang Y, Jones L, Gibbs T, Feher T, Angerer C, Steinegger M, Bhowmik D, Rost B (2021) ProtTrans: towards cracking the language of lifes code through self-supervised deep learning and high performance computing. In: IEEE Transactions on pattern analysis and machine intelligence, p 1. https://doi.org/10.1109/TPAMI.2021.3095381
13. Steinegger M, Mirdita M, Söding J (2019) Protein-level assembly increases protein sequence recovery from metagenomic samples manyfold. Nature methods 16(7):603–606

# Chapter 17

# Predicting Peptide Bioactivity Using the Unified Model Architecture UniDL4BioPep

**Zhenjiao Du, Nandan Kumar, and Yonghui Li**

## Abstract

The unified prediction model architecture, UniDL4BioPep, offers a significant advancement in applying machine learning approaches to bioactive peptide discovery. By streamlining the model development process, this architecture reduces the effort required to create custom models, allowing wet-lab researchers to accelerate scientific discovery by easily tailoring models to their specific needs. UniDL4BioPep leverages protein language models, specifically evolutionary scale modeling (ESM), and simplifies model preparation to a single click, making it both accessible and efficient for users. This chapter provides the technical details and practical operation guide for utilizing this unified architecture and demonstrates its effectiveness in binary classification tasks.

**Key words** Bioactive peptide, Machine learning, Transfer learning, Large language model, Deep learning, Protein sequence annotation

## 1 Introduction

Identifying and characterizing a new bioactive peptide experimentally is resource-intensive in the field of bioactive peptide discovery [1, 2]. Therefore, as an alternative approach to accelerate the discovery and reduce the monetary and time costs associated with trial-and-error experiments, machine learning-based prediction approaches are gaining prominence in bioactive peptide discovery across the field of food, nutraceuticals, cosmetics, and pharmaceuticals [1].

Building a prediction model generally involves three steps: (1) dataset construction, (2) peptide representation/embedding/encoding (i.e., converting a peptide sequence into a numerical vector/matrix), and (3) mode development [1]. The most challenging aspect is the peptide representation. Traditional methods, such as

Dukka B. KC (ed.), *Large Language Models (LLMs) in Protein Bioinformatics*, Methods in Molecular Biology, vol. 2941,
https://doi.org/10.1007/978-1-0716-4623-6_17,

those based on physicochemical and biochemical properties of amino acids (e.g., AAIndex), amino acid descriptors, composition, molecular descriptors, peptide fingerprints, etc., have been widely explored over the past decades [1]. The key idea is to maximize the information contained in the numerical vector for the representation of a given peptide. Those traditional approaches are generally underrepresented when used alone but can become overrepresented when most of them are combined, resulting in significant efforts in feature selection and model development. This complicates the development of high-quality models. To overcome these challenges and make machine learning more accessible, a unified model architecture, without the need for intricate feature selection and tedious model selection and hyperparameter tuning, is critical to popularizing the machine learning models among biochemistry researchers who are interested in building their own customized models for specific applications.

Deep learning-based language models are a new class of machine learning approaches for learning representation in natural language processing (NLP), where the semantic properties and linguistic relationships between words are autonomously learnt from millions or billions of text data. This is done by conditioning on predicting the next word in a sentence given the previous context or by reconstructing masked text [3, 4]. Protein sequences, represented by single-letter amino acid codes, are highly similar to human natural language, where each single letter (amino acid) corresponds to a word in a sentence. Similarly, protein language models (PLMs) are trained on millions of protein sequences to internalize the information encrypted in protein sequences and generate numerical vectors/matrices for a given protein sequence [4]. Most recently, some PLMs, such as evolutionary scale modeling (ESM), ProtTrans, have been released and have achieved state-of-the-art performance in various downstream tasks with a simple classifier [5–7].

To make this approach more accessible and user-friendly, we built this unified architecture for researchers in biochemistry who are interested in employing this cutting-edge approach for their own bioactive peptide prediction model development. We provide an application example, along with a detailed explanation and annotation of each line and function in the code. By following the protocol, the interested users only need to prepare their dataset for model development and then the trained mode can be used for future prediction tasks. Additionally, we will provide guidance on how to extend the usage of the protocol, when modifications are needed.

## 2 Prerequisites Before Model Development

In this protocol, we use Google Colab for the demonstration. This allows users to run the protocol anywhere with internet access, a Google Account, and any browser. Colab is a hosted Jupyter Notebook service that requires no setup to use and provides free access to computing resources (e.g., GPU). We highly recommend using Colab for the running if not familiar with Python and the environment setup. For local execution of this protocol, the necessary Python packages are listed in "requirements.txt" and can be found at https://github.com/dzjxzyd/LLM_book_chapter.

## 3 Protocol for Application in Binary Classification Tasks

The protocol is based on the UniDL4BioPep paper [5], with a few modifications. Users can find it at https://github.com/dzjxzyd/UniDL4BioPep, where additional benchmark datasets for practice, as well as the original references are also available.

### 3.1 Dataset Construction

To begin, we create a bioactive peptide dataset for embedding and model development. Generally, the peptide dataset size is relatively small (<10,000 sequences). To simplify the dataset construction process, we use Microsoft Excel (xls or xlsx format). Example files are provided at https://github.com/dzjxzyd/LLM_book_chapter.

- **Step 1:** Create a Microsoft Excel file with the first column named "sequence" and the second column named "label," for example, "bitter_single_dataset_example.xlsx."
- **Step 2:** Fill out the two columns with available bioactive peptide data, labeling positive samples as 1 and negative samples as 0 (*see* **Note 1**).
- **Step 3 (optional):** Dataset splitting. If you are building a model from scratch, it is recommended to save all your data into a single dataset file and randomly split it for evaluation. If you are comparing model performance with existing models whose datasets have already been divided, it is suggested to use their training and test datasets directly to ensure a fair comparison.

**Common Issues**

1. Peptide sequences must be fully represented using single-letter amino acid code in uppercase; otherwise, a "KeyError" will occur.
2. You must use the column name same as the example files; otherwise, the code will not recognize the two columns.

### 3.2 Peptide Embedding

Peptide embedding, also known as peptide representation, involves converting the peptide sequences in the dataset into numerical vectors or matrices (in our case, we will transform those sequences into vectors). In this protocol, we will use the ESM language model (esm2_t6_8M_UR50D) for the peptide embeddings. Before proceeding, please load Jupyter Notebook in Colab, for example, "Book_chapter_example_binary_classification_320_dimension. ipynb", upload your dataset file "bitter_single_dataset_example. xlsx" to Google Colab (using the left sidebar) (Fig. 1a–c).

**Step 1** ESM language model loading and embedding function construction. The language model can be easily changed to other ESM2 language models (*see* **Note 2**).

```
def esm_embeddings(esm2, esm2_alphabet, peptide_sequence_list):
  # load the python packages
  import torch
  import esm
  import collections
  import pandas as pd
  import gc
  # Choose cuda for acceleration if available
  if torch.cuda.is_available():
    device = torch.device("cuda")
  else:
    device = torch.device("cpu")
  esm2.eval()
  esm2 = esm2.to(device)
  batch_converter = esm2_alphabet.get_batch_converter() # this function is to tokenize your peptide
sequences from amino acid sequence into numbers
  # load the peptide sequence list into the bach_converter
  batch_labels, batch_strs, batch_tokens = batch_converter(peptide_sequence_list)
  batch_lens = (batch_tokens != alphabet.padding_idx).sum(1)
  ## batch tokens are the tokenized results of the whole data set
  batch_tokens = batch_tokens.to(device)
  # Extract per-residue representations (on CPU)
  with torch.no_grad():
      # Here we export the last layer of the EMS model output as the representation of the peptides
      # model'esm2_t6_8M_UR50D' only has 6 layers, and therefore repr_layers parameters are equal to 6
      # If you want to change the esm model, for example: esm2_t12_35M_UR50D, which has 12 layers, they
you will need to change the 6 to 12 at the following two line
      results = esm2(batch_tokens, repr_layers=[6], return_contacts=False)
  token_representations = results["representations"][6].cpu() # the representation is generated for each
amino acid residue; for example, if you have 3 residues, then the ourput shape is 3 * 320, 3 is the number of
residues and the 320 is the vector for a single residue
```

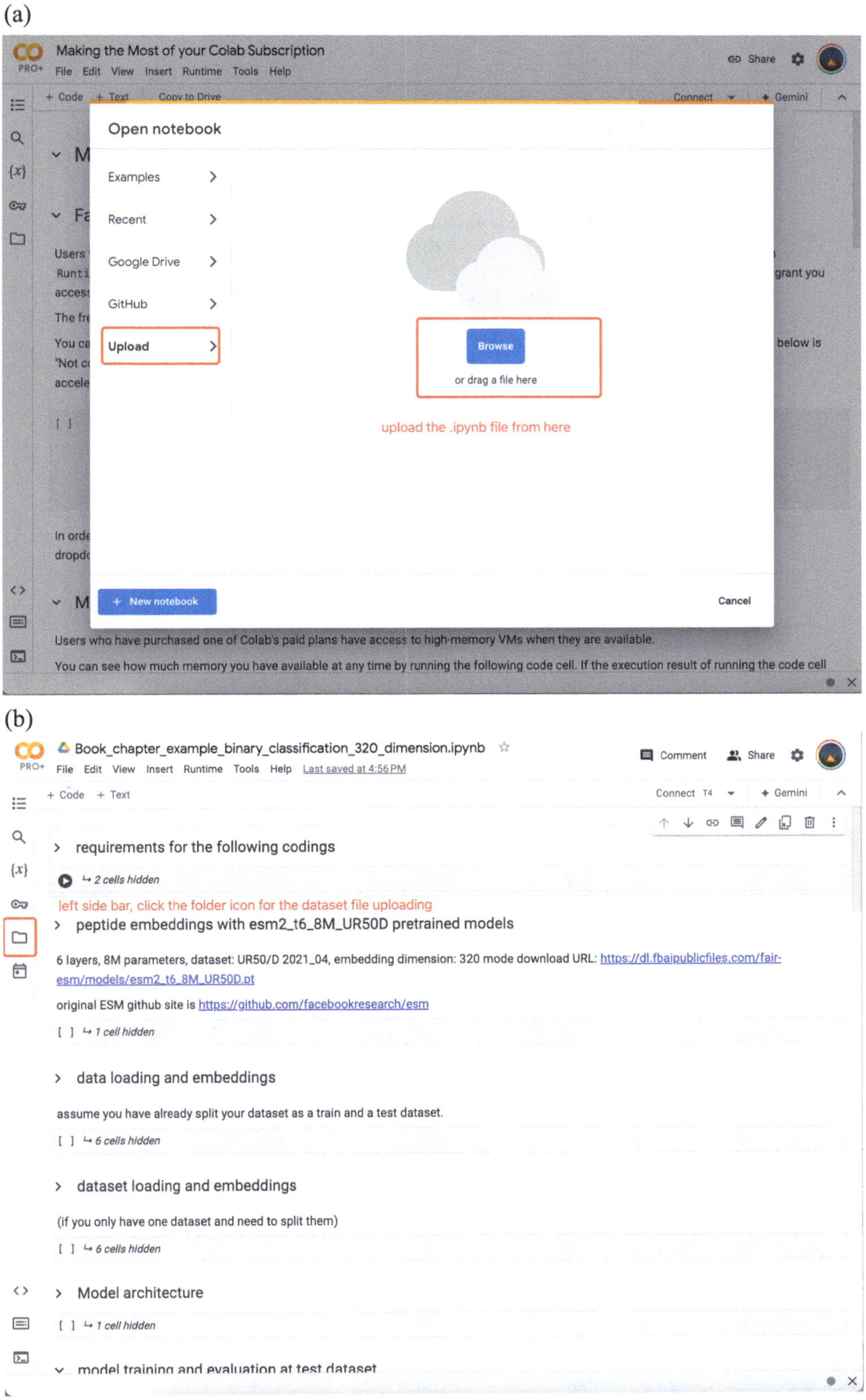


**Fig. 1** Guidelines for using Google Colab with our protocol from (**a**) step 1 to (**b**) step 2, and (**c**) step 3

(c)

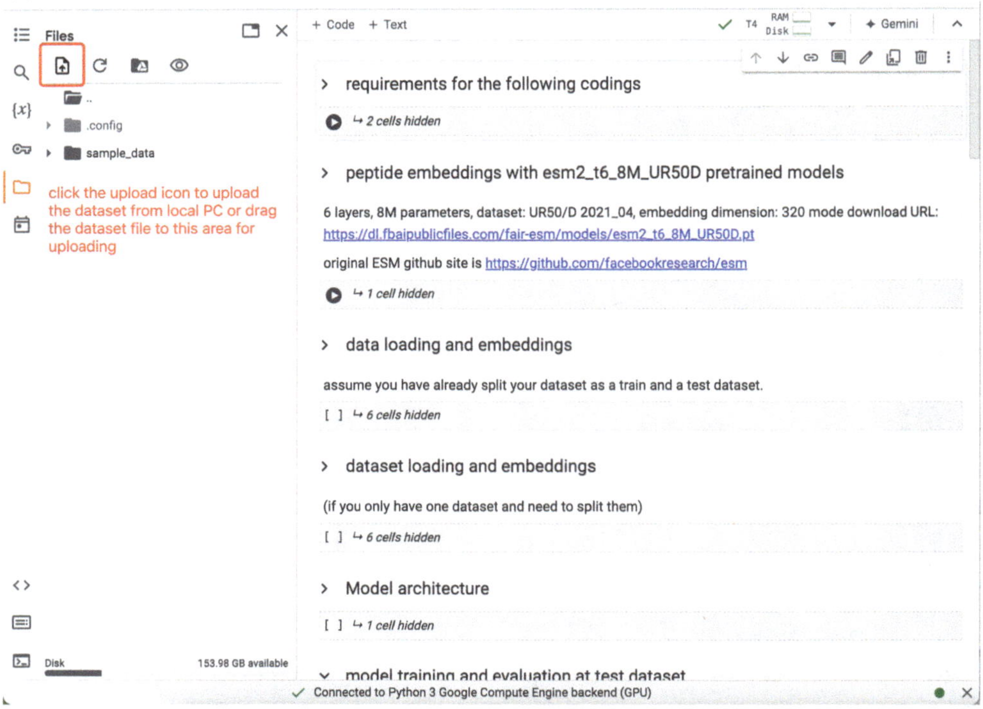


**Fig. 1** (continued)

```
# Generate per-sequence representations via averaging
# NOTE: token 0 is always a beginning-of-sequence token, so we will discard it and start from token 1
sequence_representations = []
for i, tokens_len in enumerate(batch_lens):
    sequence_representations.append(token_representations[i, 1 : tokens_len - 1].mean(0))
# sequence_representations is a list and each element is a tensor
embeddings_results = collections.defaultdict(list)
for i in range(len(sequence_representations)):
    # tensor format can be transformed as numpy format sequence_representations[0].numpy() or
sequence_representations[0].to_list
    each_seq_rep = sequence_representations[i].tolist()
    for each_element in each_seq_rep:
        embeddings_results[i].append(each_element)
embeddings_results = pd.DataFrame(embeddings_results).T
del  batch_labels, batch_strs, batch_tokens, results, token_representations # delete those variables to
save GPU memory
gc.collect() # release the GPU memory
return embeddings_results
```

```
# load the esm2_t6_8M_UR50D
model, alphabet = esm.pretrained.esm2_t6_8M_UR50D()
```

**Step 2** Dataset loading and division. This protocol includes two sections: one for using a single dataset (the current example) and the other for using separate train and test datasets. If you are using your own dataset, be sure to update the dataset name in the scripts or adjust the dataset filename accordingly. In this step, you will generate three datasets at the end: train, validation, and test.

```
# One dataset
Dataset = pd.read_excel('bitter_single_dataset_example.xlsx',header=0, index_col = None)
# Two datasets
# train dataset loading
dataset = pd.read_excel('bitter_train.xlsx',header=0, index_col = None)
# test dataset loading
dataset = pd.read_excel('bitter_test.xlsx',header=0, index_col = None)
```

```
# one dataset: we will split the dataset two times to get the train, validation and test, respectively.
X_train, X_test, y_train, y_test = train_test_split(X, y, test_size=0.2, random_state=123, shuffle= True)
X_train, X_valid, y_train, y_valid = train_test_split(X_train, y_train, test_size=0.2, random_state=234,
shuffle= True)
# two datasets: we only need to split the train and validation dataset
# random state is to make sure the dataset division is repeatable
X_train, X_valid, y_train, y_valid = train_test_split(X_train, y_train, test_size=0.2, random_state=111,
shuffle= True)
```

**Note** The ratio for dataset splitting is adjustable. In this protocol, the training dataset is further divided randomly into training and validation datasets. The validation dataset is used to save the model checkpoints. The "random_state" parameter can be adjusted to control the randomness of the split. If you prefer a repeatable dataset division, you should use the same "random_state" value each time you run the protocol.

**Step 3** Peptide embeddings. Embeddings are computed only for peptide sequences. To save the GPU memory or RAM consumption, in our demo, we conduct the embedding process one by one. We will load the peptide sequences and make sure the format of the input sequence meets the input format requirement of ESM2. Finally, we save the embedding results into a csv file for further usage (*see* **Note 3**).

```
# training dataset loading
dataset = pd.read_excel('bitter_single_dataset_example.xlsx',header=0, index_col = None)
# generate embedding for seqeunces
sequence_list = dataset['sequence']
embeddings_results = pd.DataFrame()
for seq in sequence_list:
   # the setting is just following the input format setting in ESM model, [name,sequence]
   tuple_sequence = tuple([seq,seq])
   peptide_sequence_list = []
   peptide_sequence_list.append(tuple_sequence) # build a summarize list variable including all the
sequence information
   # employ ESM model for converting and save the converted data in csv format
   one_seq_embeddings = esm_embeddings(model, alphabet, peptide_sequence_list)
   embeddings_results= pd.concat([embeddings_results,one_seq_embeddings])
embeddings_results.to_csv(whole_sample_dataset_esm2_t6_8M_UR50D_unified_320_dimension.csv')
```

**Potential Bug** This protocol can also be used for protein-related prediction. If your protein sequence is too long (more than 2000 residues), it might surpass the maximum RAM or GPU memory limits of a free Tesla T4 GPU. In such a case, the notebook may crash and automatically restart.

**Step 4** Normalization (optional). A min–max normalization is conducted using the training dataset to scale the features in the range (0, 1). In this context, each peptide is represented by a 320-dimension vector, and normalization ensures that each dimension is scaled within this range (0, 1). The scaler for normalization learnt from training dataset will be applied to normalize the validation and test datasets as well.

```
# normalize the X data range
from sklearn.preprocessing import MinMaxScaler
scaler = MinMaxScaler()
scaler.fit(X_train)
X_train = scaler.transform(X_train) # normalize X to 0-1 range
X_valid = scaler.transform(X_valid) # normalize X to 0-1 range
X_test = scaler.transform(X_test)
```

### 3.3 Model Training and Performance Evaluation at Test Dataset

Keras library is used for the model architecture design since it is more readable. In the demo notebook, we use the same architecture as that in UniDL4BioPep with a few modifications. The first layer is the input layer, where the peptide sequence information is represented by a numerical vector generated by the ESM model. Then, the next layer is a 1D convolutional layer with filter sizes of 32, kernel size of 3 and ReLU activation, and then a 1D max pooling layer was added to reduce the dimensionality. In addition, batch normalization and dropout (rate = 0.15) are added to avoid overfitting. Subsequently, the output of the convolutional layer is flattened and followed by a dense layer containing 64 hidden neurons with ReLU activation function and a dropout rate of 0.15. The last layer is also a dense layer with two neurons with a SoftMax activation function for binary classification. The Adam optimizer is chosen to accelerate the model fitting and maximize its performance. The best model will be stored under the consideration of accuracy at the validation dataset. The maximum epoch time is 200, but in practice, the best model is obtained around 100 epochs, which takes around 5 min on the Google Colab platform with T4 GPU.

**Step 1** Model architecture construction.

```
def ESM_CNN(X_train, y_train, X_test, y_test):
  inputShape=(320,1) # input feature size
  input = Input(inputShape)
  x = Conv1D(32,(3),strides = (1),name='layer_conv1',padding='same')(input)
  x = BatchNormalization()(x)
  x = Activation('relu')(x)
  x = MaxPooling1D((2), name='MaxPool1',padding="same")(x)
  x = Dropout(0.15)(x)
  x = Flatten()(x)
  x = Dense(64,activation = 'relu',name='fc1')(x)
  x = Dropout(0.15)(x)
  x = Dense(2,activation = 'softmax',name='fc2')(x)
  model = Model(inputs = input,outputs = x,name='Predict')
  adam = Adam(learning_rate=0.001)
  # compile the model
  model.compile(loss='sparse_categorical_crossentropy',optimizer=adam, metrics=['accuracy'])
  # set checkpoint and save the best model
  mc = ModelCheckpoint('best_model.keras', monitor='val_accuracy', mode='max', verbose=1,
save_best_only=True, save_weights_only=False)
  # summary the callbacks_list
  callbacks_list = [mc]
  model_history = model.fit(X_train, y_train, validation_data=(X_test, y_test),
epochs=100,callbacks=callbacks_list,batch_size = 8, verbose=1)
  return model, model_history
```

**Step 2** Train the model and save the best model during training process.

```
# train the model & best model checkpoint will be save as 'best_model.keras'
model, model_history = ESM_CNN(X_train, y_train, X_valid, y_valid)
```

**Step 3** Load the save model for performance evaluation at test dataset.

```
# load the saved best model for performance evaluation at test dataset
saved_model = load_model('best_model.keras')
# result collection list
ACC_collecton = []
BACC_collecton = []
Sn_collecton = []
Sp_collecton = []
recall_collecton = []
precision_collecton = []
f1_collecton = []
MCC_collecton = []
AUC_collecton = []
# confusion matrix
predicted_class= []
predicted_protability = saved_model.predict(X_test,batch_size=1)
for i in range(predicted_protability.shape[0]):
  index = np.where(predicted_protability[i] == np.amax(predicted_protability[i]))[0][0] # the class with higher
protability was treated as the prediction class
  predicted_class.append(index)
predicted_class = np.array(predicted_class) # transformed into a numpy for performance evaluation
y_true = y_test
from sklearn.metrics import confusion_matrix
import math
# np.ravel() return a flatten 1D array
TN, FP, FN, TP = confusion_matrix(y_true, predicted_class).ravel() # shape [ [True-Positive, False-positive],
[False-negative, True-negative] ]
ACC = (TP+TN)/(TP+TN+FP+FN)
ACC_collecton.append(ACC)
Sn_collecton.append(TP/(TP+FN))
Sp_collecton.append(TN/(TN+FP))
recall_collecton.append(TP/(TP+FN))
precision_collecton.append(TP/(TP+FP))
f1_collecton.append(2*TP/(TP+FP)*TP/(TP+FN)/ (TP/(TP+FP)+TP/(TP+FN)))
MCC = (TP*TN-FP*FN)/math.pow(((TP+FP)*(TP+FN)*(TN+FP)*(TN+FN)),0.5)
MCC_collecton.append(MCC)
BACC_collecton.append(0.5*TP/(TP+FN)+0.5*TN/(TN+FP))
# print the performance in the test dataset
from sklearn.metrics import roc_auc_score
AUC = roc_auc_score(y_test, predicted_protability[:,1])
AUC_collecton.append(AUC)
print(ACC_collecton[0])
print(BACC_collecton[0])
print(Sn_collecton[0])
print(Sp_collecton[0])
print(recall_collecton[0])
print(precision_collecton[0])
print(f1_collecton[0])
print(MCC_collecton[0])
print(AUC_collecton[0])
```

### 3.4 Model Application on New Dataset

Once the model is developed, you can use it to make predictions on new data.

**Step 1** Create a dataset for your new data similar to the dataset construction process, but leave the label column empty, for example, "bitter_new_data.xlsx."

**Step 2** Generate embeddings for the sequences in the new dataset.

```
model, alphabet = esm.pretrained.esm2_t6_8M_UR50D()
# new dataset loading
dataset = pd.read_excel('bitter_new_data.xlsx',header=0, index_col = None)
# generate embedding for seqeunces
sequence_list = dataset['sequence']
embeddings_results = pd.DataFrame()
for seq in sequence_list:
    # the setting is just following the input format setting in ESM model, [name,sequence]
    tuple_sequence = tuple([seq,seq])
    peptide_sequence_list = []
    peptide_sequence_list.append(tuple_sequence) # build a summarize list variable including all the
sequence information
    # employ ESM model for converting and save the converted data in csv format
    one_seq_embeddings = esm_embeddings(model, alphabet, peptide_sequence_list)
    embeddings_results= pd.concat([embeddings_results,one_seq_embeddings])
embeddings_results.to_csv('new_data_esm2_t6_8M_UR50D_unified_320_dimension.csv')
```

**Note** Ensure that the filename is updated to match your new dataset file name.

**Step 3 (Optional)** If you normalize your dataset during the training process, apply the same normalization to the new dataset using the scaler learned from the training dataset.

```
# read the peptide embeddings
X_data =
pd.read_csv('whole_sample_dataset_esm2_t6_8M_UR50D_unified_320_dimension.csv',header=0,
index_col = 0,delimiter=',')
X = np.array(X_data)

# please make sure the following numbers for random_state is consistant with the above data division
section.
from sklearn.model_selection import train_test_split
X_train, X_test, y_train, y_test = train_test_split( X, y, test_size=0.2, random_state=123, shuffle= True,
stratify =y)
X_train, X_valid, y_train, y_valid = train_test_split(X_train, y_train, test_size=0.2, random_state=234,
shuffle= True, stratify =y_train)
# loading the peptide embddings
X_new_embeddings =
pd.read_csv('new_data_esm2_t6_8M_UR50D_unified_320_dimension.csv',header=0, index_col =
0,delimiter=',')
# normalize the X data range
from sklearn.preprocessing import MinMaxScaler
scaler = MinMaxScaler()
scaler.fit(X_train)
X_new = np.array(X_new_embeddings)
X_new = scaler.transform(X_new) # normalize X to 0-1 range
```

**Step 4** Load the saved model to make predictions and generate an output file in Microsoft Excel.

```
# load the saved best model for performance evaluation at test dataset
saved_model = load_model('best_model.keras')
# confusion matrix
predicted_class= []
predicted_protability = saved_model.predict(X_new,batch_size=1)
for i in range(predicted_protability.shape[0]):
  index = np.where(predicted_protability[i]==np.amax(predicted_protability[i]))[0][0]
  predicted_class.append(index)
predicted_class = np.array(predicted_class)
for i in range(dataset.shape[0]):
  dataset.iloc[i,1] = predicted_class[i]
dataset.to_excel('new_data_prediction_result.xlsx')
```

## 4 Notes

1. There is no specific requirement for the order of samples in the Excel file, as the dataset will be randomized prior to model training.
2. In this protocol, we use ESM language model (esm2_t6_8-M_UR50D) for peptide embeddings. Five additional ESM models varying in sizes are available at https://github.com/facebookresearch/esm. Our previous study indicated that larger ESM models generally provide better performance in peptide representation [8]. For dataset containing over 1000 sequences, it is recommended to use a larger ESM model, as it may provide slight performance improvement compared to the esm2_t6_8M_UR50D model. When selecting a different ESM model, hyperparameter adjustments may be necessary to optimize model performance. For guidance on hyperparameter tuning across different ESM2 models, the users may refer to the supplementary documents of pLM4Alg, where we conducted a comprehensive hyperparameter tunning (https://github.com/dzjxzyd/pLM4Alg) [8].
3. We separate the embedding and training processes by saving the embedding results first and then loading them for training. This approach improves training efficiency, as embedding is time-intensive. Training, in contrast, is an iterative process, where each peptide sequence is used repeatedly, leading to redundant embedding operations of the same sequence if done concurrently. By directly using the precomputed embeddings for training, we avoid the repetitive time cost of embedding the same peptide sequences multiple times.

## Acknowledgments

This is contribution No. 25-037-B from the Kansas Agricultural Experimental Station.

### References

1. Du Z, Comer J, Li Y (2023) Bioinformatics approaches to discovering food-derived bioactive peptides: reviews and perspectives. TrAC Trends Anal Chem:117051. https://doi.org/10.1016/j.trac.2023.117051
2. Szymczak P, Możejko M, Grzegorzek T et al (2023) Discovering highly potent antimicrobial peptides with deep generative model HydrAMP. Nat Commun 14:1453. https://doi.org/10.1038/s41467-023-36994-z
3. Dallago C, Schütze K, Heinzinger M et al (2021) Learned Embeddings from deep learning to visualize and predict protein sets. Curr Protoc 1:e113. https://doi.org/10.1002/cpz1.113
4. Rives A, Meier J, Sercu T et al (2021) Biological structure and function emerge from scaling unsupervised learning to 250 million protein sequences. Proc Natl Acad Sci USA 118:

e2016239118. https://doi.org/10.1073/pnas.2016239118

5. Du Z, Ding X, Xu Y, Li Y (2023) UniDL4BioPep: a universal deep learning architecture for binary classification in peptide bioactivity. Brief Bioinform 24:bbad135. https://doi.org/10.1093/bib/bbad135
6. Elnaggar A, Heinzinger M, Dallago C et al (2022) ProtTrans: toward understanding the language of life through self-supervised learning. IEEE Trans Pattern Anal Mach Intell 44:7112–7127. https://doi.org/10.1109/TPAMI.2021.3095381
7. Lin Z, Akin H, Rao R et al (2023) Evolutionary-scale prediction of atomic-level protein structure with a language model. Science 379:1123–1130. https://doi.org/10.1126/science.ade2574
8. Du Z, Xu Y, Liu C, Li Y (2023) pLM4Alg: protein language model-based predictors for allergenic proteins and peptides. J Agric Food Chem:752. https://doi.org/10.1021/acs.jafc.3c07143

# Chapter 18

# CLAPE: Protein–Ligand Binding Site Prediction via Protein Language Models

**Yufan Liu and Boxue Tian**

## Abstract

Protein–ligand interactions are fundamental to critical biological processes such as transcription, translation, and drug–target interaction. The precise identification of protein–ligand binding residues is essential for accurately modeling these interactions and facilitating subsequent research. However, the development of computational methods that are both efficient and accurate for this purpose remains a significant challenge. Progress in this field holds the potential to drive substantial advancements in biotechnology and drug discovery. In response to this need, we have recently developed a method known as Contrastive Learning And Pretrained Encoder (CLAPE). This approach integrates a pretrained protein language model with contrastive learning to predict ligand-binding residues. In this chapter, we provided a detailed overview of the CLAPE framework, including instructions for using the associated Python package and command-line tools. Additionally, we outline the methods employed for result visualization and describe the comprehensive processes involved in model training.

**Key words** Protein–ligand interaction, Protein language model, Contrastive learning

## 1 Introduction

The interaction between proteins and ligands is fundamental to numerous biological processes, including protein–protein, protein–small molecule, and protein–nucleic acid interactions. Protein–DNA interactions, in particular, are essential to fundamental cellular processes such as DNA replication, transcription, signal transduction, and metabolic regulation. DNA-binding proteins, including transcription factors (TFs), are critical regulators of gene expression, influencing processes like immune response and stem cell pluripotency [1, 2]. Aberrations in TF function are associated with the pathogenesis of various diseases, including cancer and neurodegenerative disorders [3]. These proteins recognize and bind to conserved sequence motifs in DNA, which are preserved across evolutionary lineages [4]. Furthermore, other DNA-binding proteins, such as histones, DNA polymerases, and DNA

Dukka B. KC (ed.), *Large Language Models (LLMs) in Protein Bioinformatics*, Methods in Molecular Biology, vol. 2941, https://doi.org/10.1007/978-1-0716-4623-6_18,

topoisomerases, are integral to chromatin structure, DNA replication, and transcriptional regulation, with implications for human health and disease. Protein–RNA interactions are primarily involved in post-transcriptional regulation, where RNA-binding proteins (RBPs) play pivotal roles in processes such as RNA splicing, mRNA transport, translation, and degradation [5, 6]. The interaction between proteins and small molecules forms the basis of drug discovery, focusing on drug–target interactions and binding affinities [7]. Protein–protein interactions (PPIs) represent one of the most critical molecular events within organisms. The capacity of proteins to form large complexes is closely associated with various human diseases, and the intricate PPI network is vital for cellular communication, cell cycle control, and signal transduction pathways [8, 9]. Given the significance of protein–ligand interactions, their identification has emerged as a prominent area of research in recent decades.

The identification of ligand-binding sites on a protein represents a crucial initial step in modeling the binding properties of protein–ligand interactions. Various experimental techniques have been developed to elucidate protein–ligand interactions either in vivo or in vitro. For instance, chromatin immunoprecipitation (ChIP) [10] is employed to study protein–DNA interactions, cross-linking, and immunoprecipitation (CLIP) [11] is used for protein-RNA interactions, and the yeast two-hybrid (Y2H) [12] system is a common method for investigating protein–protein interactions. Structural biology approaches, including X-ray crystallography and nuclear magnetic resonance (NMR) spectroscopy, have also been extensively utilized to identify ligand-binding residues and domains within proteins. While these molecular biology-based experimental methods have provided significant insights over the past few decades, they are often time-consuming and require substantial resources. Consequently, the computational prediction of ligand-binding residues using machine learning techniques has emerged as an appealing alternative, offering the potential for faster and more efficient identification of key interaction sites.

The success of these models is largely dependent on the effectiveness of the representation, typically leveraging evolutionary information from protein sequences through methods like PSI-BLAST [13] and HHblits [14]. Additionally, structural features, including protein secondary structure predicted by tools like DSSP [15] and PSIPRED [16], contribute valuable information. Both sequence-based models, such as BindN [17], iDRNA-ITF [18], and PPINet [19], and structure-based models, including NucleicNet [20], HIGH-PPI [21], P2Rank [22], and GraphBind [23], have been developed to enhance the accuracy of protein–ligand interaction predictions. Notably, the application of deep learning techniques, including Convolutional Neural Networks (CNNs), Recurrent Neural Networks (RNNs), and Graph Neural

Networks (GNNs), has demonstrated significant promise in improving the predictive capabilities of these models.

In general, structure-based models tend to outperform sequence-based models due to the comprehensive information in protein structures. Nonetheless, accurate protein structures are often required [24], making sequence-based prediction of ligand-binding sites a critical research area. Current sequence-based models still fall short of practical application, often relying on manually designed features that lack refinement. Thus, there is a need for an end-to-end model that bypasses the dependency on handcrafted features. Techniques like pretraining and contrastive learning hold significant promise in this regard, with pretraining leveraging large-scale unlabeled data for unsupervised learning and contrastive learning enhancing the discriminative ability of protein embeddings.

To address these challenges, we developed the Contrastive Learning and Pretrained Encoder (CLAPE) in our previous study [25] to predict ligand-binding sites directly from raw protein sequence data. CLAPE presents a versatile framework for predicting ligand-binding sites solely based on protein sequences, contributing to the advancement of feature extraction and model development in this domain.

In this chapter, we overview the architecture, model performance, and datasets utilized in the training of CLAPE. We also offer detailed instructions on how to use the CLAPE Python package and command-line interface for predicting ligand-binding sites from a given protein sequence. Additionally, we demonstrate how to visualize the prediction results based on available protein structures. Finally, we outline the process for training a CLAPE model from scratch, including data preprocessing and model component design, with the aim of benchmarking on provided datasets or applying the CLAPE model to various protein–ligand binding tasks.

## 2 Materials

### 2.1 The Model Architecture of CLAPE

To provide a comprehensive framework for protein–ligand binding site prediction, we developed the CLAPE model. The overall architecture of CLAPE, illustrated in Fig. 1, consists of three core modules: the sequence embedding module, the backbone network module, and the loss computation module. In the sequence embedding module, ProtBert [26], a pretrained protein language model, was utilized to encode protein sequences and extract residue-level features. These features were then processed by the backbone network, which in the CLAPE model was composed of a four-layer 1D convolutional neural network (CNN). The loss computation module employed a contrastive loss function guided by a

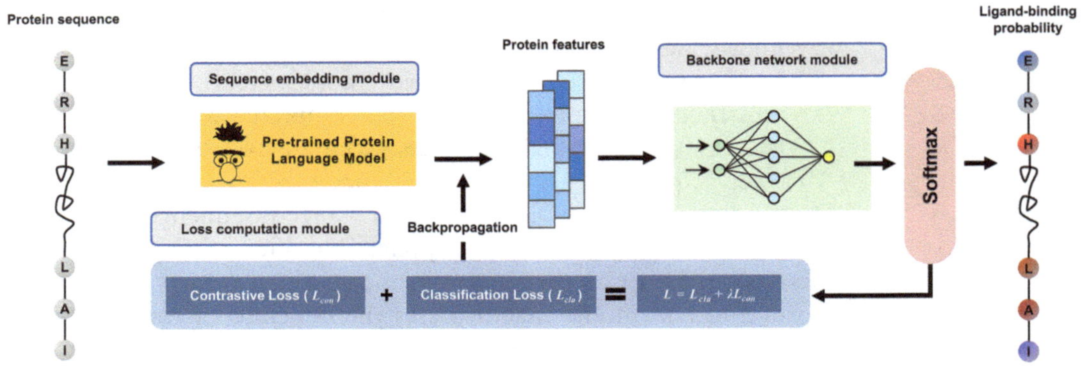


**Fig. 1** Schematic representation of the CLAPE model architecture. The sequence embedding module utilizes the pretrained ProtBert model to generate representations of protein sequences. The backbone network module processes these embeddings through flexible implementations of various neural networks. The loss computation module is responsible for calculating classification loss and contrastive loss, enabling backpropagation to refine the backbone model parameters. A classification head is used to produce classification scores ranging from 0 to 1

binary classification loss. Finally, the prediction scores generated by the backbone network are converted into classification probabilities using a Softmax function in the classification head. The pretrained protein language model was utilized as a feature extractor to eliminate the need for laborious manual feature extraction processes. Additionally, various pretrained protein language models can be employed within the sequence embedding module. In the subsequent sections, the training scheme of CLAPE will be introduced, which has been designed to allow for modifications to different components of the model.

### *2.2 Prediction Quality of CLAPE*

The CLAPE series models, denoted as CLAPE-DB (DNA-binding), CLAPE-RB (RNA-binding), and CLAPE-AB (antibody–antigen binding) in our original research paper, have demonstrated superior performance across all three ligand binding tasks. On the DNA-binding dataset, the CLAPE-DB model exceeded the performance of all sequence-based models and achieved a substantial improvement over the second-best model, with a 9.6% increase in AUC. Remarkably, CLAPE-DB outperformed most structure-based models despite not incorporating any structural information. Furthermore, it was shown that CLAPE-DB outperformed structure-based models that utilized predicted structures of lower quality, underscoring the advantages of protein language models in the absence of high-quality structural data. Additionally, the CLAPE-DB model exhibited superior generalization ability in a cross-prediction setting. In RNA-binding and antibody–antigen binding tasks, the CLAPE-RB and CLAPE-AB models also achieved leading performances compared to existing models. Notably, the CLAPE-AB model attained an AUC of 0.920, which is relatively high and suggests its pratical application in accurately predicting the paratope of a given antibody sequence.

### 2.3 Datasets

In our original research manuscript, the CLAPE model was trained and evaluated across three different ligand datasets, including DNA, RNA, and antibody–antigen, and a detailed analysis was conducted on the task of predicting protein-DNA binding sites.

For the DNA-binding site prediction, two benchmark datasets were employed, which have been widely used in previous studies, specifically from the DBPred [27] and GraphBind [23] models. To mitigate the potential issue of information leakage and enhance the robustness of the models, both datasets underwent similar preprocessing steps, including the reduction of sequence similarity to 30% using CD-HIT [28]. This preprocessing ensured that the training and testing datasets shared less than 30% sequence similarity, thereby enabling the model to predict binding sites in dissimilar proteins. Binding residues were defined as those with a distance of less than 0.5 Å plus the sum of the Van der Waals radii of the nearest atoms between the residue and the nucleic acid molecule, according to the BioLIP [29] database. The two datasets comprised 1219 sequences for training and 175 sequences for testing, with a total of 33,320 binding residues and 489,093 non-binding residues. Additionally, an independent test set generated by the GraphSite [24] model was utilized for further analysis of model performance. All datasets, including those for RNA-binding and antibody–antigen binding tasks, faced issues related to data imbalance. The protein-RNA binding site dataset was collected from GraphBind [23] and pprint2 [30], while the antibody–antigen binding site dataset was sourced from the SAbDaB [31] database. The same data preprocessing procedures were applied to the RNA-binding and antibody–antigen binding site tasks as used for the protein–DNA binding site task.

## 3 Methods

To facilitate the scientific community in predicting protein–ligand binding sites via CLAPE, the framework has been made publicly available on GitHub, including the model weights, which is hosted at https://github.com/YAndrewL/CLAPE. Additionally, a Python package and a corresponding command-line interface (CLI) have been developed, both of which are accessible for download and installation from the Python Package Index (PyPI) at https://pypi.org/project/clape/.

In this section, we will provide a detailed guide on the usage of the Python package and the command-line interface. Furthermore, we will explain how to visualize results when protein structural data is available and, importantly, outline the process for training a

CLAPE-like model from scratch. Unless otherwise stated, we will primarily use protein-DNA binding site prediction as an example in the following sections as in our original research paper [25].

### 3.1 Predicting with CLAPE Python Package

To utilize the CLAPE Python package for predicting ligand binding properties within Python code, it is necessary to prepare a coding environment, which can be set up using Conda or Mamba. This step includes installing essential dependencies. Although the CLAPE package offers considerable flexibility, it primarily relies on a pretrained large language model, specifically ProtBert, and its implementation is mainly built on PyTorch and the Transformers library from HuggingFace [32]. The environment can be manually configured according to the provided description to fit the user's specific system requirements best. Alternatively, a YAML file has been provided to facilitate the creation of a Conda environment.

The CLAPE can be installed using the PyPi command line utility as follows:

```
$ wget https://github.com/YAndrewL/CLAPE/blob/main/environment.yaml

$ conda env create -f environment.yaml

$ conda activate clape

$ pip install clape
```

Alternatively, CLAPE can be installed from the source to access the latest version by executing the following commands:

```
$ git clone https://github.com/YAndrewL/CLAPE.git

$ cd CLAPE

$ pip instal .
```

After installing, users can utilize CLAPE-related functions within Python code. CLAPE receives protein sequences in FASTA format with extensions as .fasta or .fa, which is convenient for batch predictions. Here, we will use the prediction of DNA-binding sites as an example. A sample file has been provided to demonstrate the use of CLAPE, which can be obtained from the CLAPE GitHub repository. The command line and the format of the sample data file are as follows:

```
$ wget https://github.com/YAndrewL/CLAPE/blob/main/example.fa


# the content of example.fa (DNA-binding proteins)

>seq1

TSDLFNEIIPLGRLIHMVNQKKDRLLNEYLSPLDITAAQFKVLCSIRCAACITPV

ELKKV

LSVDLGALTRMLDRLVSKGWVERLPNPNDKRGVLVKLTTGGAAICEQCHQL

VGQDLHQEL

TKNLTADEVATLEYLLKKVLP

>seq2

SELKMGELSELLGYALKRAQLRVFEDFLHCVAPVQLTPAQFSVLLLLDANPG

RNQTEIAT

TLGILRPNFVAMLDALEGRGLCVRTRSPSDRRSHILMLTDKGRATLARAKKL

VATRHEDR

LTELLGRDNRDALLSMLATIAR
```

Each protein sequence is processed by the ProtBert model to extract features, which are then fed into a backbone network to generate binding scores. Below, we present a basic use case in Python and provide an explaination of the code snippets.

```
from clape import Clape, get_weights



get_weights("your/model/path", "DNA")

clape_model = Clape(model_path="your/model/path", ligand="DNA")

results = clape_model.predict(input_file="example.fa")
```

The results variable contains the predicted residue-level binding sites for the given protein sequences, which is a Python list structured as: ['0000000111100', '0011100011101100001110000']. In this list, "0" denotes a binding site, while "1" represents a non-binding site. To retain the original prediction scores, which range between 0 and 1 and indicate the probability of a residue being a binding site, users can set the keep_score parameter to True in the predict method.

The get_weights function is employed to download the pre-trained model weights, which are then loaded by CLAPE (*see* **Notes 1** and **2**). Additionally, users have the option to train their own models from scratch using the CLAPE-like framework, with the training loop detailed in the following section.

Moreover, the switch_ligand method, encapsulated within the Clape class, enables users to switch between different ligand prediction tasks. The threshold parameter can be used to specify the classification threshold: residues with scores above this threshold are classified as positive (binding sites), while those with scores below are classified as negative (non-binding sites).

If a feasible protein crystal structure is available, CLAPE also provides an interface that allows users to predict the binding site directly from PDB files. Predictions from PDB files can be made after defining the model, as previously described.

```
from clape import Clape, get_weights


get_weights("your/model/path", "DNA")

clape_model = Clape(model_path="your/model/path", ligand="DNA")

results = clape_model.predict_from_pdb(pdb_file, chain)


# or predict by a known PDB file

results = clape_model.predict_from_pdb(pdb_id, chain)
```

In the predict_from_pdb method, we offer a method to predict binding sites directly from PDB files. The user should specify the ID of the protein chain to be analyzed. The method first extracts the protein sequences from the PDB file and then generates embeddings using the model. Additionally, this method can be called by providing a known PDB entry in four letters, such as "1A0A," in which case the corresponding PDB file will be automatically retrieved from the PDB website. As with sequence-based predictions, the keep_score parameter can be set to True to retain the prediction scores, reflecting the probability of each residue being a binding site.

In most scenarios, visualizing the prediction results is crucial for identifying potential binding sites and binding pockets within the 3-dimensional structure of the protein. To facilitate this, we provide a visualize parameter that allows users to generate visualizations directly from the input PDB file. This feature highlights the predicted binding sites within the protein's 3D structure. For greater flexibility, an independent function is also provided to visualize the prediction results within protein structures, enabling users to customize their visualization needs. This function will be described in detail later in this section.

### 3.2 Predicting with CLAPE Command-Line Interface

Apart from performing predictions in a Python console, we further provide a command-line interface, which provides a more straightforward and faster way to generate prediction results in a Unix-like operating system. The command-line interface is installed automatically when the CLAPE package is installed using the pip installer. Below is a simple use case demonstrating the command-line interface of CLAPE:

```
$ wget https://github.com/YAndrewL/CLAPE/blob/main/weights/DNA.pth

$ clape --input example.fa --output out.txt --lignd DNA --model

/path/to/downloaded/model
```

The first command line is employed to download pretrained model weights from GitHub. The subsequent command line generates prediction results, which are stored in a predefined file named out.txt. The parameters largely correspond to those used in the CLAPE Python package, as described earlier, with slight differences in the output file format, which are more suitable for terminal usage. Specifically, the parameter—input specifies the input protein sequences in FASTA format,—output defines the path for the output file,—ligand indicates the ligand whose binding is to be predicted, and—model designates the path to the downloaded model weights, as illustrated in the example command line. In addition to the parameters shown in the usage example,—threshold is used to set the classification threshold, and—cache specifies the download path for the pretrained protein language model weights, such as ProtBert for DNA-binding site predictions.

The prediction results generated by the command-line interface are similar to the FASTA format but include an additional line that indicates the classification labels. An example instance is provided below. It is noteworthy that the training data shares the same format as the output file generated by the command-line interface. The training process will be detailed in the following section. Overall, the CLAPE command-line interface facilitates the rapid identification of potential ligand-binding sites, which can be utilized for downstream biological analyses and experimental applications.

```
# The content of a sample DNA-binding site results from file out.txt

>seq1

WVERLPNPNDKRGVLVKLTTGGAAICEQCHQLVGQDLHQELTKNLTADEVA

TLEYLLKKVLP

0001000000111110000000000000000000000000000000000000000000000000000000

>seq2

RGLCVRTRSPSDRRSHILMLTDKGRATLARAKKLVATRHEDRLTELLGRDN

RDALLSMLATI

0000000000011111000000000000000000000000000000000000000000000000000000
```

### 3.3 Visualizing the Results with CLAPE

The protein–ligand binding sites can be determined through the aforementioned steps using various methods, including Python package and command-line interface. However, there is a prevalent need to explore these binding sites from a structural perspective, which is advantageous for pocket discovery and ligand design. To address this, we have provided functions that enable visualization of the results for a given PDB file. As previously discussed, visualization can be generated by setting the visualize parameter to True within the predict_from_pdb method. Additionally, the visualization can be executed via an independent function, allowing for flexible integration with the Python package for analysis. This visualization function relies on PyMOL (https://www.pymol.org/), a widely used molecular visualization software, which can be installed through Conda for use in a Python console (*see* **Note 3**).

```
# Need a license for activation

$ conda install -c conda-forge -c schrodinger pymol-bundle
```

The installation can be tested by running the command from pymol import cmd in a Python console. Additionally, the CLAPE package offers utilities for visualizing the results. Below, we provide an example of how to predict DNA-binding sites using a PDB structure and subsequently visualize the results. The visualize function will handle all necessary processes to generate the visual representation of the results.

```
from clape import Clape, visualize


clape_model = Clape(model_path="your/model/path", ligand="DNA")

score, result = clape_model.predict_from_pdb(pdb_id="1A0A",

                                             chain="A",

                                             keep_score=True)

result = list(map(lambda x: int(x), list(result)))


visualize(pdb_file="1A0A.pdb", chain="A", result=result,

          out_file="out.pse", select_only=False, ligand_id='C,D')


# or color the PDB structure by predicted scores

visualize(pdb_file="1A0A.pdb", chain="A", result=score,

          out_file="out.pse", select_only=False, ligand_id='C,D')
```

Firstly, the predict_from_pdb method will predict DNA-binding sites using a specified PDB entry, with the resulting PDB file stored in the current directory (*see* **Note 4**). This file will be used for subsequent visualization, although this behavior can be adjusted by using the pdb_cache parameter to specify a different path. It is important to note that the prediction results need to be converted to integer type to meet the visualization requirements, as the original output is in string format.

The visualize function will then color the identified positive residues according to the specified color, with the default being red. This function accepts the predicted results, provided as a Python list, along with the PDB file as its primary inputs. The output will be a PyMOL session file in .pse format, which can be easily loaded into PyMOL. Users can opt to retain only the target protein chain by setting the select_only parameter to True. However, this may conflict with the ligand_id parameter, which is intended to retain ligand chains in the visualization (*see* **Note 5**). The default display mode is "surface," but it can be modified by specifying the shown_as parameter, with options such as "cartoon" and "sticks" being supported.

Additionally, the function can take predicted scores as inputs and color the structure based on the binding probability, offering a more intuitive representation of potential binding sites. This is achieved by altering the B-factor values of the PDB file and applying gradient coloring according to these values.

In summary, the visualize function seamlessly integrates prediction results with structural data. Furthermore, this visualization function is compatible with predicted protein structures generated by advanced tools such as AlphaFold3 [33] and ESMFold [34], thereby enhancing sequence-to-function analysis. Figure 2 illustrates an example of visualizations produced by CLAPE, showcasing both binary results and binding probabilities.

### *3.4 Training CLAPE Model from Scratch*

In the previous sections, we explored the use cases, application scenarios, and visualization tools associated with CLAPE. Furthermore, CLAPE is a highly adaptable framework that integrates a protein language model, a backbone network, and contrastive learning modules, all of which can be replaced with various state-of-the-art methods. In this section, we will outline the training process for CLAPE, which can be utilized to train models on diverse datasets and architectures for protein–ligand binding tasks. We will continue to use the DNA-binding site prediction task as a representative example.

To ensure a fair comparison when benchmarking models on CLAPE, we have provided the original training data, with the data collection process detailed in the previous section. The dataset can be downloaded as follows:

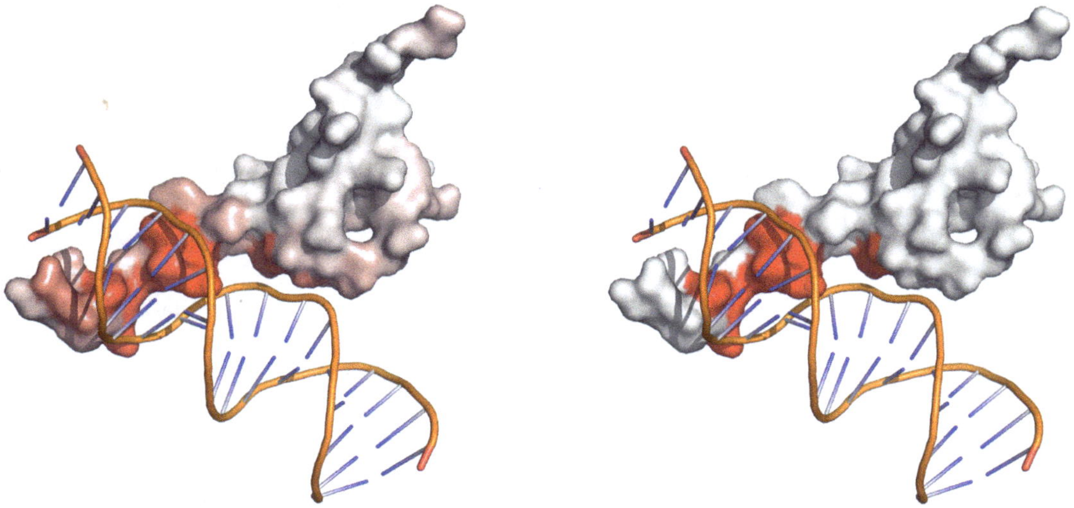

**Fig. 2** The prediction results were visualized using the CLAPE visualize function and subsequently rendered in PyMOL. In the left panel, the protein structure and DNA ligand are displayed, with a gradient color scheme indicating binding probability. The right panel presents the structure colored according to binary binding outcomes. The protein structure is referenced by the PDB entry 1A0A

```
$ wget https://github.com/YAndrewL/CLAPE/blob/main/raw_data/DNA.tar.gz

$ tar -zxvf DNA.tar.gz
```

Alternatively, the dataset can be downloaded directly in a Python console using the following command:

```
from clape import get_data


get_weights("your/data/path", "DNA")
```

The DNA directory contains training and testing data in a FASTA-like format, with an additional line indicating binary labels for residues, consistent with the output file format of the CLAPE command-line interface. Given the rapid increase in available protein structure data, the dataset can be expanded using various database sources, such as BioLIP [29], which is an actively updated resource for protein–ligand binding data. The binding labels can be generated based on specific criteria. In the training process of CLAPE, binding sites are defined as residues that lie within a distance of 0.5 Å plus the sum of the Van der Waals radii of the two nearest atoms between the residue and the nucleic acid molecule.

Initially, protein sequences are passed into a protein language model to extract the initial protein features. In CLAPE's training, ProtBert [26], a model from the Prottrans protein language model family, was utilized. However, users can employ different pretrained protein feature extractors, such as ESM [34], which was compared in our original research paper, or SaProt [35], which uses a predefined structural vocabulary.

Below, we present two examples of feature extraction using ProtBert and ESM, respectively. In these models, the first and last tokens represent the start and end tokens, while the middle tokens serve as embeddings for the residues in the protein sequences. For batched training, the generated features and labels usually need to be padded to the same length, which is typically the maximum length within a batch. This padding process can be customized in a collation function.

```
# Extract features using ProtBert
from transformers import BertModel, BertTokenizer
import re

tokenizer = BertTokenizer.from_pretrained("Rostlab/prot_bert",
do_lower_case=False)
pretrain_model = BertModel.from_pretrained("Rostlab/prot_bert")

sequence = ' '.join(seq)    # seq is the input protein sequences
sequence = re.sub(r"[UZOB]", "X", sequence)
encoded_input = tokenizer(sequence, return_tensors='pt')
features = pretrain_model(**encoded_input).last_hidden_state.squeeze(0)[1:-
1,:]
```

```
# Extract features using ESM
```

```
import esm

model, alphabet = esm.pretrained.esm2_t33_650M_UR50D()
batch_converter = alphabet.get_batch_converter()
model.to('cuda').eval()

batch_labels, batch_strs, batch_tokens = batch_converter(seq)
with torch.no_grad():
    results = model(batch_tokens.to('cuda'), repr_layers=[33],
return_contacts=False)
token_representations = results["representations"][33]
features = token_representations.squeeze(0)[1:-1]
```

```
# Collate sequence embedding into the same length
class BatchCollate(object):
    def __call__(self, data):
        # ([bz, length, dim], label)
        # (feature, label)
        tensors = pad_sequence([t[0] for t in data], batch_first=True)  # ([bz,
length, dim])
        labels = pad_sequence([t[1]) for t in data], batch_first=True,
padding_value=-1)
        return tensors, labels
```

Second, the generated features are passed into a backbone network, which can be chosen flexibly based on comparative analysis. In the case of CLAPE, we found that a 1D Convolutional Neural Network (1DCNN) model outperformed other mainstream backbone models, such as MLPs, RNNs, or Transformer-like models. The training process can either start from randomly initialized weights or utilize pretrained weights. In CLAPE training, backpropagation is typically halted at the feature extractor. However, our experiments indicate that fine-tuning the protein language model can enhance model performance, albeit with a potential risk of overfitting.

The backbone model outputs the binding probability for each residue. The final step involves defining a loss function, which in CLAPE consists of two components: classification loss and contrastive loss. For CLAPE training, we employed a triplet center loss [36], but other loss functions, such as SupCon [37] for supervised settings and InfoNCE [38] for unsupervised settings, can also be applied. We evaluated the performance of different contrastive loss functions in our original research paper. The total loss function is a weighted sum of the contrastive loss and classification loss.

In summary, the CLAPE framework is highly flexible, allowing for customized training at each step. Common training loop practices can be applied once the sequence feature extractor, backbone network, and loss functions are configured. We utilized PyTorch Lightning to structure the training process, making it easy to scale up to large datasets and model sizes. To provide clarity on the training processes of CLAPE, we offer an implementation hosted at https://github.com/JueWangTHU/CLAPE-SMB, which is a CLAPE model designed for small-molecule binding site prediction tasks [39] and can serve as a general template for other ligand types.

## 4 Notes

1. The model_path or a similar parameter should refer to the parent directory of the model weights, excluding the name of the weight file itself. The file name will be matched based on the specified ligand type to ensure the correct model weights are loaded.
2. If not explicitly specified, the download function for weights and data will retrieve all ligand types mentioned in the original paper, including DNA, RNA, and antibody–antigen datasets.
3. The PyMOL API in the Python console needs to be activated, and an educational license can be obtained at https://pymol.org/edu/.
4. When attempting to visualize a customized PDB file, ensure the file name is in lowercase, as this is purposely designed for predictions made from a PDB entry.
5. The parameters select_only and ligand_id in the visualize function cannot be set to True simultaneously, as this will cause a conflict error.

### References


1. Lambert SA, Jolma A, Campitelli LF, Das PK, Yin Y, Albu M, Chen X, Taipale J, Hughes TR, Weirauch MT (2018) The human transcription factors. Cell 172(4):650–665
2. Dillon SC, Dorman CJ (2010) Bacterial nucleoid-associated proteins, nucleoid structure and gene expression. Nat Rev Microbiol 8(3):185–195
3. Spitz F, Furlong EE (2012) Transcription factors: from enhancer binding to developmental control. Nat Rev Genet 13(9):613–626
4. Yang L, Zhou T, Dror I, Mathelier A, Wasserman WW, Gordân R, Rohs R (2014) TFBSshape: a motif database for DNA shape features of transcription factor binding sites. Nucleic Acids Res 42(D1):D148–D155
5. Oliveira C, Faoro H, Alves LR, Goldenberg S (2017) RNA-binding proteins and their role in the regulation of gene expression in Trypanosoma cruzi and Saccharomyces cerevisiae. Genet Mol Biol 40(01):22–30
6. Gebauer F, Schwarzl T, Valcárcel J, Hentze MW (2021) RNA-binding proteins in human genetic disease. Nat Rev Genet 22(3):185–198
7. Zhao L, Zhu Y, Wang J, Wen N, Wang C, Cheng L (2022) A brief review of protein–ligand interaction prediction. Comput Struct Biotechnol J 20:2831–2838
8. Athanasios A, Charalampos V, Vasileios T, Md Ashraf G (2017) Protein-protein interaction (PPI) network: recent advances in drug discovery. Curr Drug Metab 18(1):5–10
9. Safari-Alighiarloo N, Taghizadeh M, Rezaei-Tavirani M, Goliaei B, Peyvandi AA (2014) Protein-protein interaction networks (PPI) and complex diseases. GHFBB 7(1):17

10. Park PJ (2009) ChIP–seq: advantages and challenges of a maturing technology. Nat Rev Genet 10(10):669–680
11. Ramanathan M, Porter DF, Khavari PA (2019) Methods to study RNA–protein interactions. Nat Methods 16(3):225–234
12. Brückner A, Polge C, Lentze N, Auerbach D, Schlattner U (2009) Yeast two-hybrid, a powerful tool for systems biology. Int J Mol Sci 10(6):2763–2788
13. Altschul SF, Madden TL, Schäffer AA, Zhang J, Zhang Z, Miller W, Lipman DJ (1997) Gapped BLAST and PSI-BLAST: a new generation of protein database search programs. Nucleic Acids Res 25(17):3389–3402
14. Remmert M, Biegert A, Hauser A, Söding J (2012) HHblits: lightning-fast iterative protein sequence searching by HMM-HMM alignment. Nat Methods 9(2):173–175
15. Joosten RP, Te Beek TA, Krieger E, Hekkelman ML, Hooft RW, Schneider R, Sander C, Vriend G (2010) A series of PDB related databases for everyday needs. Nucleic Acids Res 39 (Suppl_1):D411–D419
16. McGuffin LJ, Bryson K, Jones DT (2000) The PSIPRED protein structure prediction server. Bioinformatics 16(4):404–405
17. Wang L, Brown SJ (2006) BindN: a web-based tool for efficient prediction of DNA and RNA binding sites in amino acid sequences. Nucleic Acids Res 34(suppl_2):W243–W248
18. Wang N, Yan K, Zhang J, Liu B (2022) iDRNA-ITF: identifying DNA-and RNA--binding residues in proteins based on induction and transfer framework. Brief Bioinform 23(4): bbac236
19. Cong H, Liu H, Cao Y, Liang C, Chen Y (2023) Protein–protein interaction site prediction by model ensembling with hybrid feature and self-attention. BMC Bioinformatics 24(1): 456
20. Lam JH, Li Y, Zhu L, Umarov R, Jiang H, Héliou A, Sheong FK, Liu T, Long Y, Li Y (2019) A deep learning framework to predict binding preference of RNA constituents on protein surface. Nat Commun 10(1):4941
21. Gao Z, Jiang C, Zhang J, Jiang X, Li L, Zhao P, Yang H, Huang Y, Li J (2023) Hierarchical graph learning for protein–protein interaction. Nat Commun 14(1):1093
22. Krivák R, Hoksza D (2018) P2Rank: machine learning based tool for rapid and accurate prediction of ligand binding sites from protein structure. J Chem 10:1–12
23. Xia Y, Xia C-Q, Pan X, Shen H-B (2021) GraphBind: protein structural context embedded rules learned by hierarchical graph neural networks for recognizing nucleic-acid-binding residues. Nucleic Acids Res 49(9):e51–e51
24. Yuan Q, Chen S, Rao J, Zheng S, Zhao H, Yang Y (2022) AlphaFold2-aware protein–DNA binding site prediction using graph transformer. Brief Bioinform 23(2):bbab564
25. Liu Y, Tian B (2024) Protein–DNA binding sites prediction based on pre-trained protein language model and contrastive learning. Brief Bioinform 25(1):bbad488
26. Elnaggar A, Heinzinger M, Dallago C, Rehawi G, Wang Y, Jones L, Gibbs T, Feher T, Angerer C, Steinegger M (2021) Prottrans: toward understanding the language of life through self-supervised learning. IEEE Trans Pattern Anal Mach Intell 44(10): 7112–7127
27. Patiyal S, Dhall A, Raghava GP (2021) DBpred: a deep learning method for the prediction of DNA interacting residues in protein sequences. BioRxiv:2021.2008. 2005.455224
28. Fu L, Niu B, Zhu Z, Wu S, Li W (2012) CD-HIT: accelerated for clustering the next-generation sequencing data. Bioinformatics 28(23):3150–3152
29. Yang J, Roy A, Zhang Y (2012) BioLiP: a semi-manually curated database for biologically relevant ligand–protein interactions. Nucleic Acids Res 41(D1):D1096–D1103
30. Patiyal S, Dhall A, Bajaj K, Sahu H, Raghava GP (2023) Prediction of RNA-interacting residues in a protein using CNN and evolutionary profile. Brief Bioinform 24(1):bbac538
31. Dunbar J, Krawczyk K, Leem J, Baker T, Fuchs A, Georges G, Shi J, Deane CM (2014) SAbDab: the structural antibody database. Nucleic Acids Res 42(D1):D1140–D1146
32. Wolf T (2019) Huggingface's transformers: state-of-the-art natural language processing. arXiv preprint arXiv:191003771
33. Abramson J, Adler J, Dunger J, Evans R, Green T, Pritzel A, Ronneberger O, Willmore L, Ballard AJ, Bambrick J (2024) Accurate structure prediction of biomolecular interactions with AlphaFold 3. Nature 630:1–3
34. Lin Z, Akin H, Rao R, Hie B, Zhu Z, Lu W, Smetanin N, Verkuil R, Kabeli O, Shmueli Y (2023) Evolutionary-scale prediction of atomic-level protein structure with a language model. Science 379(6637):1123–1130
35. Su J, Han C, Zhou Y, Shan J, Zhou X, Yuan F (2023) Saprot: protein language modeling with structure-aware vocabulary. bioRxiv:2023.2010. 2001.560349
36. He X, Zhou Y, Zhou Z, Bai S (2018) Bai X Triplet-center loss for multi-view 3d object retrieval. In: Proceedings of the IEEE

conference on computer vision and pattern recognition, pp 1945–1954

37. Khosla P, Teterwak P, Wang C, Sarna A, Tian Y, Isola P, Maschinot A, Liu C, Krishnan D (2020) Supervised contrastive learning. Adv Neural Inf Proces Syst 33:18661–18673

38. Oord A, Li Y, Vinyals O (2018) Representation learning with contrastive predictive coding. arXiv preprint arXiv:180703748

39. Wang J, Liu Y, Tian B (2024) Protein-small molecule binding site prediction based on a pre-trained protein language model with contrastive learning. J Chem 16(1):125

# Chapter 19

# Large Language Model (LLM)-Based Advances in Prediction of Post-translational Modification Sites in Proteins

**Pawel Pratyush, Suresh Pokharel, Stefan Schulze, Lisa Bramer, Robert H. Newman, and Dukka B. KC**

## Abstract

Post-translational modifications (PTMs) are vital regulators of protein function, influencing a myriad of cellular processes and disease mechanisms. Traditional experimental methods for PTM identification are both costly and labor-intensive, underlining the pressing need for efficient computational approaches. Early computational strategies predominantly relied on primary amino acid sequences and handcrafted features, which often lacked the contextual and structural understanding necessary for precise PTM site prediction. The emergence of transformer-based large language models (LLMs), particularly protein language models (pLMs), has revolutionized PTM prediction by producing context-aware embeddings that capture functional and structural intra-sequence dependencies. In this chapter, we provide a comprehensive review of recent advancements in leveraging LLMs (or, pLMs) for PTM site prediction, an important residue-level task in protein research. We identify emerging trends in the field, including the application of fine-tuning techniques, the integration of embeddings from multiple pLMs, and the incorporation of multiple modalities such as codon-aware embeddings, 3D structural data, and conventional representations. Additionally, we discuss tools that employ graph-based representations, the mamba architecture, and contrastive learning paradigms to further refine pLM-powered PTM site prediction models. We finally explore the interpretability and explainability aspects of the embeddings used in various tools. Despite the significant progress made, persistent limitations remain, and we outline these challenges while proposing directions for future research.

**Key words** Post-translational modification, Large language model, Protein language model, Mamba, Graph, GPT, Fine-tuning, AlphaFold, Explainability, Contrastive learning

## 1 Introduction

The functional complexity of proteomes is driven not only by the number of protein-coding gene sequences, transcriptional regulation, or alternative splicing of mRNA transcripts but also by post-translational modifications (PTMs) that result in a multitude of proteoforms [1, 2]. These modifications can expand the proteomic complexity by several orders of magnitude and can alter a protein's

Dukka B. KC (ed.), *Large Language Models (LLMs) in Protein Bioinformatics*, Methods in Molecular Biology, vol. 2941, https://doi.org/10.1007/978-1-0716-4623-6_19,

structure and function, or serve as signals for interactions, localization, or degradation [3–5]. Since many PTMs can be dynamically added or removed, they serve crucial functions in numerous cellular processes, signaling pathways, and regulatory networks [6–8]. Thus, PTMs are crucial for understanding cell biology, health, and disease [9, 10].

The addition and removal of PTMs is a non-template-driven process and hence cannot directly be inferred from a protein's sequence. Existing methods for the experimental identification of PTMs, such as mass spectrometry-driven proteomics, are time-consuming, expensive, and often yield an incomplete overview of the proteins and sites that can be modified [11, 12]. Therefore, computational approaches are needed to accurately predict PTM sites across whole proteomes.

Computational methods often rely on encoding amino acid sequences into numerical representations known as feature vectors [13]. These vectors act as proxies for the sequences, allowing them to serve as input for various inference models. The accuracy of these models depends heavily on the quality of the feature vectors, among other factors. In residue-level tasks such as PTM site prediction, the focus narrows to the context of specific residues within the overall sequence. For example, in crotonylation site prediction, lysine residues are of particular interest. Within a single protein sequence, some lysines may be crotonylated (positive sites) while others are not (negative sites). Therefore, feature vectors must distinguish between lysines belonging to different classes (i.e., "between" positives and negatives). This is known as inter-class differentiation. Simultaneously, they should differentiate lysines within the same class (i.e., "within" positives and "within" negatives), which is referred to as intra-class differentiation. These concepts lead to the notion of *contextualization*. Simply put, the feature vector for each residue should reflect its unique context within the sequence.

Based on the approach to contextualization, computational tools can be divided into two groups. The first group represents the *pre-contextualization* era of tools, where contextualizing vectors was challenging. To address this challenge, researchers introduced the concept of sequence windows. Instead of considering only the residue-of-interest, a predefined window of amino acids around it is selected [14]. The target residue is positioned at the center, with upstream and downstream flanking residues on either side. These window sequences, which are sometimes referred to as peptides due to their shorter length, are extracted for each residue-of-interest (i.e., lysines, in the case of crotonylation prediction). Since the local environment of each residue differs, the corresponding window sequences capture different contexts. For instance, the window sequence around a lysine residue at position 4 will differ from that at position 2, as the neighboring residues are different at these positions. This approach imposes a level of contextualization because the likelihood of two window sequences being identical is relatively low. Furthermore, homology removal tools, such as CD-HIT [15] and MMseqs [16], are applied to these window sequences to eliminate similar sequences, ensuring the uniqueness of the contextual representation for the residues. These window sequences are then encoded into feature vectors. Initially, researchers used the physicochemical properties of amino acids to encode these sequences [14, 17]. The major drawback of this class of features was that the optimal set of properties to use was not known a priori. Researchers had to use some heuristics or feature selection algorithms [18–20] to create the optimal feature set for their tasks. This posed extra overhead and introduced the possibility of missing important properties since it was challenging to exhaustively explore all features. These types of features were commonly known as "handcrafted" features due to the constant interventions required to optimize the feature set. Natural Language Processing (NLP) encoding techniques have greatly streamlined feature extraction by removing the need for manual intervention [13]. With these techniques, features can be automatically obtained by simply inputting window sequences into NLP models, ranging from simple binary encoding to more advanced embeddings (via language models). For a comprehensive review of tools from this era and the encodings used (including various NLP encodings), please refer to the literature by Pakhrin et al [14] and Pokharel et al [13]. However, the tools in this era had limitations. For instance, they primarily used linear sequence data (i.e., the primary amino acid structure), which assumes proteins are linear and fails to capture three-dimensional spatial relationships. Likewise, window sequences represent isolated fragments, limiting the encoding's scope to local information and omitting important global context. This is significant because, in the 3D structure, distant amino acids in the primary sequence can be spatially close to the residue-of-interest, an aspect not captured by primary sequence-based methods alone. To overcome these limitations, there is a need for encoders that are capable of processing the entire protein sequence and generating global context-aware embeddings. Ideally, such an encoder would represent each residue with a vector embedding that is aware of all other residues in the protein sequence, capturing essential long-range interactions and structural context.

The early advancements in NLP saw a surge in models capable of generating contextualized embeddings, which capture the meaning of words based on their context within a sequence [21]. These embeddings were initially driven by sequential models like Recurrent Neural Networks (RNNs) and Long Short-Term Memory (LSTM) networks, which were designed to process sequential data by retaining information across token positions [22, 23]. Recognizing the sequential nature of protein sequences, researchers began adopting these models for protein-related tasks, enabling contextualized representations of amino acids within protein sequences [24]. However, the inherent sequential nature of RNNs and LSTMs introduced several critical limitations [25, 26]. For instance, these models process tokens one at a time, which makes it challenging to retain information over long sequences. As they progress sequentially, earlier contextual information may gradually diminish, especially in lengthy sequences. As a consequence, RNN and LSTM-based models are often limited in their ability to capture long-range dependencies. Additionally, sequential processing restricts parallel computation, making training and inference slower, especially for large datasets. These limitations highlighted the need for alternative architectures that could handle entire sequences at once to efficiently capture complex dependencies across long sequences and facilitate faster computations. This need eventually catalyzed the development of attention-based architectures, with the transformer model [27] marking a revolutionary shift. Unlike RNNs and LSTMs, transformers employ self-attention mechanisms, allowing them to capture relationships between words across an entire sequence simultaneously. The self-attention mechanism processes all tokens at once in parallel, efficiently computing the context over a sentence or even a passage "in a single shot." As a result, transformers excel at capturing long-range dependencies and are highly parallelizable, making them ideal for large-scale training. This initial model architecture laid the foundation for scaling. Researchers rapidly increased both the depth (layers) and width (parameters) of transformers, training these larger models on vast corpora to create what we now call large language models (LLMs). These LLMs, like BERT (Bidirectional Encoder Representations from Transformers) [28], GPT (Generative Pretrained Transformers) [29], and their successors, which are typically pretrained on a large corpus of data, leveraged immense datasets and billions of parameters that produced models capable of performing a wide array of complex language tasks, often without task-specific fine-tuning. This scalability of the transformer model has enabled state-of-the-art performance across various domains, fundamentally changing the landscape of NLP and driving advancements in fields such as machine translation, text generation, and question-answering [30].

Inspired by the success of NLP in processing text, researchers have applied transformers to protein sequences, where each amino acid (or, group of amino acids or motif in some cases) is treated as a "word" and each protein sequence as a "sentence" [13]. This approach led to the development of protein language models (pLMs), LLMs pretrained on large protein datasets like UniRef50 and UniRef100 using self-supervised techniques usually based on masked language modeling (MLM) or casual language modeling (CLM). These pLMs leverage extensive pretraining datasets to capture complex relationships between residues (i.e., "grammar" of protein language), creating context-aware embeddings that encode both local and global dependencies essential for understanding protein functions and structural relationships inherent in the sequences. Some prominent examples of existing pLMs include ProtTrans [31] by Rostlab, which incorporates models like Prot-BERT (based on BERT) and ProtT5 (based on T5) trained on UniRef100 for tasks like secondary structure prediction, subcellular localization, and membrane protein classification; ESM (Evolutionary Scale Modeling) [32] series by Meta, which captures evolutionary information for structure prediction, function prediction, and mutation effect prediction; AlphaFold's Evoformer [33], which integrates sequence and evolutionary data for robust 3D structure predictions; and TAPE [34], a benchmark suite pretrained on the Pfam dataset, which is designed for tasks like secondary structure prediction and homology detection. For an in-depth exploration of various pLMs and their architectures and applications, please refer to Chapter 1.

This brings us to the current generation of tools, which we refer to as the *post-contextualization era*, where pLMs and LLMs have transformed many downstream tasks including PTM site prediction. New approaches utilizing these models have demonstrated superior performance compared to earlier, pre-contextualized encoding methods [35–40]. With pLMs, we now have encoders that generate embeddings with full sequence context and awareness of protein functions and properties. Building on these advancements, this chapter explores how LLMs and pLMs are being utilized to address various PTM site prediction tasks. We will start with an overview of tools that leverage LLMs/pLMs for PTM site prediction, and then identify trends and advancements in these tools. Finally, we will discuss their limitations and areas for future research directions. This chapter aims to provide readers with a comprehensive understanding of the current landscape of PTM site prediction tools and how these advancements contribute to more accurate and contextually aware predictions. We hope the chapter will be especially useful for researchers looking to apply or develop pLM-based approaches, offering insights into state-of-the-art methodologies and highlighting areas where further development is possible.

Subheading 2 provides a brief overview of single PTM and multi-PTM site prediction tools that utilize LLMs/pLMs, and custom transformers. Subheading 3 discusses the key trends in the field and is further divided into seven subsections (Subheadings 3.1, 3.2, 3.3, 3.4, 3.5, 3.6, and 3.7), each highlighting the key trends observed in these tools. Subheading 4 discusses the limitations of these tools through five subsections (Subheadings 4.1, 4.2, 4.3, 4.4, and 4.5) and outlines future directions inspired by these shortcomings. We conclude the chapter with a discussion and final remarks in Subheading 5.

## 2 Brief Review of Tools Utilizing pLMs/Custom Transformer for PTM Site Prediction

While this chapter primarily focuses on discussing trends and advancements in LLM-based approaches for PTM site prediction, this section aims to familiarize readers with existing state-of-the-art LLM/pLM-based tools developed in recent years for some well-studied PTMs. It is important to note that we provide only a high-level overview of these tools and encourage readers to consult the original published literature for a more in-depth description of each tool.

Table 1 presents a categorized list of tools designed for the prediction of sites of some well-studied PTM types, separating those that utilize pretrained pLMs from those that incorporate custom transformers into their workflows. The subsequent subsections (Subheadings 2.1, 2.2, 2.3, 2.4, 2.5, 2.6, 2.7, 2.8, 2.9, 2.10, and 2.11) provide a brief description of these tools, organized by PTM type.

**Table 1**
**Recently developed tools for predicting PTMs using pretrained LLMs/pLMs and custom transformers**

| PTM(s) | Target residue | Pretrained LLM/pLM-based tools | Custom transformer-based tools |
|---|---|---|---|
| Phosphorylation | S, T, Y | Phosformer [41], LMPhosSite [42], CaLMPhosKAN [43] | TransPhos [44], DL-SPhos [45] |
| Glycosylation (N-linked, O-linked, and O-GlcNAcylation) | N, S, T | *Glycosylation (N: N-linked, O: O-Linked)*: LMGlyPred [46] (*N*), EMNGLy [47] (*N*), HOTGpred [48] (*O*), CoNglyPred [49] (*N*), Stack-OglyPred-PLM [50] (*O*)<br>*O-GlcNAcylation*: LM-OGlcNAc-Site [51] | – |
| Ubiquitination | K | Work of Pourmirzaei et al. [52] | – |
| Acetylation | K | MSTL-Kace [53], TransPTM [54] | – |
| S-Nitrosylation | C | pLMSNOSite [55] | – |
| Succinylation | K | LMsuccSite [39] | – |
| SUMOylation | K | SumoPred-PLM [56], iSUMO-RsFPN [57] | GPS-SUMO 2.0 [58] |
| Crotonylation | K | BERT-Kcr [59], LMCrot [38] | – |
| Glutarylation | K | Deepro-Glu [60], ProtTrans-Glutar [61] | – |
| Glycation | K | BERT-Kgly [62] | – |
| Multi-PTM approaches | Multiple residues | PTMGPT2 [36], PTG-PLM [63], PTM-Mamba [64], UniPTM [65] | MIND-S [66], PTM-CMGMS [35] |

As can be observed from Table 1, existing tools for PTM site prediction mostly employ pretrained pLMs to extract protein sequence representations, which are then used as input features for downstream PTM site prediction tasks. Additionally, we discuss tools based on custom transformer architectures, highlighting innovations in design and functionality. Generative models, representing a novel paradigm in PTM site prediction, are also reviewed for their unique contributions. Lastly, multi-PTM site prediction approaches are explored, showcasing tools that expand predictive capabilities across diverse modification types and residues.

### 2.1 Phosphorylation Site Prediction

Phosphorylation is a PTM in which protein kinases catalyze the addition of the gamma phosphate of ATP to specific amino acid residues, primarily serine (S), threonine (T), and tyrosine (Y) [67]. Several custom transformer-based and pretrained protein language model (pLM)-based tools have been developed to predict phosphorylation sites, including Phosformer [41], TransPhos [44], DL-SPhos [45], LMPhosSite [42], and CaLMPhosKAN [43].

Phosformer [41] is a deep learning (DL) model where a domain-specific protein language model is trained specifically for predicting kinase-specific phosphorylation sites in proteins given the kinase domain sequence and the substrate peptide sequence. Phosformer is trained in two phases. In the pretraining phase, a transformer model consisting of an encoder and decoder is trained, which learns from the overall sequence using the masked language modeling (MLM) objective (the goal of training is to predict the masked token). The pretraining phase enables the model to understand the structural and functional properties of proteins. This phase of the model is trained on 295,320 diverse kinase domain sequences and 5538 substrate proteins. The encoder consists of 6 attention layers, with each layer having 12 attention heads. Together, these layers produce an output tensor with 768 embedding dimensions. In the second phase, the model is fine-tuned to predict kinase-specific phosphorylation. The fine-tuning step takes two inputs: (1) an 11-mer window sequence and (2) an unaligned kinase domain sequence. Using this information, it predicts the likelihood of phosphorylation for the middle residues ("S," "T," and "Y") in the peptide by the corresponding kinase.

Similarly, TransPhos [44] and DL-SPhos [45] are other custom transformer-based tools tailored for phosphorylation site prediction. TransPhos focuses on general phosphorylation site prediction, using a transformer encoder integrated with densely connected convolutional neural network blocks. On the other hand, DL-SPhos specializes in serine phosphorylation site prediction that uses five transformer blocks equipped with multi-headed attention and causal mask layers. Interestingly, DL-SPhos not only identifies conserved motifs linked to serine phosphorylation but also excels in kinase-specific predictions.

LMPhosSite [42] takes a different approach by utilizing a pretrained pLM to predict general phosphorylation sites. The architecture of LMPhosSite is stack-based consisting of two components and a meta-classifier. The first component is a supervised embedding layer that encodes local sequence information from a window of amino acids surrounding the site-of-interest. The local encodings are then trained using a one-dimensional convolutional neural network (1D-CNN). Parallelly, the second component first extracts embeddings from ProtT5 pLM using the full sequence as an input and subsequently uses a feed-forward neural network on top of the embedding corresponding to the site-of-interest ("S," "T," and "Y"). The prediction probabilities from these two components are fused using a multi-layer perceptron (MLP) as a meta-classifier, allowing the model to incorporate local and global information to enhance prediction accuracy.

CaLMPhosKAN [43] further extends the capabilities of pLM-based models for general phosphorylation site predictor by integrating codon-aware embeddings from CaLM [68] (a pLM pretrained on coding DNA sequences) with amino acid-aware embeddings from ProtT5 (a pLM pretrained on amino acid sequences). The architecture of CaLMPhosKAN consists of three components (or modules) in succession. The first component, the *Embedding Extraction and Fusion Module*, combines the codon-aware and amino acid-aware window embeddings (generated from full sequence) using early fusion strategy. The second component, the *Spatiotemporal Feature Extraction Module*, utilizes a ConvBiGRU (Convolutional Bi-directional Gated Recurrent Unit Network) that extracts features from the fused embeddings to capture spatial and sequential correlations among neighboring residues. The last component, the *Wav-KAN Module*, consists of a classification head based on wavelet-based Kolmogorov–Arnold Network (KAN) [69] which produces the classification inference of the site-of-interest. The novelty of this work lies in the use of complementary information contained in the codon space.

### 2.2 Glycosylation Site Prediction

In glycosylation, glycotransferases catalyze the attachment of carbohydrate chains (glycans) to specific amino acid residues [70]. It is broadly categorized into N-linked glycosylation, which occurs on asparagine (N) residues within a consensus sequence (N-X-S/T, where X is any amino acid except proline), and O-linked glycosylation, where glycans attach to serine (S) or threonine (T) residues. A specific type of O-linked glycosylation is O-GlcNAcylation, in which a single N-acetylglucosamine (GlcNAc) moiety is added to serine or threonine residues [71].

LM-OGlcNAc-Site [51] is a tool for predicting O-GlcNAcylation sites in proteins by integrating embeddings from multiple pretrained pLMs. Specifically, it utilized embeddings corresponding to the site-of-interest, extracted using full sequence as an input from three pLMs: Ankh, ESM-2, and ProtT5, which potentially capture complementary sequence representations. Feed-forward neural networks (FFNNs) were trained on embeddings from each pLM independently and integrated via decision-level fusion for improved performance. Decision-level fusion, using a hard-voting mechanism, outperformed individual models and other ensemble approaches, demonstrating superior performance on benchmark datasets. The findings emphasize the advantages of combining embeddings from multiple pLMs for PTM site prediction.

LMNglyPred [46] is an N-linked glycosylation prediction tool, trained using two datasets: N-GlyDE and N-GlycositeAtlas. Using full sequence as an input to ProtT5 pLM, embeddings corresponding to the site-of-interest were extracted. Thus obtained per-residue embeddings were trained using the MLP model for classification purposes.

EMNGly [47] further advances N-linked glycosylation prediction by integrating embeddings from ESM-1b [32] pLM with structure-aware embeddings from the Masked Inverse Folding (MIF) model [72]. ESM-1b captures both global and local sequence dependencies, while the MIF model provides local 3D structural information through structure-aware embeddings. For prediction, sequence embeddings are generated as 1024-dimensional vectors specific to the site-of-interest, supplemented by a window-level embedding spanning 41 residues. Structural embeddings, encoded as 256-dimensional vectors, represent the spatial properties of the glycosylation site environment. These combined features are then input into a Support Vector Machine (SVM) classifier for prediction. This work exemplifies the integration of sequence and structure to improve the accuracy of N-linked glycosylation site prediction.

CoNglyPred [49] is another tool designed for the prediction of N-linked glycosylation sites by integrating sequence and structural features using pLMs and graph-based approaches. It uses ESM-2 to extract sequence representations that capture contextual and interactional information across amino acid sequences. Specifically, it uses window embeddings of length 41 to extract sequential representations from ESM-2 using a full-length input sequence. For structural features, CoNglyPred employs AlphaFold2 [33] predicted 3D structures to construct protein graphs, where residues are represented as nodes and their spatial relationships as edges, processed through a Graph Transformer Network (GTN). The GTN dynamically captures spatial relationships between residues using the attention mechanism. To effectively integrate sequence and structural information, CoNglyPred uses a co-attention mechanism, allowing mutual enhancement of features derived from ESM-2 and GTN. The fused features are then processed through an MLP classifier to make the final predictions.

Stack-OglyPred-PLM [50] is a stacked generalization framework designed to predict O-linked glycosylation sites by integrating per-residue embeddings from two pLMs: ProtT5 and Ankh. These embeddings, corresponding to the site-of-interest ("S" or "T"), are used to train two independent base models based on MLP. The outputs of these base models are then combined using a meta-MLP classifier, creating a stacked architecture that leverages complementary features from both pLMs to enhance predictive performance.

Similarly, HOTGpred [48] introduces a novel hybrid approach to predicting O-linked glycosylation sites by leveraging features from 14 pretrained pLMs, such as ESM 1v, FastText, Seq2Vec, ProtTransT5UniRef50, and ProtTransT5XLU50, alongside 11 conventional feature descriptors. The top five pLM-based encodings, selected for their superior discriminative ability, were linearly integrated into high-dimensional hybrid features, multiple tree-based algorithms were trained to generate feature importance scores (FIS), and optimal feature sets were identified using these scores. The final architecture combines the selected 220-dimensional feature set with an extreme gradient boosting (XGBT) classifier.

### 2.3 Ubiquitination Site Prediction

In Ubiquitination PTM, ubiquitin ligases catalyze the covalent attachment of ubiquitin to lysine residues of target proteins [73]. The work of Pourmirzaei et al. [52] presents a comprehensive study on ubiquitination site prediction in human proteins, comparing three paradigms: feature-based classical ML methods, end-to-end sequence-based DL models, and hybrid strategies combining engineered features with raw sequence embeddings. The models utilize architectures such as LSTMs and variants of BERT variants (BERT-small, BERT-tiny, Nystromformer, SqueezeBERT), which effectively capture long-range dependencies and contextual information in protein sequences. Additionally, one-hot encoding and word embeddings were employed for sequence-based inputs, while engineered features like position-specific scoring matrices (PSSM), amino acid composition (AAC), and physicochemical properties (PCP) provided valuable biological insights. Sliding window techniques ensured that fixed-length subsequences were created with the target residue at the center. Among the models tested, the final model, an LSTM-based hybrid approach, outperformed others (even BERT) by combining sequence embeddings and engineered features, such as PSSM and PCP.

### 2.4 Acetylation Site Prediction

In acetylation, acetyltransferases catalyze the addition of an acetyl group from acetyl-CoA to the ε-amino group of lysine residues [74]. Recently, pLM-based computational tools such as MSTL-Kace [53] and TransPTM [54] have been developed to predict acetylation sites, leveraging advances in deep learning and protein sequence embeddings.

MSTL-Kace [53] model architecture is built upon the Bidirectional Encoder Representations from Transformers (BERT) [28], adapted to encode protein sequences as natural language sentences. Specifically, the model begins with a domain-specific BERT pre-trained on protein sequences from six prokaryotic species, ensuring that embeddings capture biological context. Fine-tuning is performed in two stages to address the small sample size issue: the first stage uses larger datasets to create an intermediate model and the second stage further fine-tunes this model using species-specific data. The architecture uses the embeddings of the "CLS" token, which aggregates sequence-level features, and incorporates embeddings from the last six encoder layers for optimal feature fusion followed by a downstream classification.

The TransPTM [54] is a transformer-based model designed for non-histone acetylation site prediction, leveraging the power of pLMs and graph neural networks (GNNs). TransPTM uses ProtT5 pLM to generate contextualized embeddings of amino acid sequences. These embeddings encapsulate residue-level dependencies and serve as node features in a complete graph, where each amino acid is a node and edges represent interactions between residues. The model's architecture incorporates three TransformerConv layers, which use self-attention enabling the detection of critical relationships among residues. The encoded graph features are followed by a downstream classification MLP that predicts the probability of acetylation at specific lysine residues.

### 2.5 S-nitrosylation Site Prediction

S-nitrosylation is a PTM in which a nitric oxide (NO) group is covalently attached to the thiol side chain of cysteine ("C") residues in proteins [75]. pLMSNOSite [55] is a tool designed to predict S-nitrosylation sites in proteins by stack generalization of two models trained using supervised word embeddings and embeddings from ProtT5 pLM. The word embedding module uses a 2D CNN to capture local amino acid interactions around a window sequence centered on the "C" residue-of-interest. The second module uses the per-residue ProtT5 embeddings, corresponding to the site-of-interest, to capture the global context of the full protein sequence. These two separately trained modules are then fused using a stacked generalization approach, where a meta-classifier combines features from both models to improve prediction performance outperforming existing SNO site prediction tools.

### 2.6 Succinylation Site Prediction

Succinylation PTM involves covalent attachment of succinyl group to the ε-amino group of lysine (K) residues in proteins [76]. LMSuccSite [39] is a pLM-driven tool designed to predict succinylation sites, utilizing both supervised word embeddings and unsupervised pLM embeddings. Its architecture is built upon a stack generalization of two modules. The first module encodes protein sequence fragments into vector representations through a supervised word embedding layer with a 2D CNN, which learns local sequence patterns by analyzing interactions within a windowed sequence centered on the target "K" residue. In parallel, the second module employs per-residue ProtT5 embeddings, focusing specifically on the site-of-interest to capture the global context of the full protein sequence. The integrated model effectively combines local and global contexts, making LMSuccSite highly proficient in accurately identifying succinylation sites and outperforming conventional sequence-based prediction methods.

### 2.7 SUMOylation Site Prediction

SUMOylation is a PTM where Small Ubiquitin-like Modifier (SUMO) proteins are covalently attached to lysine residues on target proteins by ligases [77]. Recent advancements include the development of tools like SumoPred-PLM [56], iSUMO-RsFPN [57], and GPS-SUMO 2.0 [58] (non-LLM/custom transformer-based), which leverage pLM and custom transformer to predict sumoylation sites with improved accuracy and efficiency.

SumoPred-PLM [56] utilizes pLMs for predicting SUMOylation and SUMO2/3 sites in proteins. It uses ProtT5 pLM to extract contextualized embeddings of lysine residues in protein sequences. These embeddings are 1024-dimensional vectors acting as a high-level feature essential for site-specific predictions. The extracted embeddings serve as inputs to the downstream classification model.

iSUMO-RsFPN [57] is another lysine SUMOylation site prediction tool that integrates physicochemical properties features: AAindex and ZScale encodings and ProteinBert, a pretrained protein language model, which generates 128-dimensional embeddings capturing contextual sequence information. Specifically, an input protein sequence fragment of length 39 is fed to ProteinBERT [78] which outputs a representation of size (39, 1024) which is fed to a Residual Pyramid Network (RsFPN). A similar modal is trained using AAindex and ZScale features and results are combined using an ensemble approach.

GPS-SUMO 2.0 [58] is built on the combined architecture of transformers, dense neural networks (DNNs), and penalized logistic regression (PLR) to predict SUMOylation sites and SUMO-interacting motifs (SIMs) with high accuracy. At the core, a custom transformer model is pretrained on 145,545 lysine modification sites, capturing rich contextual information within sequence windows, and is fine-tuned on a focused dataset of 52,404 SUMOylation sites and 163 SIMs. To enhance prediction accuracy, the system integrates seven sequence-based features, such as k-spaced amino acid pairs (CKSAAPs) and position-specific scoring matrices (PSSMs), along with three structural features, including secondary structure (SS), accessible surface area (ASA), and backbone torsion angles (BTAs). Each feature is scored through a DNN model, and the outputs, along with transformer-derived contextual representations, are consolidated using PLR with an L2 penalty for final prediction.

### 2.8 Crotonylation Site Prediction

Crotonylation is a PTM wherein a crotonyl group is covalently attached to the ε-amino group of lysine (K) residues in proteins [79]. In this section, we discuss language model-based computation tools: BERT-Kcr [59] and LMCrot [38] to predict crotonylation sites.

BERT-Kcr [59] utilizes a pretrained BERT model, originally developed for NLP tasks, by treating window sequences as "sentences" and individual amino acids as "words." Raw protein sequences are directly input into the BERT-Mini pretrained model to extract high-dimensional feature representations from the final encoder layer, which are then passed to a Bidirectional Long Short-Term Memory (BiLSTM)-based classifier for lysine crotonylation (Kcr) site prediction. The authors also fine-tuned the BERT model with protein sequences to adapt it specifically for this task. Interestingly, the original natural language BERT outperformed the protein-specific fine-tuned version, highlighting its inherent ability to capture meaningful contextual information applicable to protein sequences.

LMCrot [38] improves predictive performance by integrating embeddings from the ProtT5 pLM with local window-based representations. Its architecture employs intermediate fusion-based stacked generalization across three base models. The first base model, ResConvBiLSTM (Residual Convolutional BiLSTM), is trained on window-level embeddings of length 33, derived from ProtT5 using the full sequence as input, referred to as FSWE (Full Sequence-based Window Embeddings). The second and third base models leverage local representations from a supervised embedding layer and physicochemical properties of window sequences. The second base is a 2D-CNN, while the third is a DNN. Latent representations from these models are merged via stacked generalization and trained on a DNN-based meta-classifier to produce the final output.

### 2.9 Glutarylation Site Prediction

Glutarylation is a PTM in which a glutaryl group, derived from glutaryl-CoA, is covalently attached to the ε-amino group of lysine residues in proteins [76]. Deepro-Glu [60] and ProtTrans-Glutar [61] are two pLM-based glutarylation site prediction tools.

Deepro-Glu [60] is an ensemble prediction model for glutarylation site prediction developed by integrating features from ProtBERT [31] with handcrafted biochemical and evolutionary features. The architecture consists of two parallel sub-models: one processes embeddings obtained for a fragment of protein sequence (length 41) generated by ProtBert using a 1D-CNN to capture local sequence semantics, while the other utilizes handcrafted features including physicochemical properties, evolutionary profiles, and sequence patterns fed into a BiLSTM network to learn sequence dependencies. The two feature sets are combined using an attention mechanism, which assigns weights to fused features to focus on important information for prediction. The final predictions are generated using an MLP model.

ProtTrans-Glutar [61] is another model designed to predict lysine glutarylation sites by leveraging both traditional physicochemical features and pLM-based embeddings. At its core, the model incorporates the per-residue embeddings derived from ProtT5, alongside amino acid composition (EAAC) and Composition/Transition/Distribution (CTDD) features. The combined features are then used to train a downstream classification model consisting of a XGBoost classifier.

### 2.10 Glycation Site Prediction

Glycation is a non-enzymatic modification in which reducing sugars, such as glucose, covalently attach to lysine or arginine residues in proteins [80]. BERT-Kgly [62] is a pLM-based model to predict lysine glycation (Kgly) sites in proteins. The model combines the representations from three language models: BERT-Base, TAPE (trained on 31 million Pfam protein domains), and BERT-Prot [31] (trained on 556,603 UniProt sequences). Specifically, the "CLS" token embedding of the protein fragment is extracted from the pLMs. It is further processed through the 1D-CNN and BiLSTM to train the downstream classification model. This architecture combines the strengths of the natural language model and protein language models offering an effective approach for lysine glycation prediction.

### 2.11 Multi-PTM Site Prediction

In this section, we will discuss tools that utilize the same architecture for predicting more than one type of PTM site. These include pretrained LLM/pLM-based tools such as PTMGPT2 [36], PTG-PLM [63], PTM-Mamba [64], and UniPTM [65] and two tools that train their own custom transformer architectures: MIND-S [66] and PTM-CMGMS [35].

PTMGPT2 [36] is a multiple PTM site prediction framework capable of predicting 19 different types of PTMs. It is built upon the ProtGPT2 [81] (a GPT-based decoder-only pLM) and applies a prompt-based fine-tuning method tailored to PTM identification. Here, each prompt is crafted to embed both the protein sequence segment of interest (window sequence) and a classification label, guiding the model to learn patterns associated with PTM sites. To achieve this, PTMGPT2 constructs prompts by framing the input as a sequence paired with a "fill-in-the-blank" label: the prompt includes tokens that specify "SEQUENCE:" followed by the amino acid segment, and "LABEL:" with either a "POSITIVE" or "NEGATIVE" indicator for modified or unmodified regions, respectively. During fine-tuning, the ProtGPT2 model, which was originally trained on large protein sequence data, learns to associate these prompt labels with specific PTM sites. This prompt-based setup allows PTMGPT2 to operate without a specialized classification head; instead, the decoder predicts PTM sites by interpreting each sequence within the structured prompt.

Similarly, PTG-PLM [63] is a CNN-based model developed to predict glycosylation and glycation sites utilizing features from six pLMs: ProtBERT-BFD, ProtBERT, ProtALBERT, ProtXLNet, ESM-1b, and TAPE. The features are extracted using WSWE (Window Sequence Window Embedding), where protein sequence fragments are fed to pLMs to extract window-level embeddings. These representations from the pLMs are combined and fed as input to the CNN classification model.

UniPTM [65] is another tool designed for multi-PTM site prediction, accommodating nine distinct modifications. It uses full-length protein sequences for training and testing, bypassing the limitations of traditional sliding window approaches. It utilizes embeddings generated by state-of-the-art pLMs, including ESM-2, ProtBert, and ProtT5, which capture contextual relationships between amino acids. The model architecture integrates these per-residue embeddings with a convolutional layer and an eight-head transformer encoder for feature extraction. This design enables the model to learn long-range dependencies across the protein sequence, which is essential for understanding the nuanced processes influencing PTM formation.

PTM-Mamba [64] is a PTM-specific pLM designed to explicitly represent PTMs within protein sequences. Existing pLMs such as ESM-2 and ProtT5 effectively capture the structural and functional properties of proteins but often overlook the critical role of PTMs in proteins. To address this gap, PTM-Mamba incorporates a PTM-specific token into its vocabulary, enabling the model to process both unmodified and modified protein sequences. It is built upon the Mamba model architecture, a structured state space model (SSM) known for its efficiency in handling long sequences. By utilizing hardware-aware primitives, Mamba overcomes the quadratic time complexity associated with traditional transformer models [27]. PTM-Mamba employs bidirectional gated Mamba blocks that integrate outputs with ESM-2 embeddings through a novel gating mechanism. This fusion allows the model to effectively incorporate PTM information, enhancing its predictive capabilities. PTM-Mamba exhibits improved performance over ESM-2 in various PTM-specific tasks, including phosphorylation and acetylation site prediction, disease association studies, and druggability assessments. As the first pLM capable of uniquely representing both unmodified and PTM-modified sequences, PTM-Mamba offers a significant advancement in proteomics, facilitating more accurate modeling and design of proteins with PTMs.

MIND-S [66] is designed to predict 26 types of PTM sites leveraging both sequence and structural information. The model processes protein sequences using one-hot encodings, transforming them into dense feature representations through a feedforward neural network. These embeddings are further refined using a BiLSTM layer to encode positional information, which acts as a foundation for two key components: multi-head self-attention and graph attention layers. The multi-head self-attention captures sequential relationships within the protein, while the graph attention layer integrates spatial proximity insights derived from protein contact maps. By combining outputs from these layers, MIND-S generates PTM site prediction scores for each amino acid residue. Its multi-label training strategy enables the simultaneous prediction of multiple PTM types, even for those with limited data, by sharing learned representations across PTMs.

PTM-CMGMS [35] predicts three types of PTM sites (crotonylation, succinylation, and S-nitrosylation) by integrating multiscale sequence data and structural information. Specifically, the model employs a novel architecture that incorporates structure-aware representation that captures structural features at the atom, amino acid, and whole protein levels, leveraging AlphaFold-predicted protein structures. These representations are optimized using contrastive learning, which enhances their discriminative capacity. Additionally, the model extracts multi-scale sequence representations through a combination of BiLSTM, transformer encoders, and multi-scale CNNs, capturing local and contextual sequence patterns. A motif-based approach aligns sequences to identify recurring PTM-related patterns, further refining predictions of crotonylation, succinylation, and S-nitrosylation modifications.

## 3 Trends in LLM-Based Approaches for PTM Site Prediction

In the previous section, we briefly reviewed existing tools for predicting PTMs that use LLMs, pLMs, or custom transformer-based architectures. In this section, we will outline the key trends identified within these tools. A high-level overview of these trends is depicted in Fig. 1.

### 3.1 Use of Foundation Models

LLMs/pLMs are often referred to as "foundation" models as they are pretrained on large datasets in an unsupervised manner causing them to be task agnostic [82]. In PTM site prediction, foundation models are either used as a feature extractor (i.e., feature-based approach) or fine-tuned to adapt to a specific task. This section will elucidate these two approaches in detail.

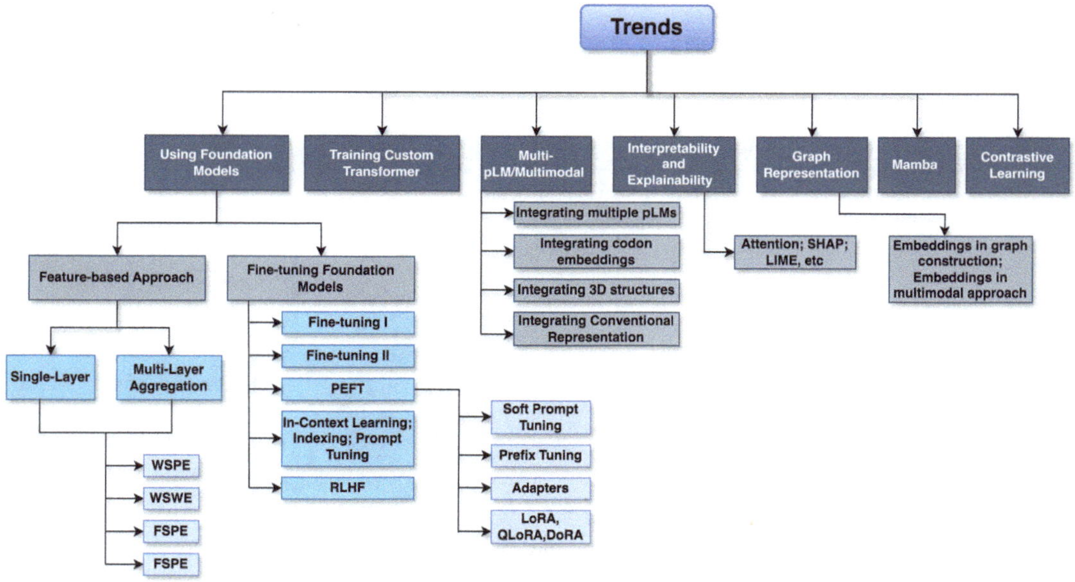


**Fig. 1** Schematic diagram of the trends discussed in this chapter

*3.1.1 Using Feature-Based Approach (LLM/pLM as a Feature Extractor)*

Pretrained embeddings are tensors obtained from the final hidden layer of a foundation model, in our case, a pLM. These embeddings are often referred to as "last hidden states." The dimension of the embedding is determined by the size of the hidden layer from which they are extracted. For example, the final hidden layer of GPT2-small has 768 dimensions, meaning each token is represented by a tensor of size $768 \times 1$. These embeddings can be fine-tuned for any downstream protein task by adding task-specific heads while keeping the parameters of the foundation model frozen. This approach is therefore also called *Feature-based fine-tuning* [83].

Depending on the task, the input to these prediction heads must be adjusted to provide an optimal representation that addresses the problem. Protein-level tasks (e.g., DNA binding protein prediction) typically employ one of three representation techniques [31, 82]: pooling, attention-based pooling, or using the CLS token. In pooling, either mean pooling is applied, where all residue embeddings across the sequence are averaged, or max pooling, where only the maximum value for each embedding dimension is used. Alternatively, attention-based pooling calculates a weighted average of the embeddings through an attention mechanism [84].

PTM site prediction is a residue-level task, meaning the local environment of the specific residue or site-of-interest should be incorporated to obtain the site-specific representation. Since the advent of pLMs, researchers have adopted several strategies to derive site-specific representations [39, 47, 53, 55, 59]. Notably,

**Table 2**
**Comparison of representation strategies in protein language models for residue-level tasks**

| Strategy | Input type | Input dim. | Embedding used | Representation dim. | Comments |
|---|---|---|---|---|---|
| WSPE | Window sequence | $W \times 1$ | Per-residue corresponding to the site of interest | $L \times 1$ | No global context<br>Limited capture of association between adjacent residue embeddings |
| WSWE | Window sequence | $W \times 1$ | Window residues | $L \times W$ (entire window) or $L \times 1$ (after pooling or using [CLS] token) | No global context<br>Better capture of association between adjacent residue embeddings |
| FSPE | Full sequence | $N \times 1$ | Per-residue corresponding to the site of interest | $L \times 1$ | Captures global context<br>Limited capture of association between adjacent residue embeddings |
| FSWE | Full sequence | $N \times 1$ | Window residues | $L \times W$ (entire window) or $L \times 1$ (after pooling or using [CLS] token) | Captures global context and association between adjacent residue embeddings |

Each strategy (WSPE, WSWE, FSPE, FSWE) differs by input type, input dimension, embedding usage, and representation dimension. Comments highlight their strengths and limitations in capturing global context and relationships between adjacent residues embedding. $W$ is the length of the window, $L$ is the embedding dimension, and $N$ is the length of the full sequence. [CLS] stands for classification token which is prepended in the input sequence in some transformers architecture (e.g., BERT)

the work of Pratyush et al. [38] categorizes these strategies into four main types: Window Sequence-based Per-Residue Embedding (WSPE), Window Sequence-based Window Embedding (WSWE), Full Sequence-based Per-Residue Embedding (FSPE), and Full Sequence-based Window Embedding (FSWE). In WSPE and FSPE, the embedding is derived only for the site-of-interest, but they differ in the input provided to the pLM. WSPE uses a window of residues around the site as input, while FSPE takes the entire sequence. In contrast, WSWE and FSWE consider the embeddings of all residues within a window surrounding the site-of-interest. WSWE uses the window as input to the pLM, while FSWE employs the full sequence. The choice of input, whether a window or the full sequence, determines whether the obtained embeddings capture the global context. Furthermore, the use of per-residue or window-level embeddings influences whether the relationships between adjacent embeddings are captured. Table 2 summarizes the strategies above, including input sequence type, input dimension, embeddings used for representation, the corresponding representation dimension, and comments highlighting the benefits and drawbacks of these approaches. It should be noted that these four approaches are mentioned because they are among the most commonly adopted in the field, and this is by no means an exhaustive list of all possible methods for representing site-specific representation.

#### Variants of Feature-Bsed Representation

While the prevailing trend in the field is to use the last hidden states (i.e., tensors from the final hidden layer) for obtaining latent representations, the original BERT paper [28] also proposed using hidden states from intermediate layers. This approach has been adopted by MSTL-Kace [53], an acetylation site prediction tool, which uses the mean of embeddings (corresponding to [CLS] token) from the last six layers instead of obtaining the final hidden states of pretrained BERT. Similarly, other variations can be applied, and these approaches generally fall into two main categories listed below [28, 53]. These approaches can be incorporated with any of the aforementioned representation strategies (WSPE, WSWE, FSPE, FSWE).

- *Single-layer extraction:* Embeddings are extracted from either the final hidden layer or intermediate layers. As previously mentioned, extraction from the final hidden layer is the most common approach. However, in certain cases, intermediate layers may prove more useful, depending on the task. For example, earlier layers may capture more syntactic information (higher order arrangement of residues), while later layers tend to capture more semantic information (protein function). If the goal is to focus on a particular aspect of the sequence, embeddings from specific intermediate layers can be extracted.
- *Multi-layer aggregation:* In this approach, tensors from multiple layers are aggregated, either by concatenating or averaging them (e.g., MSTL-Kace). As mentioned in the single-layer extraction approach, each layer captures a different range of information (from syntactic to semantic to task-specific). Aggregating embeddings from multiple layers can help preserve this diverse information, some of which might be lost by the time the final layer is reached.

In Table 3, we report the different PTM site prediction tools, their respective representation strategies, and layer extraction methods, highlighting how each tool is optimized for specific PTMs by leveraging different embedding approaches and layer extraction techniques.

**Table 3**
**Summary of PTM site LLM/pLM-based prediction tools, representation strategies (WSPE, WSWE, FSPE, and FSWE), and layer extraction methods (single-layer and multi-layer aggregation)**

| Tools | PTM | Representation strategy for pLM/transformer model | Layer extraction | Comments |
|---|---|---|---|---|
| CaLMPhosKAN [43] | Phosphorylation | FSWE | Single layer (final hidden layer layer) | Combines FSWE embeddings from CaLM and ProtTrans |
| MSTL-Kace [53] | Acetylation | WSWE | Multi-layer aggregation | The embeddings are the mean of [CLS] token of the last six layers of BERT |
| LMCrot [38] | Crotonylation | FSWE | Single layer (final hidden layer layer) | LMCrot also explores WSPE, WSWE, and FSPE |
| LMSuccSite [39], pLMSNOSite [55], LMPhosSite [42] | Succinylation, S-Nitrosylation, and phosphorylation, respectively | FSPE | Single layer (final hidden layer layer) | These tools have similar architectures |
| LM-O-GlcNAc-Site [51], LMGlyPred [46] | N-linked and O-linked glycosylation | FSPE | Single layer (final hidden layer layer) | LM-O-GlcNAc-site is for O-linked while LMGlyPred is for N-linked |
| EMNGLy [47] | N-linked glycosylation | WSWE + WSPE | Single layer (final hidden layer layer) | Combines WSWE (via [CLS] token) and WSPE for sequence embeddings |
| CoNglyPred [49] | N-linked glycosylation | FSWE | Single layer (final hidden layer layer) | – |
| SumoPred-PLM [56] | SUMOylation | FSPE | Single layer (final hidden layer layer) | – |
| iSUMO-RsFPN [57] | SUMOylation | WSWE | Single layer (final hidden layer layer) | – |
| BERT-Kgly [62] | Glycation | WSWE | Single layer (final hidden layer layer) | WSWE using [CLS] token was optimal. Also explores WSWE via averaging and WSPE |
| BERT-Kcr [59] | Crotonylation | WSWE | Single layer (final hidden layer layer) | – |

(continued)

**Table 3 (continued)**

| Tools | PTM | Representation strategy for pLM/transformer model | Layer extraction | Comments |
|---|---|---|---|---|
| ProtTrans-Glutar [61] | Glutarylation | WSWE | Single layer (final hidden layer layer) | WSWE via averaging |
| Deepro-Glu [60] | Glutarylation | WSWE | Single layer (final hidden layer layer) | WSWE via averaging |
| TransPTM [54] | Acetylation | WSWE | Single layer (final hidden layer layer) | WSWE via averaging |
| PTG-PLM [63] | N-linked glycosylation and glycation | WSWE | Single layer (final hidden layer layer) | WSWE |
| UniPTM [65] | Multiple PTMs | FSPE | Single layer (final hidden layer layer) | – |

### 3.1.2 Fine-Tuning of Foundation Models

An alternative to the feature-based approach is adapting pretrained foundation models (i.e., LLMs or pLMs) to a downstream task using a labeled dataset. This process involves fine-tuning the model's weights while using task-specific labels to guide learning. The fine-tuning of pLMs can be simplified into three steps:

*Step 1. Pretraining:* The model is first pretrained on a large corpus of protein sequences, learning general features of sequences through self-supervised learning, typically using a Masked Language Modeling (MLM) objective.

*Step 2. Constructing the labeled dataset:* A labeled dataset is then created, represented as pairs <I, O>, where *I* refers to the input (in this case, protein sequences, as used during pretraining), and *O* represents task-specific labels (i.e., the outputs).

*Step 3. Fine-tuning:* The final step involves updating the weights of the pretrained model using the task-specific labels via backpropagation.

It is important to note that not all the fine-tuning approaches need to adhere to these three steps. There are more ways to perform fine-tuning for adapting pretrained models to domain-specific tasks. Raschka [83] classifies these approaches into six categories, one of which is the feature-based approach that we have already discussed in the previous section. In subheadings "Fine-tuning I (updating output layers only)", "Fine-tuning II or full fine-tuning (updating all layers)", "Parameter-efficient fine-tuning (PEFT)", "In-context learning, indexing, and prompt tuning", and "Reinforcement learning with human feedback (RLHF)", we will briefly describe the remaining five categories to equip the readers with foundational knowledge although we recommend going through the excellent book by Raschka [83] for comprehensive explanations. Following this, in subheading "Fine-tuning in PTM site prediction context", we will focus on fine-tuning approaches specifically in the context of PTM site prediction.

#### Fine-Tuning I (Updating Output Layers Only)

This method updates only the output layer of the pretrained model, leaving the core layers frozen. It is efficient in memory and computation, and suitable when minor task-specific adaptations are sufficient. This approach should not be conflated with the feature-based approach (or, feature-based fine-tuning) that we discussed in the Subheading 3.1.1. The latter involves adding a classification head to the pretrained model without tuning any of its layers, while this method updates the output layers within the existing pretrained model itself.

#### Fine-Tuning II or Full Fine-Tuning (Updating All Layers)

Fine-Tuning II updates all model layers, fully adapting them to the target task. Although computationally demanding, this strategy often achieves higher performance for domain-specific applications than updating output layers only.

#### Parameter-Efficient Fine-Tuning (PEFT)

These methods update only select parts of the parameter space of the model, thereby saving resources. PEFT can be further classified into four sub-components:

1. *Soft prompt tuning:* Soft prompt tuning adds learnable prompts to the input layer, tuning these without altering core model layers.
2. *Prefix tuning:* Prefix tuning inserts trainable prompts within transformer blocks to enable efficient tuning across layers.
3. *Adapters:* In this method, small layers are added to transformer blocks, with only adapters being updated while keeping the main layers frozen.
4. *Low-rank adaptation (LoRA)*: LoRA [85] decomposes the weight-updated matrix into trainable low-rank matrices that are injected into the original model. Thus, the number of parameters participating in the fine-tuning is considerably reduced. Popular variants of LoRA include quantization LoRA (QLoRA), which combines LoRA with quantization to reduce memory usage and enable fine-tuning of large models on limited hardware [86]. Similarly, Weight-Decomposed Low-Rank Adaptation (DoRA) decomposes the pretrained weights into magnitude as well as direction and finetunes these components separately [87].

#### In-context Learning, Indexing, and Prompt Tuning

These techniques adapt the model via input manipulation rather than parameter updates:

- *In-context learning*: In-context learning embeds examples directly in the prompt to guide the model on the task. This method is ideal for zero-shot and few-shot tasks, where labeled data for training may be limited.
- *Indexing*: Indexing retrieves relevant information from external sources based on query similarity, aiding response generation. This strategy is commonly used in retrieval-augmented generation (RAG) applications, which enables LLMs to synthesize responses using both model knowledge and external resources.
- *Prompt tuning*: Prompt tuning optimizes specific prompts for task performance without changing model parameters. Unlike in-context learning, prompt tuning often relies on a small labeled dataset to identify the optimal prompt formulation.

#### Reinforcement Learning with Human Feedback (RLHF)

RLHF combines human feedback with reinforcement learning, training a reward model to guide model responses. Common in applications like ChatGPT, RLHF aligns the model's behavior with human preferences through reinforcement techniques.

#### Fine-Tuning in PTM Site Prediction Context

Various fine-tuning techniques have been adopted for the task of PTM site prediction. For instance, BERT-Kcr [59] utilizes task-adaptive pretraining (TAPT) by fully fine-tuning (Fine-tuning II) the BERT to adapt the model to domain knowledge by pretraining with protein sequences. Similarly, MSTL-Kace [53] fully fine-tunes BERT across multiple species to create species-specific embeddings for acetylation site prediction. Meanwhile, Phosformer follows a two-stage process for phosphosite prediction. Specifically, it first pretrains a custom transformer on broad protein sequence data to capture general protein relationships. The second stage then applies full fine-tuning on kinase-substrate pairs using a multi-level negative sampling strategy that is framed as a context-based question-answering (CBQA) task. Here, the kinase sequence acts as the "question," the peptide sequence as the "context," and the phosphorylation probability as the "answer."

PTMGPT2 [36], a multi-PTM site prediction tool, fine-tunes the ProtGPT2 [81] pLM (decoder-only model) via prompt tuning by constructing tailored prompts where the window sequence and its inference are embedded within the prompt. The model is then fine-tuned based on these tailored prompts, adapting ProtGPT2 to the specific PTM site prediction task through prompt-based fine-tuning.

While PEFT methods have not been explored in PTM site prediction, the experiments done by Schreiber [88] show promising results in PTMs using LoRA and QLoRA.

At the time of writing (Oct. 2024), we have not identified other fine-tuning approaches applied in the PTM domain. Moreover, the tools using feature-based fine-tuning (using pLM as a feature extractor) for PTM site prediction have already been addressed in Subheading 3.1.1, so we will refrain from reiterating that discussion here. It is worth noting, however, that this approach is relatively straightforward compared to other fine-tuning methods, as it involves merely adding a classification head on top of the pretrained embeddings without altering the foundation model itself. A study conducted by Sledzieski et al. [82] has observed that this method can even outperform both full fine-tuning and PEFT techniques in the context of protein-protein interaction prediction. This suggests that, for certain tasks, a classification head alone may be sufficient without necessitating further task-specific modifications to the pretrained model.

In Table 4, we summarize the tools that use the aforementioned fine-tuning strategies for PTM site prediction, highlighting each tool's specific approach and application.

### 3.2 Training Custom Transformer

While the use of foundation models (e.g., pretrained LLMs) and their fine-tuning has become prevalent in the field, some tools opt

**Table 4**
**Tools utilizing fine-tuning approaches for PTM site prediction, with comments providing further insights into the specific fine-tuning strategies adopted by each tool**

| Fine-tuning type | Tools (target PTM) | Comments |
|---|---|---|
| Full-fine tuning (fine-tuning II) | BERT-Kcr [59] (crotonylation), MSTL-Kace [53] (acetylation), phosformer [41] (phosphorylation) | The BERT model undergoes full fine-tuning for BERT-Kcr and MSTL-Kace. Phosformer uses a custom transformer pretrained in the first stage on protein data, followed by full fine-tuning |
| Feature-based fine-tuning (using pLM as a feature extractor) | LMCrot [38] (crotonylation), LMSuccsite [39] (succinylation), EMNGLy [47] (N-linked glycosylation), ProtTrans-Glutar [61] (glutarylation), etc. | All the tools mentioned in the previous subsection (i.e., Subheading 3.1.1) belong to this category. Here, the pretrained models are used as a feature extractor |
| Prompt-based fine-tuning | PTMGPT2 [36] (multiple PTMs) | Pretrained ProtGPT2 [81] is fine-tuned using customized prompts |
| Parameter efficient fine-tuning (PEFT) | Analysis by Schreiber [88] (PTM type not specified) | Shows application of LoRA and QLoRA in PTM site prediction |
| Other fine-tuning approaches (fine-tuning I, prefix tuning, indexing RLHF, etc.) | N/A | None found at the time of writing (Oct. 2024) |

to train their own transformers on sequences. These transformers are often lightweight compared to LLMs and can adapt to tasks with supervised training using fewer computational resources. Overall, training custom transformers offers several advantages, such as task-specific optimization, reduced computational cost, flexibility in architecture design, better alignment with domain-specific data, and improved adaptability for emerging tasks [89, 90]. In Subsection 2, we briefly reviewed tools that train their own transformers to learn representations for PTM site prediction. These tools were listed in Table 1 as a separate column. Therefore, we will not reiterate the descriptions of these tools again in this subsection.

### 3.3 Multi-pLM Integration and Multimodal/Multifacted Approaches

While sequence-aware pLMs have demonstrated impressive performance across various downstream tasks, their predictive capabilities can be limited when relying solely on a single pLM or unimodal embeddings. To address these limitations, approaches that integrate multiple pLMs, embeddings from other modes, or even conventional representation have been developed.

#### 3.3.1 Integrating Embeddings from Multiple pLMs

Each pLM differs in aspects such as architecture, pretraining datasets, and training objectives. These differences enable pLMs to capture unique characteristics of the same sequence. By combining embeddings from different pLMs, we can harness their complementary strengths to capture a richer representation of the sequence. The integration of embeddings from multiple pLMs was first explored in LM-OGlcNAc-Site [51], where embeddings from three pLMs—Ankh [91], ESM-2 [32], and ProtT5 [31] were combined at the decision level to predict O-GlcNAc modification sites. Similarly, Stack-OglyPred-PLM [50] similarly combines embeddings from Ankh and ProtT5 using stacked generalization for O-linked glycosylation prediction. Additionally, HOTGpred [48] employs an early fusion strategy, integrating embeddings from ESM-1 [32], ProtT5 [31], and Seq2Vec [92] for the same task. Likewise, PTM-Mamba [64] integrates embeddings from ESM-2 with its own PTM-specific embeddings, generated using the Mamba architecture, through a gated fusion mechanism, enabling multiple PTM site predictions, including phosphorylation and acetylation.

#### 3.3.2 Integrating Codon-Aware Embeddings

Due to the degeneracy of the genetic code, each amino acid can be encoded by up to six different codons [68]. This suggests that codon sequences (coding DNA) may contain additional information beyond the amino acid sequence alone. Codon usage is known to influence protein synthesis rates and co-translational protein folding [93]. This indicates codon usage can have an influence on the occurrence of PTMs. Recently, a pretrained pLM called CaLM

[68], which was trained on codon alphabets (coding DNA), outperformed amino acid-based pLMs like ProtTrans and the ESM family in several protein-level tasks.

The recently developed phosphosite prediction tool, CaLMPhosKAN [43], is the first work to utilize codon-aware embeddings in the domain of PTM site prediction and residue-level tasks in general. Interestingly, while CaLM embeddings alone did not outperform ProtTrans embeddings for this specific task, integrating CaLM embeddings (codon-aware) with ProtTrans embeddings (amino acid-aware) yielded the best results. This indicates that although CaLM may not surpass amino acid-based pLMs individually, it provides complementary representations that improve performance when combined.

It should be noted that translating sequences into the most reliable coding sequences, commonly referred to as "reverse translation," plays a crucial role. One approach is to predict codons using models like LinearDesign [94]. Another method involves mapping protein sequences to actual coding DNAs [95]. For instance, CaLMPhosKAN adopted a process where protein sequences were translated into actual coding DNA by mapping UniProt identifiers to non-redundant RefSeq nucleotide sequences [43]. Specifically, CaLMPhosKAN retrieved coding sequences from the NCBI Nucleotide database [96] and verified the sequence identity with UniProt proteins using pairwise alignment via the Needleman–Wunsch (NW) algorithm. Only sequences with 100% identity were included in the dataset, resulting in a reduced but highly reliable dataset.

#### *3.3.3 Integrating 3D Structure-Aware Embeddings or 3D Structures*

Protein function and interactions are intrinsically linked to their three-dimensional (3D) spatial conformations. While sequence-based pretrained pLMs capture linear amino acid sequences, they may overlook critical spatial relationships and structural motifs that are only apparent in the folded protein structure [97]. To address this limitation, integrating embeddings derived from 3D structures or from structure-aware models can provide additional context. The advent of the AlphaFold model has enabled noteworthy advancements in protein bioinformatics by providing high-confidence predicted 3D structures [33]. This has facilitated the development of tools that utilize structural information for PTM site prediction. This chapter discusses four tools that leverage 3D structures for PTM site prediction: MIND-S [66] for predicting 26 types of PTMs, EMNGLy [47], and CoNglyPred [49] for N-linked glycosylation prediction, and PTM-CMGMS [35] for predicting succinylation, S-nitrosylation, and crotonylation. MIND-S and PTM-CMGMS utilize 3D structures by learning latent representations through graph attention layers and contrastive learning, respectively, while EMNGLy employs structure-aware embeddings from a Masked Inverse Folding (MIF) model [72]. In all three methods, the 3D structural information serves as a complementary representation to sequence-based features rather than relying on 3D structures alone.

### 3.3.4 Integrating with Conventional Representation

Integrating conventional features such as physicochemical properties and evolutionary information can complement pLM embeddings. Additionally, conventional NLP encoders learned on local window sequences can be integrated. A notable example is combining encodings from the supervised embedding layer (SEL) learned from window sequences with pLM embeddings, providing both global and local views of sites [38, 51]. Tools like pLMSNOSite [55], LMSuccSite [39], and LMPhosSite [42] utilize stacked generalization of two models: one trained on FSPE-based (Full Sequence Per-Residue Embedding) pLM embeddings and another on SEL features from window sequences. Similarly, LMCrot [38] integrates latent representations of FSWE-based (Full Sequence Window Embedding) pLM embeddings with those from SEL and physicochemical properties. Likewise, other tools that combine pLM embeddings with conventional representations are presented in Table 5.

It is worthwhile to note that the integration (or, fusion) of other modalities with sequence modality can be achieved through four primary approaches [37]: *early fusion*, where sequence and modality-specific features are combined at the input level before being fed into the model; *intermediate fusion*, where each input type is first processed independently to extract high-level features, which are then combined at a hidden layer within the model; *late fusion*, where sequence and modality-specific data are processed separately through separate models or network branches, with their outputs (probability or inference) combined at the decision level; and *attention based fusion* (e.g., cross-attention, co-attention), which utilizes attention mechanisms to dynamically model interactions between sequence and modality-specific features, allowing the model to focus on relevant parts of one modality based on information from the other. While these four fusion methods are the most prevalent in PTM site prediction, other strategies also exist (e.g., fusion via gating mechanisms). In Table 5, we present tools that employ multi-pLM and multimodal strategies, along with the specific fusion methods used.

## 3.4 Interpretability and Explainability of LLM or LLM-Derived Embeddings

One of the drawbacks of deep learning models is their "black-box" nature [98]. Explaining and interpreting predictions by correlating inputs with outputs can help us understand the behavior of these models. The core mechanism of LLMs, namely the multi-head attention mechanism, computes attention maps that enable interpretation of the parts of the sequence that were focused on by the model. Each value in these maps, called an attention weight,

**Table 5**
**Overview of PTM site prediction tools integrating multiple pLMs and complementary modalities**

| Tool | PTM | Sequence modality (pLM-based) | Complementary modality | Type of integration | Fusion method |
|---|---|---|---|---|---|
| LM-OGlcNAc-Site [51] | O-GlcNAcylation | (Ankh + ESM-2 + ProtT5) embeddings | – | Multiple pLM integration | Late fusion |
| Stack-OglyPred-PLM [50] | O-linked glycosylation | (Ankh + ProtT5) embeddings | – | Multiple pLM integration | Late fusion |
| PTM-Mamba [64] | Multiple PTMs | (ESM-2 + mamba) embeddings | – | Multiple pLM integration | Gating based |
| HOTGpred [48] | O-linked glycosylation | (ESM-1 + ProtT5 + Seq2Vec) embeddings | FastText | Multiple pLMs as well as conventional representation integration | Early fusion |
| CaLMPhosKAN [43] | Phosphorylation | ProtT5 embeddings | Codon-aware embeddings from CaLM | Integration with codons | Early fusion |
| MIND-S [66] | 26 PTMs | Custom transformer embeddings | 3D structure + graph attention layer | Integration with 3D structure | Intermediate fusion |
| EMNGLy [47] | N-linked glycosylation | ESM-1 embeddings | 3D-structure aware embeddings from MIF | Integration with 3D structure | Early fusion |
| CoNGlyPred [49] | N-linked glycosylation | ESM-2 | 3D structure + graph transformer | Integration with 3D structure | Attention-based fusion |
| PTM-CMGMS [35] | S-Nitrosylation, succinylation, and crotonylation | Custom transformer embeddings | 3D structure + contrastive learning | Integration with 3D structure | Intermediate fusion |
| LMSuccSite [39], pLMSNOSite [55], and LMPhosSite [42] | Succinylation, S-nitrosylation, and phosphorylation, respectively | ProtT5 embeddings | Supervised embedding layer (SEL) | Integration with conventional representation | Intermediate fusion |

| | | | | | |
|---|---|---|---|---|---|
| iSUMO-RsFPN [57] | SUMOylation | ProtBERT embeddings | Physicochemical properties | Integration with conventional representation | Late fusion |
| GPS-SUMO 2.0 [58] | SUMOylation | Custom transformer embeddings | Physicochemical properties and evolutionary information | Integration with conventional representation | Late fusion |
| LMCrot [38] | Crotonylation | ProtT5 embeddings | SEL and physicochemical properties | Integration with conventional representation | Intermediate fusion |
| Deepro-Glu [60] | Glutarylation | ProtBERT embeddings | Physicochemical properties | Integration with conventional representation | Attention based |
| TransPTM [54] | Acetylation | ProtT5 embeddings + graph transformer | One-hot encoding of window | Integration with conventional representation | Early fusion |

Note that entries in the sequence modality and complementary modality columns show pairings with methods such as graph transformers (e.g., 3D structure + graph attention) and contrastive learning (e.g., 3D structure + contrastive learning) in some places, indicating that latent embeddings are learned from their respective modalities by applying these techniques indicates the degree of correlation between two tokens in the sequence. This can help derive important associations between tokens (amino acids) that the model has learned.

The lysine crotonylation (Kcr) site prediction tool, BERT-Kcr [59], derives average attention weights over the heads from different layers of BERT to obtain the distribution of weights over each position in the input window sequence. This enabled the authors to infer that higher layers capture more complex sequence relationships. Another Kcr prediction tool, LMCrot [38], also examines the attention map from the final encoder layer of ProtT5 to analyze the associations of other amino acids with the site of interest. They also reference the corresponding 3D structure of the sequence to extract insights into the relationship between the attention weights and the actual structure. This analysis enabled them to understand why both full-context and window embeddings are needed, and why FSWE (Full Sequence Window Embedding) provides the optimal result. Similar to BERT-Kcr, MSTL-Kace [53], an acetylation site predictor, examines averaged attention maps from all layers of BERT to observe the interactions between acetylation sites and other residues. Based on these maps, they also inferred why aggregated embeddings across layers are more effective than single-layer embeddings. Additionally, CaLMPhosKAN [43], a phosphorylation site predictor, utilizes attention maps from the final encoder layers of both ProtT5 [31] and CaLM [68]. By contrasting the associations captured by ProtT5 (an amino acid-based pLM) and CaLM (a codon-based pLM) with the site-of-interest, they provided insights into how codon-level information complements amino acid–level information in predicting phosphorylation sites. Similarly, PTMGPT2 [36], a multiple PTM site predictor, extracts attention scores from the last decoder layer averaged across the heads to generate Position Specific Probability Matrices (PSPMs). These PSPMs reveal amino acid patterns and preferences, providing insights into important sequence determinants for protein modifications, such as lysine acetylation and various kinase family motifs. Likewise, Phosformer [41] uses attention mechanisms within its model architecture to identify which positions in peptide sequences are most influential for kinase-specific phosphosite prediction. By examining attention weights from the final encoder layers, Phosformer reveals how the model focuses on substrate specificity determinants; critical amino acid residues for kinase recognition.

Some existing predictors have also utilized established explainable algorithms to interpret their models or derived embeddings. For instance, LMCrot [38] uses the model-agnostic method SHAP (SHapley Additive exPlanations) [99] to determine which residue positions have the most impact on the prediction of crotonylation sites. It was observed that the site-of-interest had the most significant impact, as expected, while adjacent residues also contributed notably. The SHAP values (mean absolute per position in their case) gradually decreased as the linear distance from the site-of-interest increased. This observation corroborated the findings from the attention map examination and provided additional support for using FSWE embeddings. Similarly, MIND-S [66] employs the Integrated Gradients [100] method to interpret its predictions. By calculating the contribution of each amino acid residue to the final prediction, MIND-S was able to identify important residues and known motifs associated with PTMs. For example, it successfully highlighted the N-X-S/T motif in N-linked glycosylation and distinguished different kinase-specific patterns in phosphorylation sites. Additionally, HOTGpred [48], which integrates multiple pLM embeddings for O-linked glycosylation, uses SHAP to determine the most influential pLM embeddings impacting the predictions. Tools that train their custom transformers have also applied these algorithms; for instance, DL-SPhos [45], a serine phosphorylation site predictor, employs the LIME (Local Interpretable Model-agnostic Explanations) [101] method to highlight the amino acid residues most critical to its predictions and to identify conserved motifs associated with serine phosphorylation. Table 6 summarizes the interpretable and explainable approaches utilized by the tools discussed above.

### 3.5 Use of Graph Representation

While sequence and structural-based representation have been extensively used in the field, graph-based representations offer several advantages by capturing complex, non-linear relationships essential for understanding protein structure and function. Unlike sequence-based models, graphs can represent spatial relationships by treating residues as nodes with edges that capture both close and distant interactions in 3D space. This flexibility allows the integration of diverse data types, such as evolutionary conservation and physicochemical properties, enriching the representation. Graphs also enhance interpretability by making residue interactions and dependencies more transparent and allow models to focus on biologically significant connections, such as those involved in binding or regulatory interactions. In this chapter, we will discuss three tools that leverage graph-based representations based on LLMs/pLMs/custom transformers.

The TransPTM [54] tool utilizes graph representations to predict acetylation sites. In this graph construction, each node represents a residue within a defined window sequence, with each node vectorized using ProtT5 embeddings. Edges are defined based on a complete graph assumption, meaning that every residue is connected to every other residue within the sequence (sequence adjacency). These graphs are then fed to a GNN composed of TransformerConv (TransConv) layers, which apply graph convolution with an attention mechanism to produce a graph-based latent representation of the sequence. Similarly, MIND-S [66], a tool for predicting multiple PTMs, constructs contact maps directly from

**Table 6**
**Tools that perform interpretability and explainability along with the specific strategies they used and inference obtained from it**

| Tool | PTM | Interpretable/ explainable strategy | Inference |
|---|---|---|---|
| BERT-Kcr [59] | Crotonylation | Attention weights | Higher layers capture more complex sequence relationships |
| LMCrot [38] | Crotonylation | Attention weights and SHAP | Validates the effectiveness of FSWE embeddings |
| MSTL-Kace [53] | Acetylation | Attention weights | Aggregated embeddings across layers are more effective than single-layer embeddings |
| CaLMPhosKAN [43] | Phosphorylation | Attention weights | Codon-level information complements amino acid–level information for predicting phosphorylation |
| PTMGPT2 [36] | Multiple PTMs | Attention weights | PSPMs reveal amino acid patterns and preferences critical for protein modifications |
| Phosformer [41] | Phosphorylation | Attention weights | Reveals substrate specificity determinants critical for kinase recognition in phosphosite prediction |
| MIND-S [66] | Multiple PTMs | Integrated gradients | Identifies residues contributing to predictions, highlighting motifs like N-X-S/T and kinase-specific patterns |
| HOTGpred [48] | O-linked glycosylation | SHAP | Determines most influential embeddings impacting predictions |
| DL-SPhos [45] | Phosphorylation | LIME | Highlights critical amino acids and conserved motifs for serine phosphorylation |

3D structural data. These contact maps serve as the initial graph structure, where each residue is represented as a node, and edges are defined based on physical proximity in the 3D structure rather than sequence adjacency. These contact maps are learned by a graph attention (GAT) [102] layer to create a graph-based representation of the sequence. Another tool, CoNglyPred [49], is an N-linked glycosylation predictor. Similar to MIND-S, it does not use a pLM/transformer directly to obtain the graph representation. Instead, it constructs a graph from contact maps based on 3D structures, where nodes represent residues and edges are defined by spatial proximity. Each residue node includes a combination of structural and sequence features, such as secondary structure information, physicochemical properties, evolutionary PSSM data, and torsion angles. This graph is processed by a Graph Transformer Network (GTN) [103] to create a latent graph representation, which is then fused with ESM-2 embeddings via a co-attention mechanism. These three approaches are summarized in Table 7.

**Table 7**
**Summary of graph-based tools for PTM site prediction, detailing target PTMs, node and edge definitions, and use of pLMs or transformers**

| Tool | PTM | Nodes | Edges | Use of pLMs/transformer |
|---|---|---|---|---|
| TransPTM [54] | Acetylation | Each residue in a defined window sequence | Complete graph (sequence adjacency) | ProtT5 embeddings as node features |
| MIND-S [66] | Multiple PTMs | Each residue in a 3D structure | Contact map (spatial proximity in 3D space) | Graphs are processed by the graph attention layer |
| CoNglyPred [49] | N-linked glycosylation | Each residue in a 3D structure | Contact map structure (spatial proximity in 3D space) | Graphs are processed by graph transformer<br>ESM-2 embeddings fused via co-attention (not used directly in the graph) |

### *3.6 Mamba Meets PTM (PTM-Mamba): Alternative to Transformers*

Recent advances in sequence modeling have explored alternatives to transformers to address their inherent computational inefficiencies, particularly their quadratic scaling with sequence length. Mamba [104], a novel state space model (SSM), has emerged as a promising alternative by leveraging hardware-aware primitives that enable efficient modeling of long-range dependencies in sequential data. Unlike transformers, which rely on attention mechanisms to capture relationships across tokens, Mamba employs structured state spaces to achieve sub-quadratic scaling, making it computationally efficient for large-scale datasets without sacrificing modeling capacity. Its streamlined design is especially suited for tasks requiring the analysis of extensive sequences, such as protein modeling.

Building on this innovation, PTM-Mamba's [64] extends the capabilities of Mamba to explicitly account for PTMs within protein sequences. PTM-Mamba integrates bidirectional Mamba blocks, processing sequences both forward and backward, thereby capturing context and dependencies that span the entire sequence. This is critical for PTMs, which often influence regions distant from the site of modification. Additionally, PTM-Mamba introduces a gated fusion mechanism to combine embeddings from the ESM-2 model with its own PTM-specific embeddings. By explicitly incorporating PTM information, PTM-Mamba has shown to outperform traditional transformer-based models PTM site prediction, disease association, and druggability analysis.

### 3.7 Use of Contrastive Learning

Contrastive learning refines representations by learning an embedding space where positive samples (e.g., modified sites) are close to each other, while negative samples (e.g., non-modified sites) are pushed further apart [105]. This is achieved using a contrastive loss function that minimizes distances between similar samples (positive pairs) and maximizes distances between dissimilar samples (negative pairs). In PTM site prediction, contrastive learning can be applied by pairing samples based on their modification status, sequence similarity, or shared structural features. For example, samples with the same PTM type, such as phosphorylation sites, are treated as positive pairs, while non-phosphorylated sites serve as negative pairs. This approach allows the model to embed sites with shared modification characteristics closer together, preserving the unique biological signals of different PTM types in the feature space.

An example of this approach is PTM-CMGMS [35], a multi-PTM site prediction tool that applies contrastive learning to optimize structural representations by grouping PTM sites with similar modification characteristics. For instance, crotonylation sites are treated as positive pairs and non-crotonylation sites as negative pairs. This grouping helps the model learn to embed sites with similar PTM types closer together in the representation space while pushing dissimilar PTM types farther apart.

## 4 Limitations and Future Directions

The following limitations identified in current PTM site prediction tools leveraging LLMs/pLMs/custom transformers highlight specific gaps that could serve as promising directions for future advancements.

### 4.1 Optimal Use of 3D Structures

While 3D structures are crucial for accurately capturing spatial information in proteins, they are currently underutilized in PTM site prediction. Tools like AlphaFold have enabled access to high-confidence 3D models, but integrating 3D data into PTM models is currently limited. Although we have identified a few tools that incorporate 3D modalities in this chapter, the predictive capability of 3D-based representation alone remains suboptimal. We are likely to observe robust use of 3D structures in PTM site prediction in the future.

### 4.2 Exploration of Layers

Most models rely on embeddings from the final layer or a simple aggregation across a few layers of pretrained LLMs/pLMs. However, intermediate layers can contain useful information that can aid in capturing different aspects of protein function and structure. Conducting an exhaustive layer exploration, or layer-wise analysis, could provide deeper insights into specific sequence-structure-function relationships that are useful for PTM site prediction. Additionally, selectively aggregating embeddings from informative layers can enhance the overall representation quality.

### 4.3 Exploration of Other Fine-Tuning Approaches

The majority of PTM predictors use only limited fine-tuning methods (e.g., full fine-tuning or feature-based fine-tuning), while more approaches like selective layer fine-tuning and prefix fine-tuning remain largely underexplored. Likewise, techniques for parameter-efficient fine-tuning, such as LoRA, QLoRA, and DoRA, are also not widely applied. Exploring these approaches could optimize model performance while reducing computational costs, providing notable advancements in PTM prediction.

### 4.4 Use of Decoders

Most PTM site prediction models rely on encoder-only architectures, with decoders or the decoder components of encoder-decoder models being largely underutilized. In this chapter, we discussed PTMGPT2 [36], which is currently the only tool that employs a decoder-only model and has demonstrated impressive performance across multiple PTMs. This suggests that incorporating decoders can offer advantages in tasks requiring sequential predictions or autoregressive capabilities, and may improve the ability of PTM site prediction models to handle PTMs with dependencies across protein sequences.

### 4.5 Exploration of Other Modalities

Integrating multiple data types (e.g., sequence, 3D structure, physicochemical properties, and evolutionary information) can offer a more comprehensive view of protein behavior. Although multimodal approaches are gaining interest, few tools successfully combine different data modalities to exploit the diversity of protein characteristics needed for effective PTM site prediction. While multimodal approaches are gaining traction, relatively few tools successfully combine diverse data modalities to utilize the full spectrum of protein characteristics essential for effective PTM site prediction. Expanding the range of explored modalities could amplify predictive accuracy and insights.

In addition to commonly used geometric representations, such as graphs and contact maps, further exploration might include voxel grids and 3D density maps, which offer detailed spatial occupancy and distribution information at finer resolutions. For instance, embeddings derived from multiple sequence alignment (MSA) transformer [106] can be incorporated to reflect evolutionary conservation patterns.

### 4.6 Improvement in the Predictive Performance

Although significant advancements have been achieved in PTM site prediction with the development of LLMs, there is still ample room for improving predictive capabilities. To illustrate this, we have compiled a selection of existing tools for five well-studied PTMs,

**Table 8**
**Some existing tools for the prediction of five selected PTMs, presented as an example, along with their reported test MCC values (range: [−1,+1]) on the datasets they were trained on**

| PTM | Tools and their reported MCC |
|---|---|
| Phosphorylation | LMPhosSite [42]: 0.39<br>CaLMPhosKAN [43]: 0.41 |
| Acetylation | MSTL-Kace [53]: 0.41<br>TransPTM [54]: 0.45 |
| Glycosylation | EMNGly [47]: 0.80<br>Stack-OglyPred-PLM [50]: 0.52<br>LM-OGlcNAc-Site [51]: 0.13 |
| SUMOylation | SumoPred-PLM [56]: 0.59<br>iSUMO-RsFPN [57]: 0.48 |
| Crotonylation | BERT-Kcr [59]: 0.64<br>LMCrot [38]: 0.70 |

It is important to note that each tool (for its respective PTM) may have been evaluated on different datasets, and the reported MCC reflects performance specific to those datasets. Additionally, for tools that reported MCC values on multiple datasets or species, we have selected the highest-performing MCC observed across those datasets

along with their reported Matthew's Correlation Coefficient (MCC) values on corresponding datasets in Table 8. A similar trend can be observed across tools for other PTMs as well.

The previous subsections (Subheadings 4.1, 4.2, 4.3, and 4.4) outlined some approaches that could be integrated into current LLM-based methods to enhance their performance. Additionally, improvements can be made at the dataset level, as they directly influence the quality of downstream learning. For instance, efforts should focus on acquiring reliable negative sites to minimize potential false negatives. Similarly, positive sites should be thoroughly vetted to eliminate any false positives.

A high accuracy of predictions is crucial to translate these computational approaches into biological experiments, biomedical treatments, or biotechnological protein design. Even in relatively simple prokaryotic model systems, biological follow-up experiments to study the function of PTMs (e.g., site-directed mutagenesis of PTM sites, followed by phenotypic analyses) often take several months to perform. It is unlikely that biologists would decide to spend significant time and resources on these experiments based on predictions with an MCC $< 0.9$. Indeed, so far, we have been unable to find studies that use PTM-predictions as the sole basis for wet-lab experiments. However, reports are emerging that apply prediction tools to databases of clinical variants to assess the impact of genetic mutations on protein modifications and thus protein function [66, 107, 108]. Similarly, PTM-predictions are becoming increasingly relevant for protein design approaches [109, 110]. With an increase in prediction accuracies, functional studies and applications like these are expected to gain traction.

## 5 Discussion and Conclusion

The application of LLMs and pLMs has brought significant advances in PTM site prediction. By capturing both local and global contexts within protein sequences, these models outperform traditional sequence-based methods, which often miss key contextual interactions. Sophisticated representation strategies, such as full sequence-based embeddings (e.g., FSPE and FSWE), enable models to incorporate long-range dependencies essential for accurate PTM site prediction. Additionally, aggregating embeddings from multiple layers leverages the hierarchical structure of transformer architectures, enriching the models' understanding of protein sequences.

Fine-tuning approaches have also evolved. Full fine-tuning adapts models like BERT to specific PTMs, while feature-based fine-tuning applies task-specific classification heads to frozen models, simplifying adaptation. Prompt-based fine-tuning has expanded the use of decoder-only models like ProtGPT2 in PTM site prediction, exploiting their generative capabilities without needing specialized classification heads. However, parameter-efficient fine-tuning (PEFT) methods remain underexplored and present opportunities for optimizing model adaptation. Incorporating multiple data modalities and combining embeddings from various pLMs further enhances predictive power. Each pLM captures unique characteristics, so combining them yields richer representations. Codon-aware embeddings, for example, add biological nuance, while integrating three-dimensional structural information to address limitations in sequence-only models by capturing spatial relationships critical for protein function. Additionally, blending pLM embeddings with conventional features, like physicochemical properties, broadens the feature space and improves prediction.

Interpretability of these models, however, remains challenging due to their black-box nature. Attention mechanisms in transformer-based models offer a degree of explainability, with attention maps highlighting influential residues in predictions. Model-agnostic techniques such as SHAP values also provide insights into feature importance, but further advancements in interpretability will be essential for broader biological applications.

Recent graph-based approaches model complex protein structures by treating residues as nodes and their relationships as edges, capturing both sequential and spatial information. Contrastive learning further refines these representations by clustering modified sites together while pushing non-modified sites apart, enhancing the model's ability to distinguish PTM sites accurately.

Despite these advancements, several limitations persist. The integration of 3D structural information is not yet fully optimized, and model layer exploration is limited, potentially overlooking valuable features. Similarly, decoder architectures are underutilized, and multimodal integration could be expanded to include a broader range of biological data. Improving interpretability will be critical to making these models more transparent and broadly useful.

In conclusion, LLMs and pLMs have transformed PTM site prediction by enabling more accurate, contextually rich models. Future research should focus on addressing current limitations by exploring advanced fine-tuning techniques, utilizing decoder architectures, and integrating additional data modalities. Enhancing model transparency will also be crucial for wider acceptance in the biological community. The continued evolution of these models holds great promise for providing deeper insights into protein function and disease mechanisms, driving further innovation in computational biology.

## Acknowledgments

This work is supported in part by funds from the National Science Foundation (NSF: # 1901793 and # 2210356) and a faculty start-up grant to DBK. SS acknowledges support from the College of Science start-up grant.

## References

1. Keenan EK, Zachman DK, Hirschey MD (2021) Discovering the landscape of protein modifications. Mol Cell 81:1868–1878. https://doi.org/10.1016/j.molcel.2021.03.015
2. Aebersold R, Agar JN, Amster IJ et al (2018) How many human proteoforms are there? Nat Chem Biol 14:206–214. https://doi.org/10.1038/nchembio.2576
3. Hong X, Li N, Lv J et al (2023) PTMint database of experimentally verified PTM regulation on protein–protein interaction. Bioinformatics 39:btac823. https://doi.org/10.1093/bioinformatics/btac823
4. Lee JM, Hammarén HM, Savitski MM, Baek SH (2023) Control of protein stability by post-translational modifications. Nat Commun 14:201. https://doi.org/10.1038/s41467-023-35795-8
5. Ryšlavá H, Doubnerová V, Kavan D, Vaněk O (2013) Effect of posttranslational modifications on enzyme function and assembly. J Proteome 92:80–109. https://doi.org/10.1016/j.jprot.2013.03.025
6. Millar AH, Heazlewood JL, Giglione C et al (2019) The scope, functions, and dynamics of posttranslational protein modifications. Annu Rev Plant Biol 70:119–151. https://doi.org/10.1146/annurev-arplant-050718-100211
7. Li Z, Li S, Luo M et al (2022) dbPTM in 2022: an updated database for exploring regulatory networks and functional associations of protein post-translational modifications. Nucleic Acids Res 50:D471–D479. https://doi.org/10.1093/nar/gkab1017
8. Beltrao P, Bork P, Krogan NJ, van Noort V (2013) Evolution and functional cross-talk of protein post-translational modifications. Mol Syst Biol 9:714. https://doi.org/10.1002/msb.201304521
9. Zhong Q, Xiao X, Qiu Y et al (2023) Protein posttranslational modifications in health and diseases: functions, regulatory mechanisms,

and therapeutic implications. MedComm 4: e261. https://doi.org/10.1002/mco2.261

10. Geffen Y, Anand S, Akiyama Y et al (2023) Pan-cancer analysis of post-translational modifications reveals shared patterns of protein regulation. Cell 186:3945–3967.e26. https://doi.org/10.1016/j.cell.2023.07.013
11. Mann M, Jensen ON (2003) Proteomic analysis of post-translational modifications. Nat Biotechnol 21:255–261. https://doi.org/10.1038/nbt0303-255
12. Olsen JV, Mann M (2013) Status of large-scale analysis of post-translational modifications by mass spectrometry. Mol Cell Proteomics 12:3444–3452. https://doi.org/10.1074/mcp.O113.034181
13. Pokharel S, Sidorov E, Caragea D, KC DB (2022) NLP-based encoding techniques for prediction of post-translational modification sites and protein functions. In: Machine learning in bioinformatics of protein sequences. World Scientific, Singapore, pp 81–127
14. Pakhrin SC, Pokharel S, Saigo H, KC DB (2022) Deep learning–based advances in protein posttranslational modification site and protein cleavage prediction. In: KC DB (ed) Computational methods for predicting post-translational modification sites. Springer, New York, pp 285–322
15. Li W, Godzik A (2006) Cd-hit: a fast program for clustering and comparing large sets of protein or nucleotide sequences. Bioinformatics 22:1658–1659. https://doi.org/10.1093/bioinformatics/btl158
16. Hauser M, Steinegger M, Söding J (2016) MMseqs software suite for fast and deep clustering and searching of large protein sequence sets. Bioinformatics 32:1323–1330. https://doi.org/10.1093/bioinformatics/btw006
17. Meng L, Chan W-S, Huang L et al (2022) Mini-review: recent advances in post-translational modification site prediction based on deep learning. Comput Struct Biotechnol J 20:3522–3532. https://doi.org/10.1016/j.csbj.2022.06.045
18. Pudjihartono N, Fadason T, Kempa-Liehr AW, O'Sullivan JM (2022) A review of feature selection methods for machine learning-based disease risk prediction. Front Bioinforma 2. https://doi.org/10.3389/fbinf.2022.927312
19. Li Y, Mansmann U, Du S, Hornung R (2022) Benchmark study of feature selection strategies for multi-omics data. BMC Bioinformatics 23:412. https://doi.org/10.1186/s12859-022-04962-x
20. Miao J, Niu L (2016) A survey on feature selection. Procedia Comput Sci 91:919–926. https://doi.org/10.1016/j.procs.2016.07.111
21. Liu Q, Kusner MJ, Blunsom P (2020) A survey on contextual embeddings. arXiv 2003.07278. https://arxiv.org/abs/2003.07278
22. Sherstinsky A (2023) Fundamentals of recurrent neural network (RNN) and long short-term memory (LSTM) network. Phys D Nonlin Phenomena 404:132306. https://doi.org/10.1016/j.physd.2019.132306
23. Schmidt RM (2019) Recurrent neural networks (RNNs): a gentle introduction and overview. arXiv 1912.05911. https://arxiv.org/abs/1912.05911
24. Liu X (2017) Deep recurrent neural network for protein function prediction from sequence. arXiv 1701.08318. https://arxiv.org/abs/1701.08318
25. Nugaliyadde A, Wong KW, Sohel F, Xie H (2019) Language modeling through long term memory network. arXiv 1904.08936. https://arxiv.org/abs/1904.08936
26. Jozefowicz R, Vinyals O, Schuster M, et al (2016) Exploring the limits of language modeling. arXiv:1602.02410. https://arxiv.org/abs/1602.02410
27. Vaswani A, Shazeer N, Parmar N, et al (2017) Attention is all you need. In: Guyon I, von Luxburg U, Bengio S, Wallach H, Fergus R, Vishwanathan S, Garnett R (eds) Advances in neural information processing systems, vol 30. Curran Associates, Inc. https://proceedings.neurips.cc/paper_files/paper/2017/file/3f5ee243547dee91fbd053c1c4a845aa-Paper.pdf
28. Devlin J, Chang M-W, Lee K, Toutanova K (2019) BERT: pre-training of deep bidirectional transformers for language understanding. arXiv 1810.04805. https://arxiv.org/abs/1810.04805
29. Brown TB, Mann B, Ryder N, et al (2020) Language models are few-shot learners. arXiv 2005.14165. https://arxiv.org/abs/2005.14165
30. Hagos DH, Battle R, Rawat DB (2024) Recent advances in generative AI and large language models: current status, challenges, and perspectives. IEEE Trans Artif Intell:1–21. https://doi.org/10.1109/TAI.2024.3444742

31. Elnaggar A, Heinzinger M, Dallago C et al (2022) ProtTrans: toward understanding the language of life through self-supervised learning. IEEE Trans Pattern Anal Mach Intell 44:7112–7127. https://doi.org/10.1109/TPAMI.2021.3095381
32. Lin Z, Akin H, Rao R et al (2023) Evolutionary-scale prediction of atomic-level protein structure with a language model. Science 379:1123–1130. https://doi.org/10.1126/science.ade2574
33. Jumper J, Evans R, Pritzel A et al (2021) Highly accurate protein structure prediction with AlphaFold. Nature 596:583–589. https://doi.org/10.1038/s41586-021-03819-2
34. Rao R, Bhattacharya N, Thomas N et al (2019) Evaluating protein transfer learning with TAPE. In: Proceedings of the 33rd international conference on neural information processing systems. Curran Associates Inc., Red Hook, pp 9689–9701
35. Li Z, Li M, Zhu L, Zhang W (2024) Improving PTM site prediction by coupling of multi-granularity structure and multi-scale sequence representation. Proc AAAI Conf Artif Intell 38:188–196. https://doi.org/10.1609/aaai.v38i1.27770
36. Shrestha P, Kandel J, Tayara H, Chong KT (2024) Post-translational modification prediction via prompt-based fine-tuning of a GPT-2 model. Nat Commun 15:6699. https://doi.org/10.1038/s41467-024-51071-9
37. Pratyush P, KC DB (2025) Advances in prediction of posttranslational modification sites known to localize in protein supersecondary structures. In: Kister AE (ed) Protein supersecondary structures: methods and protocols. Springer, New York, pp 117–151
38. Pratyush P, Bahmani S, Pokharel S et al (2024) LMCrot: an enhanced protein crotonylation site predictor by leveraging an interpretable window-level embedding from a transformer-based protein language model. Bioinformatics:btae290. https://doi.org/10.1093/bioinformatics/btae290
39. Pokharel S, Pratyush P, Heinzinger M et al (2022) Improving protein succinylation sites prediction using embeddings from protein language model. Sci Rep 12:16933. https://doi.org/10.1038/s41598-022-21366-2
40. Pratyush P, Pokharel S, Ismail HD et al (2025) LMPTMSite: a platform for PTM site prediction in proteins leveraging transformer-based protein language models. In: Kloczkowski A, Kurgan L, Faraggi E (eds) Prediction of protein secondary structure. Springer, New York, pp 261–297
41. Zhou Z, Yeung W, Gravel N et al (2023) Phosformer: an explainable transformer model for protein kinase-specific phosphorylation predictions. Bioinformatics 39: btad046. https://doi.org/10.1093/bioinformatics/btad046
42. Pakhrin SC, Pokharel S, Pratyush P et al (2023) LMPhosSite: a deep learning-based approach for general protein phosphorylation site prediction using Embeddings from the local window sequence and pretrained protein language model. J Proteome Res 22:2548–2557. https://doi.org/10.1021/acs.jproteome.2c00667
43. Pratyush P, Carrier C, Pokharel S, et al (2024) CaLMPhosKAN: prediction of general phosphorylation sites in proteins via fusion of codon-aware embeddings with amino acid-aware embeddings and wavelet-based Kolmogorov–Arnold network. Bioinformatics 41(4):btaf124. https://doi.org/10.1093/bioinformatics/btaf124
44. Wang X, Zhang Z, Zhang C et al (2022) TransPhos: a deep-learning model for general phosphorylation site prediction based on transformer-encoder architecture. Int J Mol Sci 23:4263. https://doi.org/10.3390/ijms23084263
45. Shrestha P, Kandel J, Tayara H, Chong KT (2024) DL-SPhos: prediction of serine phosphorylation sites using transformer language model. Comput Biol Med 169:107925. https://doi.org/10.1016/j.compbiomed.2024.107925
46. Pakhrin SC, Pokharel S, Aoki-Kinoshita KF et al (2023) LMNglyPred: prediction of human N-linked glycosylation sites using embeddings from a pre-trained protein language model. Glycobiology 33:411–422. https://doi.org/10.1093/glycob/cwad033
47. Hou X, Wang Y, Bu D et al (2023) EMNGly: predicting N-linked glycosylation sites using the language models for feature extraction. Bioinformatics 39:btad650. https://doi.org/10.1093/bioinformatics/btad650
48. Pham NT, Zhang Y, Rakkiyappan R, Manavalan B (2024) HOTGpred: enhancing human O-linked threonine glycosylation prediction using integrated pretrained protein language model-based features and multi-stage feature selection approach. Comput Biol Med 179: 108859. https://doi.org/10.1016/j.compbiomed.2024.108859
49. Wang H, Zhao L, Yu Z, et al CoNglyPred: accurate prediction of N-linked glycosylation

sites using ESM-2 and structural features with graph network and co-attention. Proteomics 25:e202400210. https://doi.org/10.1002/pmic.202400210

50. Pakhrin SC, Chauhan N, Khan S et al (2024) Prediction of human O-linked glycosylation sites using stacked generalization and embeddings from pre-trained protein language model. Bioinformatics 40:btae643. https://doi.org/10.1093/bioinformatics/btae643
51. Pokharel S, Pratyush P, Ismail HD et al (2023) Integrating embeddings from multiple protein language models to improve protein O-GlcNAc site prediction. Int J Mol Sci 24:16000. https://doi.org/10.3390/ijms242116000
52. Pourmirzaei M, Ramazi S, Esmaili F et al (2023) Machine learning-based approaches for ubiquitination site prediction in human proteins. BMC Bioinformatics 24:449. https://doi.org/10.1186/s12859-023-05581-w
53. MSTL-Kace: prediction of prokaryotic lysine acetylation sites based on multistage transfer learning strategy. ACS Omega. https://pubs.acs.org/doi/full/10.1021/acsomega.3c07086. Accessed 7 Mar 2024
54. Meng L, Chen X, Cheng K et al (2024) TransPTM: a transformer-based model for non-histone acetylation site prediction. Brief Bioinform 25:bbae219. https://doi.org/10.1093/bib/bbae219
55. Pratyush P, Pokharel S, Saigo H, KC DB (2023) pLMSNOSite: an ensemble-based approach for predicting protein S-nitrosylation sites by integrating supervised word embedding and embedding from pre-trained protein language model. BMC Bioinformatics 24:41. https://doi.org/10.1186/s12859-023-05164-9
56. Palacios AV, Acharya P, Peidl AS et al (2024) SumoPred-PLM: human SUMOylation and SUMO2/3 sites prediction using pre-trained protein language model. NAR Genomics Bioinforma 6:lqae011. https://doi.org/10.1093/nargab/lqae011
57. Lv Z, Wei X, Hu S et al (2024) iSUMO-RsFPN: a predictor for identifying lysine SUMOylation sites based on multi-features and feature pyramid networks. Anal Biochem 687:115460. https://doi.org/10.1016/j.ab.2024.115460
58. Gou Y, Liu D, Chen M et al (2024) GPS-SUMO 2.0: an updated online service for the prediction of SUMOylation sites and SUMO-interacting motifs. Nucleic Acids Res 52:W238–W247. https://doi.org/10.1093/nar/gkae346
59. Qiao Y, Zhu X, Gong H (2022) BERT-Kcr: prediction of lysine crotonylation sites by a transfer learning method with pre-trained BERT models. Bioinformatics 38:648–654. https://doi.org/10.1093/bioinformatics/btab712
60. Wang X, Ding Z, Wang R, Lin X (2023) Deepro-Glu: combination of convolutional neural network and Bi-LSTM models using ProtBert and handcrafted features to identify lysine glutarylation sites. Brief Bioinform 24: bbac631. https://doi.org/10.1093/bib/bbac631
61. Indriani F, Mahmudah KR, Purnama B, Satou K (2022) ProtTrans-Glutar: incorporating features from pre-trained transformer-based models for predicting Glutarylation sites. Front Genet 13. https://doi.org/10.3389/fgene.2022.885929
62. Liu Y, Liu Y, Wang G-A et al (2022) BERT-Kgly: a bidirectional encoder representations from transformers (BERT)-based model for predicting lysine glycation site for homo sapiens. Front Bioinforma 2. https://doi.org/10.3389/fbinf.2022.834153
63. Alkuhlani A, Gad W, Roushdy M et al (2022) PTG-PLM: predicting post-translational glycosylation and glycation sites using protein language models and deep learning. Axioms 11:469. https://doi.org/10.3390/axioms11090469
64. Peng Z, Schussheim B, Chatterjee P (2024) PTM-mamba: a PTM-aware protein language model with bidirectional gated mamba blocks. bioRxiv 2024.02.28.581983. https://doi.org/10.1101/2024.02.28.581983
65. Meng L, Lin J, Cheng K, et al (2024) UniPTM: multiple PTM site prediction on full-length protein sequence. bioRxiv 2024.08.03.606471. https://doi.org/10.1101/2024.08.03.606471
66. Yan Y, Jiang J-Y, Fu M et al (2023) MIND-S is a deep-learning prediction model for elucidating protein post-translational modifications in human diseases. Cell Rep Methods 3:100430. https://doi.org/10.1016/j.crmeth.2023.100430
67. Szewczuk LM, Tarrant MK, Cole PA (2009) Chapter 1: Protein phosphorylation by semisynthesis: from paper to practice. In: Muir TW, Abelson JN (eds) Methods in enzymology. Academic Press, London, pp 1–24
68. Outeiral C, Deane CM (2024) Codon language embeddings provide strong signals for use in protein engineering. Nat Mach Intell 6: 170–179. https://doi.org/10.1038/s42256-024-00791-0

69. Liu Z, Wang Y, Vaidya S, et al (2024) KAN: Kolmogorov-Arnold networks. arXiv 2404.19756. https://arxiv.org/abs/2404.19756
70. He M, Zhou X, Wang X (2024) Glycosylation: mechanisms, biological functions and clinical implications. Signal Transduct Target Ther 9:1–33. https://doi.org/10.1038/s41392-024-01886-1
71. Lu Q, Zhang X, Liang T, Bai X (2022) O-GlcNAcylation: an important post-translational modification and a potential therapeutic target for cancer therapy. Mol Med 28:115. https://doi.org/10.1186/s10020-022-00544-y
72. Yang KK, Zanichelli N, Yeh H (2023) Masked inverse folding with sequence transfer for protein representation learning. Protein Eng Des Sel 36:gzad015. https://doi.org/10.1093/protein/gzad015
73. Damgaard RB (2021) The ubiquitin system: from cell signalling to disease biology and new therapeutic opportunities. Cell Death Differ 28:423–426. https://doi.org/10.1038/s41418-020-00703-w
74. Drazic A, Myklebust LM, Ree R, Arnesen T (2016) The world of protein acetylation. Biochim Biophys Acta 1864:1372–1401. https://doi.org/10.1016/j.bbapap.2016.06.007
75. Hess DT, Matsumoto A, Kim S-O et al (2005) Protein S-nitrosylation: purview and parameters. Nat Rev Mol Cell Biol 6:150–166. https://doi.org/10.1038/nrm1569
76. Weyh M, Jokisch M-L, Nguyen T-A et al (2024) Deciphering functional roles of protein succinylation and glutarylation using genetic code expansion. Nat Chem 16:913–921. https://doi.org/10.1038/s41557-024-01500-5
77. Han Z-J, Feng Y-H, Gu B-H et al (2018) The post-translational modification, SUMOylation, and cancer (review). Int J Oncol 52:1081–1094. https://doi.org/10.3892/ijo.2018.4280
78. Brandes N, Ofer D, Peleg Y et al (2022) ProteinBERT: a universal deep-learning model of protein sequence and function. Bioinformatics 38:2102–2110. https://doi.org/10.1093/bioinformatics/btac020
79. Jiang G, Li C, Lu M et al (2021) Protein lysine crotonylation: past, present, perspective. Cell Death Dis 12:1–11. https://doi.org/10.1038/s41419-021-03987-z
80. Fan X, Monnier VM (2021) Protein post-translational modification (PTM) by glycation: role in lens aging and age-related cataractogenesis. Exp Eye Res 210:108705. https://doi.org/10.1016/j.exer.2021.108705
81. Ferruz N, Schmidt S, Höcker B (2022) ProtGPT2 is a deep unsupervised language model for protein design. Nat Commun 13:4348. https://doi.org/10.1038/s41467-022-32007-7
82. Sledzieski S, Kshirsagar M, Baek M et al (2024) Democratizing protein language models with parameter-efficient fine-tuning. Proc Natl Acad Sci 121:e2405840121. https://doi.org/10.1073/pnas.2405840121
83. Raschka S (2024) Build a large language model (from scratch). Simon and Schuster, New York
84. Thumuluri V, Almagro Armenteros JJ, Johansen AR et al (2022) DeepLoc 2.0: multi-label subcellular localization prediction using protein language models. Nucleic Acids Res 50:W228–W234. https://doi.org/10.1093/nar/gkac278
85. Hu EJ, Shen Y, Wallis P, et al (2021) LoRA: low-rank adaptation of large language models. arXiv 2106.09685. https://arxiv.org/abs/2106.09685
86. Dettmers T, Pagnoni A, Holtzman A, Zettlemoyer L (2023) QLoRA: efficient finetuning of quantized LLMs. arXiv 2305.14314. https://arxiv.org/abs/2305.14314
87. Liu S-Y, Wang C-Y, Yin H, et al (2024) DoRA: weight-decomposed low-rank adaptation. arXiv 2402.09353. https://arxiv.org/abs/2402.09353
88. Schreiber A (2023) ESMBind and QBind: LoRA, QLoRA, and ESM-2 for predicting binding sites and post translational modification. bioRxiv 2023.11.13.566930. https://doi.org/10.1101/2023.11.13.566930
89. Friedman D, Wettig A, Chen D (2023) Learning transformer programs. arXiv 2306.01128. https://arxiv.org/abs/2306.01128
90. Tuli S, Dedhia B, Tuli S, Jha NK (2022) FlexiBERT: are current transformer architectures too homogeneous and rigid?. arXiv 2205.11656. https://arxiv.org/abs/2205.11656
91. Elnaggar A, Essam H, Salah-Eldin W, et al (2023) Ankh: optimized protein language model unlocks general-purpose modelling. arXiv 2301.06568. https://arxiv.org/abs/2301.06568
92. Heinzinger M, Elnaggar A, Wang Y et al (2019) Modeling aspects of the language of life through transfer-learning protein

sequences. BMC Bioinformatics 20:723. https://doi.org/10.1186/s12859-019-3220-8

93. Liu Y, Yang Q, Zhao F (2021) Synonymous but not silent: the codon usage code for gene expression and protein folding. Annu Rev Biochem 90:375–401. https://doi.org/10.1146/annurev-biochem-071320-112701
94. Zhang H, Zhang L, Lin A et al (2023) Algorithm for optimized mRNA design improves stability and immunogenicity. Nature 621:396–403. https://doi.org/10.1038/s41586-023-06127-z
95. Suyama M, Torrents D, Bork P (2006) PAL2-NAL: robust conversion of protein sequence alignments into the corresponding codon alignments. Nucleic Acids Res 34:W609–W612. https://doi.org/10.1093/nar/gkl315
96. NCBI Resource Coordinators (2016) Database resources of the National Center for Biotechnology Information. Nucleic Acids Res 44:D7–D19. https://doi.org/10.1093/nar/gkv1290
97. Heinzinger M, Weissenow K, Gomez Sanchez J, Henkel A, Mirdita M, Steinegger M, Rost B (2024) Bilingual language model for protein sequence and structure. NAR Genom Bioinform 6(4):lqae150. https://doi.org/10.1093/nargab/lqae150
98. Buhrmester V, Münch D, Arens M (2019) Analysis of explainers of black box deep neural networks for computer vision: a survey. arXiv 1911.12116. https://arxiv.org/abs/1911.12116
99. Lundberg S, Lee S-I (2017) A unified approach to interpreting model predictions. arXiv 1705.07874. https://arxiv.org/abs/1705.07874
100. Sundararajan M, Taly A, Yan Q (2017) Axiomatic attribution for deep networks. In: Precup D, Teh YW (eds) Proceedings of the 34th international conference on machine learning, vol 70. Proceedings of Machine Learning Research, pp 3319–3328. http://proceedings.mlr.press/v70/sundararajan17a.html
101. Ribeiro MT, Singh S, Guestrin C (2016) "Why should I trust you?": explaining the predictions of any classifier. In: Proceedings of the 22nd ACM SIGKDD international conference on knowledge discovery and data mining, San Francisco, California, USA, pp 1135–1144. Association for Computing Machinery. https://doi.org/10.1145/2939672.2939778
102. Veličković P, Cucurull G, Casanova A, et al (2018) Graph attention networks. arXiv 1710.10903. https://arxiv.org/abs/1710.10903
103. Yun S, Jeong M, Kim R, et al (2020) Graph transformer networks. In: Wallach H, Larochelle H, Beygelzimer A, d'Alché-Buc F, Fox E, Garnett R (eds) Advances in neural information processing systems, vol 32. Curran Associates, Inc. https://proceedings.neurips.cc/paper_files/paper/2019/file/9d63484abb477c97640154d40595a3bb-Paper.pdf
104. Gu A, Dao T (2024) Mamba: linear-time sequence modeling with selective state spaces. arXiv 2312.00752. https://arxiv.org/abs/2312.00752
105. Chen T, Kornblith S, Norouzi M, Hinton G (2020) A simple framework for contrastive learning of visual representations. In: Proceedings of the 37th international conference on machine learning (ICML'20). JMLR.org, art. no. 149
106. Rao RM, Liu J, Verkuil R, et al (2021) MSA transformer. In: Proceedings of the 38th international conference on machine learning. PMLR, pp 8844–8856
107. Kellman BP, Sandoval D, Zaytseva OO, et al (2024) Protein structure, a genetic encoding for glycosylation. bioRxiv 2024.05.15.594261. https://doi.org/10.1101/2024.05.15.594261
108. Wagih O, Galardini M, Busby BP et al (2018) A resource of variant effect predictions of single nucleotide variants in model organisms. Mol Syst Biol 14:e8430. https://doi.org/10.15252/msb.20188430
109. Strumillo M, Beltrao P (2015) Towards the computational design of protein post-translational regulation. Bioorg Med Chem 23:2877–2882. https://doi.org/10.1016/j.bmc.2015.04.056
110. Ertelt M, Mulligan VK, Maguire JB et al (2024) Combining machine learning with structure-based protein design to predict and engineer post-translational modifications of proteins. PLoS Comput Biol 20:e1011939. https://doi.org/10.1371/journal.pcbi.1011939

# INDEX

## A



## B



## C



## D



## E



## F



## G



## H



## L



Dukka B. KC (ed.), *Large Language Models (LLMs) in Protein Bioinformatics*, Methods in Molecular Biology, vol. 2941, https://doi.org/10.1007/978-1-0716-4623-6,

## M



## N



## P



## S



## T